Minerals&
Human Health

First Edition

By Larissa Dobrzhinetskaya
University of California - Riverside

cognella®
academic publishing

Bassim Hamadeh, CEO and Publisher
Michael Simpson, Vice President of Acquisitions
Jamie Giganti, Senior Managing Editor
Jess Busch, Senior Graphic Designer
Marissa Applegate, Acquisitions Editor
Gem Rabanera, Project Editor
Alexa Lucido, Licensing Coordinator
Claire Yee, Interior Designer

First published in the United States of America in 2016 by Cognella, Inc.

Trademark Notice: Product or corporate names may be trademarks or registered trademarks, and are used only for identification and explanation without intent to infringe.

Cover image copyright© 2012 Depositphotos/Nobilior.
        Copyright © 2013 Depositphotos/photominer.
        Copyright © 2011 Depositphotos/itsmejust.
        Copyright © 2010 Depositphotos/Antartis.

NOTICE
Knowledge and best practice in this field are constantly changing. As new research and experience broaden our understanding, changes in research methods, professional practices, or medical treatment may become necessary. Practitioners and researchers must rely on their own experience and knowledge in evaluating and using any information, methods, compounds, or experiments described herein. In using such information and methods, readers should be mindful of their own safety and safety of others, including parties for whom they have a professional responsibility. To the fullest extent of the law, neither the Publisher nor the author assume any liability for any injury and/or damage to persons or property as a matter of products liability, negligence or otherwise, or from any use or operation of any methods, products, instructions, or ideas contained in the materials herein. This book is not intended to be a substitute for the medical advice of a licensed physician. The reader should consult with their doctor in any matters relating to his/her health.

Printed in the United States of America

ISBN: 978-1-62661-342-3 (pbk) / 978-1-62661-343-0 (br)

cognella®
academic publishing
www.cognella.com   800-200-3908

Contents

CHAPTER 2
MINERALS AROUND US

CHAPTER 3
MEDICAL GEOLOGY

CHAPTER 4
GEOLOGICAL HAZARDS—MINERAL DUST, AEROSOLS, AND DUST STORMS

CHAPTER 5
GEOLOGICAL HAZARDS: FROM MINERALS TO CHEMICAL DISASTERS

CHAPTER 6
GEOLOGICAL AND ANTHROPOGENIC HAZARDS
FROM TOXIC METALS

CHAPTER 7
ASBESTOS—HAZARDOUS MINERALS OF THE SERPENTINE AND AMPHIBOLE GROUPS

CHAPTER 8
COAL—A FOSSILIZED GEOLOGICAL FUEL AND ITS HAZARDOUS EFFECTS

Preface

This book is written in the form of narrative reviews and a meta-analysis of existing scientific studies and published statistical data from different databases and networks which bridge together Earth's dynamics, geological and hazardous processes, minerals, rocks and environments with human and industrial activities. The connection between minerals, environments, and human health has pre-historical roots and was appreciated by all civilizations. Widely available and easy accessible rocks and minerals served as universal materials for ancient people to survive and improve their lives. Obsidian, quartz, jasper and other minerals and rocks were used by ancient humankind to make weapons to hunt and protect themselves. Clay, for example, was used for healing skin wounds and improving digestion; and clay and pigments of other minerals were used for painting and decorating bodies and religious temples.

From the Industrial Revolution to modern day, the consumption of minerals and rocks has dramatically increased, and became an important part of the industrial and economic development of nations. However, the world-wide exploration and utilization of Earth's materials for a better life has caused adverse impacts on human health. It remains mostly unknown to the public that in the modern world, the health of billions of people may be affected by geological events and materials. This book integrates elements of geology and mineralogy with topics of wider interests related to public health and the environment—from the health effects of arsenic, mercury, fibrous minerals and "geological" dusts that contribute to the diseases—to questions related to identifications of such hazards. The connection of the natural environment, particularly when both geologic processes and products of human activities are coherently involved, deserves interest from scientists of different disciplines, politicians, and society. By combining knowledge from earth sciences, life sciences, environmental, political and medical sciences, students will be able to identify hazard problems and develop solutions that can help to ease or prevent toxic effects of some minerals or hazardous geological events on human health and the environment.

The book is recommended for the broad public and students of broad specialties who are interested in understanding the earth's dynamics and its interplay with environments, public health, sustainability of human lives and technological progress. The book may be adopted by a wide range of Universities for new courses to extend the horizons of modern education in geological, environmental and societal sciences.

The book was written in the frame of activities of the Task Force IV under auspices of the International Lithosphere Program.

My sincere thanks to H. Catherine W. Skinner (Yale University), Mary Droser (University of California at Riverside), Keith D. Morrison (Arizona State University), Sara E. Henry (San Diego Community College), Junfeng Zhang (China University of Geosciences, Wuhan, China) and other colleagues and students who have shared their interesting visions and opinions regarding the preparation of this book.

—Larissa Dobrzhinetskaya

Introduction to Geological Processes

Chapter 1

How does the Earth operate?

The Earth is the largest planet within the group of terrestrial planets that includes Venus, Mars, and Mercury (Fig. 1). Its diameter (d) is 12756.3 km, while Venus has d = 12103.6 km, Mars has d = 6794 km, and Mercury is the smallest one with d = 4880 km (Fig. 1). Only the Earth is a vital planet because it has a sustainable atmosphere and liquid water, which are necessary conditions to support biological life. Although life is a "biological issue," evidence of life, starting from ancient, primitive micro-organisms, is recorded in terrestrial rocks and is assumed to be found in extraterrestrial planetary objects as well. Yet, the Earth is the source of geological materials such as mineral ore deposits, coal, oil, and clean underground water reservoirs that sustain life and provide energy. Therefore, the Earth represents a complex system that reflects dynamic interactions among the atmosphere, biosphere, hydrosphere, geosphere, cryosphere, and anthroposphere.

The Earth is 4.6 billion years old, and the geological time span and the most challenging related events are summarized in Table 1. The geological time, since the Earth's formation is characterized by six eras, is described below.

Hadean era—no rocks, no geological records. The first era, *ca.* 600 million years after accretion, is called the Hadean era, and it is a period of solidifying of the Earths continental and oceanic crusts. The International Commission on Stratigraphy does not recognize this era because no rocks are known to be formed during this time, so the characterization of the Hadean era is usually based on the lunar time scale and geological events.

FIGURE 1 Sizes of the planets in the terrestrial group

Archean era—primitive, one-celled biological life. During the Archean era, which spanned the next 1500 million years of the Earth's history, the first life developed in the form of one-celled organisms. The appearance of algae and bacteria in the oceanic waters triggered a release of free oxygen into the atmosphere.

Proterozoic era—multi-celled biological life. The Proterozoic era lasted about 1958 million years, and it is known as a bloom of life, represented by multi-celled organisms called Vendian or Ediacaran biota. Ediacaran fauna has now been found by geologists on almost all continents, and it plays an important role in our understanding of the evolution of life on the Earth. This is because it is considered to be a precursor to the appearance of hard-shelled organisms at the beginning of the Paleozoic era, e.g., 543–541 million years ago. The evolution of life started from primitive Archean forms in close proximity to the oxygen production by these micro-organisms. Eventually, as the atmosphere became enriched with oxygen and the ozone layer was formed, more complex micro-organisms evolved because the ozone layer shielded the Earth from radiation, creating conditions for new forms of life.

Paleozoic era—diversity of life and plate tectonics. During the Paleozoic era, which was a much shorter period (291 million years) than the Proterozoic era, diverse invertebrate groups evolved, including organisms with both hard skeletons (insects, fish, amphibians, reptiles, etc.) and hard shells (trilobites, brachiopods, and others). The formation of the supercontinent

TABLE 1   Geological Time and Major Biological Evolution and Geological Events

Era	Age (MYA)	Time Span (MYA)	Geological events
Cenozoic	65.5–present	65.5	Human civilization (0.011 million years ago (MYA)–present). More mammals, primates, mammoths, whales, birds, horses, dogs, etc.
Mesozoic	251 to 65.5	185.5	Extreme tectonic and volcanic activity, or gigantic asteroid impact followed by mass extinction. The supercontinent Pangaea begins to break up. Appearance of dinosaurs, reptiles, marine invertebrates, mammals, and plants.
Paleozoic	542 to 251	291	Formation of the supercontinent Pangaea and the era of active plate tectonics. Skeletal organisms and hard-shelled organisms (e.g., trilobites, fish, insects, amphibians, and reptiles).
Proterozoic	2500 to 542	1958	Multi-celled organisms (Ediacaran or Vendian biota).
Archean	4000 to 2500	1500	"Ancient life"—one-celled organisms; oxygen released from oceans into atmosphere.
Hadean	4600 to 4000	600	"Rock-less" era, from which no geological formations have been recorded.

Pangaea and the beginning of active plate tectonics are the most notable characteristics of the Paleozoic era.

Mesozoic era—the era in which dinosaurs and reptiles appeared and then became extinct. The Mesozoic era, with a time span of ~185 million years, is known as the time of reptiles, dinosaurs, and plants. The supercontinent Pangaea began to break-up, followed by the formation of six continents with configurations more or less the same as today. The Mesozoic era ended with the extinction of most of the dinosaurs and reptiles, and in most geological literature, this extinction is known as the Cretaceous-Paleocene extinction (K-T extinction). The extinction has been explained by several different hypotheses, the most popular being (i) a huge meteorite impact or (ii) extreme volcanic activity, both of which might have led to dramatic changes in the climate and environment.

Cenozoic era—age of mammals and human civilization. The Cenozoic era began 66 million years ago as a "recovery" after the K-T extinction and it continues to the present. The slow drift of

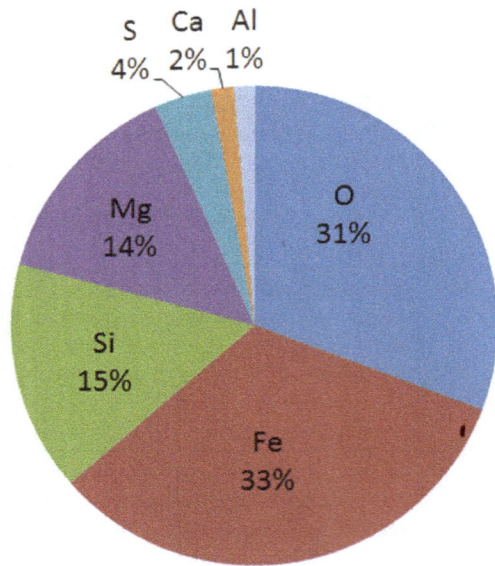

FIGURE 2 An average chemical composition of the Earth (in atomic %)

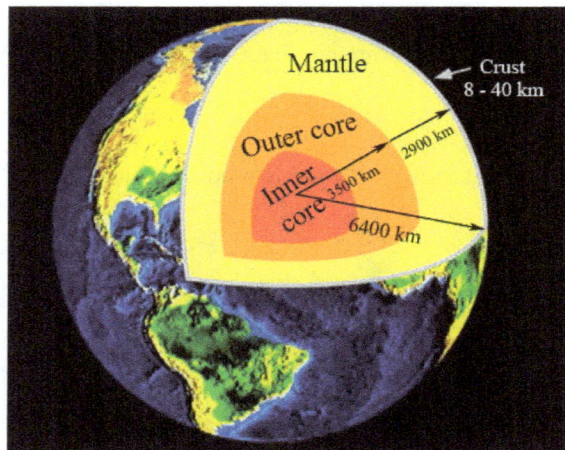

FIGURE 3 Inner layers of the Earth

continents brought them to their current positions, but the relief of the continents was dramatically changed due to closure of the huge Tethys Ocean and collisions that happened at the continental margins. Mammals became the dominant vertebrate group, and the continents became covered by forests and other vegetation, both of which eventually became similar to the present fauna and flora.

1.1.1. Composition and structure of the Earth

The Earth's mass is approximately 5.98×10^{24} kg. Although it contains all of the elements listed in the chemical periodic table (except those that have been created in the laboratory), there are only seven most-common elements that make up ~97% of the Earth's mass. They are (in weight percent, or wt.%): Fe = 32.9, O = 30.7, Si = 15.5, Mg = 14.2, S = 3.8, Ca = 1.5, and Al = 1.4 (Fig. 2).

Based on the chemical compositions and seismic properties of the rocks, the Earth's interior is divided into three major layers: crust, mantle, and core. Their structural relationships are usually presented in the form of the "onion-shell" model, which suggests radial variations of physical and chemical properties of the rocks composing the shells (Fig. 3).

Crust

Two types of crust are recognized: continental crust and oceanic crust.

The continental crust has a varying thickness of 30–70 km (the average is accepted as ~40 km) and consists of igneous, metamorphic, and sedimentary rocks diverse in composition, although

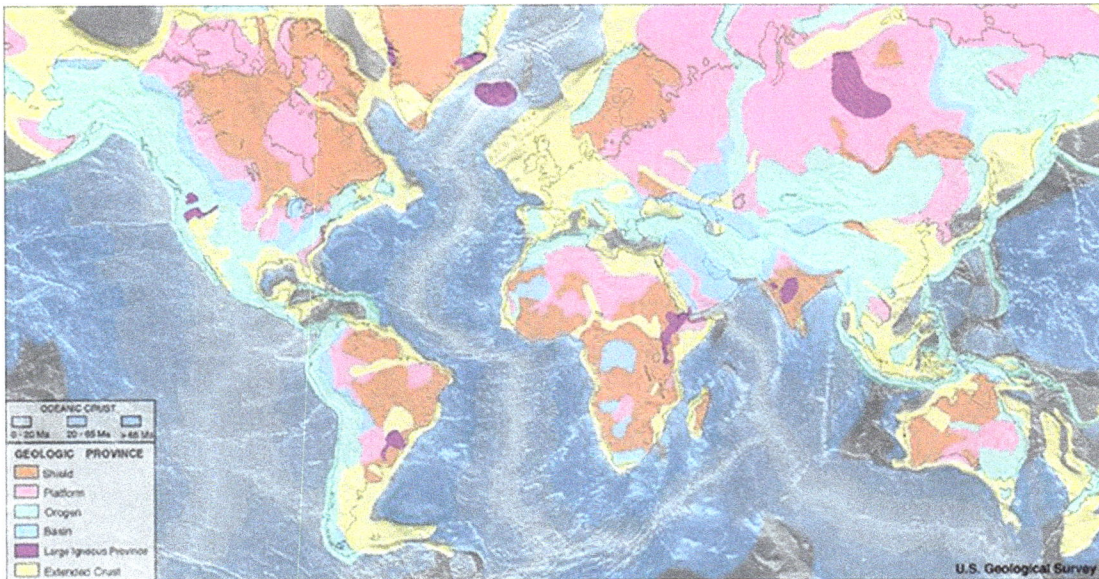

FIGURE 4 Map of the world geological provinces

in general, they are rich in SiO_2 (~60–65 wt.% of SiO_2). The thicker continental crust (~70 km) occurs beneath the Himalayas (Euro-Asian continent) and the Andes (South America). The oldest rocks that built up the stable "core" of the continents (called the platform) are nearly 4.3 billion years old. Almost all platforms are covered with younger sediments, and those areas, called shields, are where the oldest rocks are exposed on the Earth's surface. Other rocks of the continental crust were formed during different geological times in the Earth's history, and they compose the roots of mountain belts, called orogenic belts or orogens. A geological map of the world (Fig. 4) illustrates the spatial relationships between platforms, shields, orogens (continental crust), and oceanic crust. Occupying the surface of the continents, the continental crust is slightly extended into the area of the seabed close to the shoreline, and this extension is called the continental shelf. Continental margins outline and reshape the configurations of continents through long geological time. They play significant roles in the Earth's dynamic processes, and industrial mineral deposits and oil deposit reservoirs which are formed inside mountain belts and continental shelves, respectively.

An average chemical composition of the continental crust (in wt.%) is: SiO_2 = 60.6, Al_2O_3 = 15.9, FeO (total Fe^{2+} and Fe^{3+}) = 6.7, CaO = 6.4, MgO = 4.7, Na_2O = 3.1, K_2O = 1.8, TiO_2 = 0.7, and P_2O_3 = 0.1 (Rudnik and Fountain, 1995). Therefore, the most common chemical elements in the crust (in wt.%) are: oxygen (46.6), silicon (27.7), aluminum (8.1), iron (5.0), calcium (3.6), potassium (2.8), sodium (2.6), and magnesium (2.1) (Fig. 5a). The rocks that compose

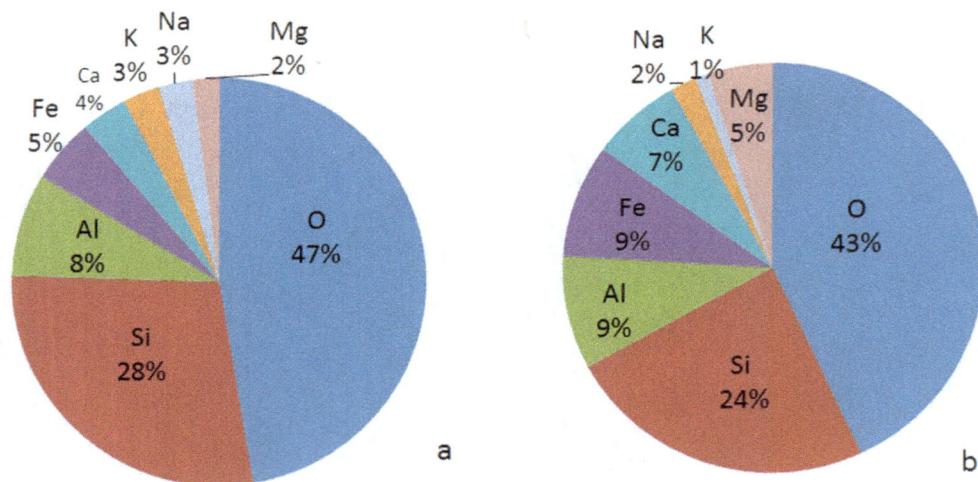

FIGURE 5 (a) An average chemical composition of the continental crust; (b) an average chemical composition of the oceanic crust

the continental crust are diverse in their mineralogical compositions, although in general, they have high SiO_2 content (~60% on average) and an average density of ~2.6–2.7 g/cm³.

The oceanic crust is ~7–8 km thick. It underlies the continental crust of the continents, and waste volume of the oceanic crust occurs at the bottoms of the oceans, which cover almost two-thirds of the Earth's surface. The oceanic crust is chemically less diverse than the continental crust, consisting mostly of basalts, which are magmatic rocks that contain ~49 ± 2 wt.% of SiO_2 and are enriched in MgO and FeO. The chemical composition of the average oceanic crust is represented by the bulk composition of the mid-oceanic-ridge basalts (MORB), which contain (in wt.%) SiO_2 = 50.45, Al_2O_3 = 15.26, FeO (total Fe^{2+} and Fe^{3+}) = 10.43, MgO = 7.58, Ca = 11.30, Na_2O = 2.68, and TiO_2 = 1.62 (Hofmann, 1988). Elements composing the oceanic crust are shown in Fig. 5b. If you compare it with the continental crust composition (Fig. 5a), you will see that the oceanic crust contains more Fe, Mg, Ca, and Na and less Si and O than the continental crust. The oceanic crust, therefore, is heavier than the continental crust, with a calculated average density of ~2.9 g/cm³ (Carlson and Raskin, 1984). Such differences in density are an important factor in plate-tectonic processes.

The oldest sea-floor formed about 180 million years ago. The world map of the ages of the oceanic crust (Fig. 6) shows in detail the spatial distribution of its youngest and oldest fragments. The ocean-floor spreading occurring at the mid-oceanic ridges is where new portions of the oceanic crust materials are generated by the upper mantle, and therefore, the youngest rocks of the oceanic crust can be found in the middle parts of the oceans. The rocks of the oceanic crust at the continental margins are being recycled back into the mantle through subduction zones.

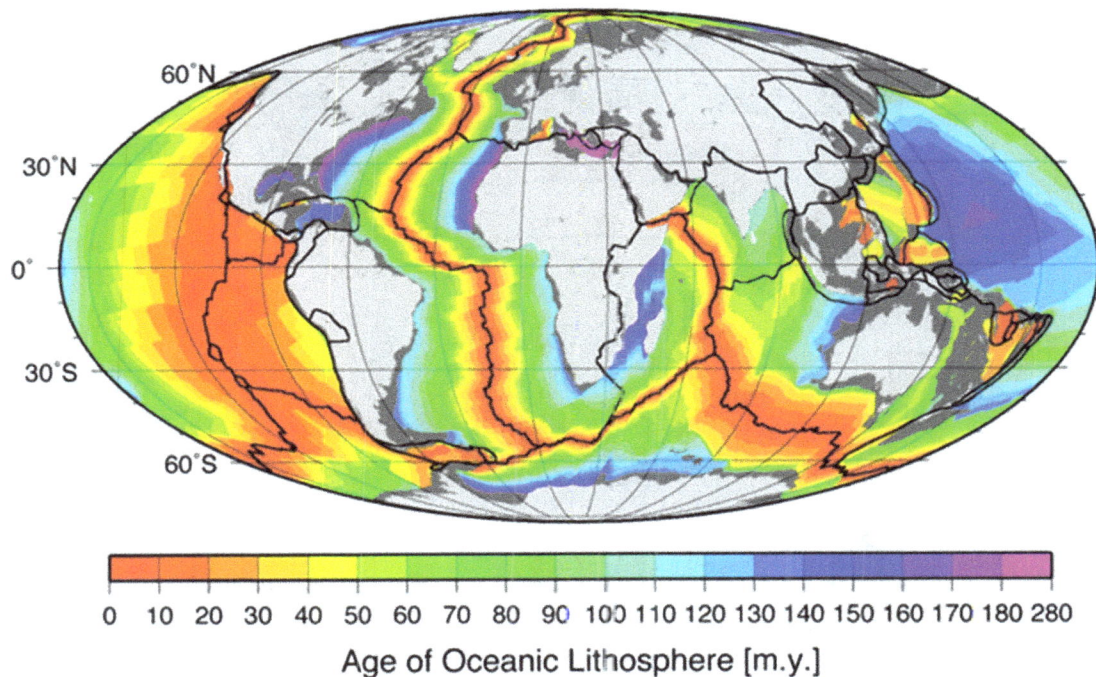

Age of Oceanic Lithosphere [m.y.]

FIGURE 6   The world map of the age of the oceanic crust

Lithospheric plates. The crust and part of the upper mantle, totaling ~100 km, form the brittle "multi-compositional layer" (Fig. 7) that is divided into several rigid lithospheric plates. A lithospheric plate may be composed entirely of oceanic or continental lithosphere, but most of them are partly oceanic and partly continental. The plates are "floating" on the top of the heavy underlying mantle.

Mantle

The crust (lithospheric plates) rests on the mantle, which has a thickness of ~2900 km (Figs. 3, 7) and a volume corresponding to two-thirds of the Earth's mass. The mantle is divided into two shells—the *upper mantle* and *lower mantle*—which are separated by a *mantle transition zone* that lies at a depth of between 410 and 660 km. The upper mantle is composed of rocks called peridotites, which are rich in Mg and capable to be ductile and flowing, thus providing a mantle convection. The lower mantle also consists of peridotites, but their main rock-forming mineral, olivine [$(Mg,Fe)_2SiO_4$], is transformed into the mineral wadsleyite, which has the same chemical composition as olivine but a denser atomic structure because the pressure and temperature increase with increasing depth (e.g., Duffy, 2008).

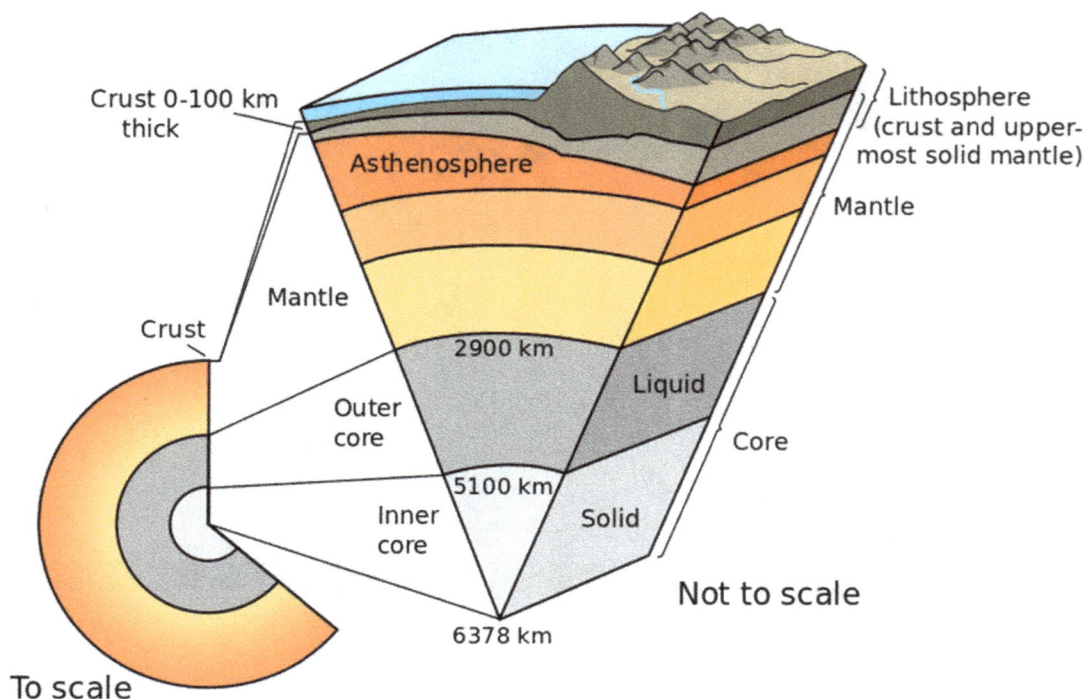

FIGURE 7 Diagram showing the internal structure of the Earth

The average composition of the mantle (in wt.%) is: SiO_2 = 46, MgO = 37.8, FeO = 7.5, Al = 4.2, CaO = 3.2, Na_2O_3 = 0.4, and K_2O_3 = 0.04 (Jackson, 2000). An average mantle composition diagram (Fig. 8a) shows that the mantle is richer in Mg (26 atomic percent—at.%) and Fe (10 at.%) and significantly poorer in Si (19 at.%) and Al (1 at.%) than the Earth's crust layers (compare Figs. 5a,b and Fig. 8a). The average density of the mantle (4.5 g/cm³) is also greater than that of the oceanic crust, with the upper-mantle density varying from 3.4 to 4.4 g/cm³ and the lower mantle consisting of rocks of still higher density (5.7 g/cm³); (Fowler, 2005). The average temperature in the upper-mantle region is ~600–900°C, and it increases with depth. In the upper part of the lower-mantle boundary region, at a depth of 620–660 km, temperatures can reach up to 1600°C, and at 2700 km depth, they can reach 2200°C (Fig. 9).

Asthenosphere. The uppermost part of the upper mantle (at a depth of ~100 km), which is called the *asthenosphere* (Fig. 7), is solid but ductile, and its low viscosity provides slow movement of the rigid lithospheric plates with respect to the mantle. The thickness of the asthenospheric layer is not perfectly defined because it depends on the temperatures, which are different beneath the oceanic and continental lithospheres. For example, the highest temperature (1350°C) is beneath the oceanic ridges, where the asthenosphere is molten (e.g., Dasgupta et al., 2013).

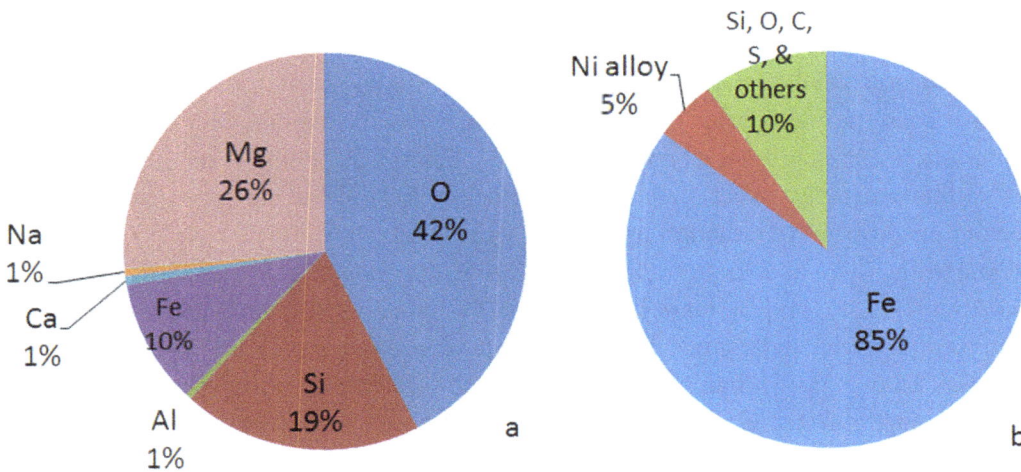

FIGURE 8 Average chemical composition of the Earth's mantle (a) and core (b)

Core

The core of the Earth is as large as 3500 km in diameter, and the average composition of the core (in at.%) is: Fe = 85, Ni = 5, and some other elements (such as S, Si, C, O) = 10 (Fig. 8b). The core includes a molten outer core attached to the lower mantle and a solid inner core isolated from the mantle (Fig. 3).

The outer core is equal to 30.8% of the Earth's mass, extending down from the lowermost mantle at 2900 km to 5100 km (Fig. 7). The outer core is a low-viscosity liquid (melt), and its temperature varies between 3200°C and 3700°C in the region close to the boundary with the solid lower mantle, and 4700°C and 5200°C at its boundary with the solid inner core (Fig. 9). The pressure at the top of the outer core is 135 GPa, and it increases to 330 GPa with increasing depth toward the outer-inner core boundary (Fig. 9).

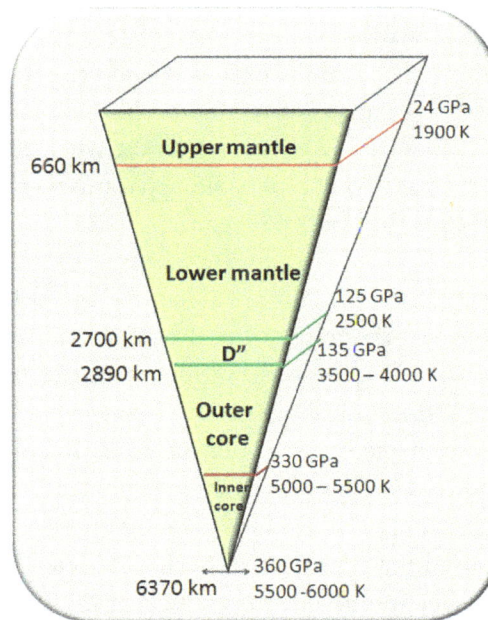

FIGURE 9 Cross-section of the Earth's interior, showing the expected range of pressures and temperatures

The boundary between the two contrasting regions—the lower mantle and the outer core (between 2700 and 2890 km depth)—is called the D" layer, which produces hot plumes (Fig. 9b) that cause partial melting in the neighboring zones. Inside the outer core, the liquid Fe-alloy produces turbulence—e.g., convective flow (Gillan et al., 1998)—and many scientists believe that this convective effect maintains the magnetic field of the Earth, which shields the surface from the destructive radiation of the solar wind (Lean, 2005).

The inner core is a solid matter suspended inside the molten outer core at a depth of 5100–6370 km (Fig. 9), and its mass is equal to 1.7% of the Earth's mass. Although the inner core has almost the same chemical composition as the outer core and has even higher temperatures (5200–5700°C) than the outer core (4700–5200°C), it is still solid. This is not a paradox—experimental studies and numerical modeling have shown that the melting point of the Fe increases with increasing pressure (Boehler, 2000; Alfè et al., 1999). Indeed, the pressure in the inner core region (max ~360 GPa) is significantly greater than the pressure in the outer core (max ~330 GPa), which suggests that the pressure effect overrides the temperature effect, causing the matter of the inner core to remain in a solid state.

1.1.2. Principles of plate tectonics—the main concept of geology

The formulation of the modern plate-tectonics theory was based on two fundamental concepts: *continental drift* and *sea-floor spreading*. In the 1920s, continental drift was the only foundation underlying the earlier version of the plate-tectonics concept. The spreading of the oceanic floor was recognized later, in the 1960s, when it was confirmed through geophysical data, direct observations by submersible apparatuses, and the lifting out of drill cores of the rocks that composed the floors of oceans.

Continental drift. Almost a century ago, Alfred Wegener, a German scientist-meteorologist, analyzing a geographic world map, recognized that the coastlines of several continents fit well together. Using this observation, as well as paleontological data showing that fossils of the same types and identical ages were found in both Brazil and Africa, he proposed that Africa was once connected with South America. In support of these paleontological data, Wegener also found that some geological formations of South Africa (e.g., the rocks of Karroo system) and South America (the Santa Catarina strata, Brazil) matched very closely to each other. He also proposed that Europe had been connected to North America and that Madagascar had been connected to India because fossils of similar types and ages were found in these now-separated land masses. Furthermore, these fossils represented ancient organisms that were not able to travel across the vast oceans that exist now.

Wegener's general idea was that all current continents had first formed a single supercontinent, called Pangaea (Fig. 10), that later split apart due to a process that he called "continental drift" (Wegener, 1924). However, at that time, Wegener's concept was met with strong

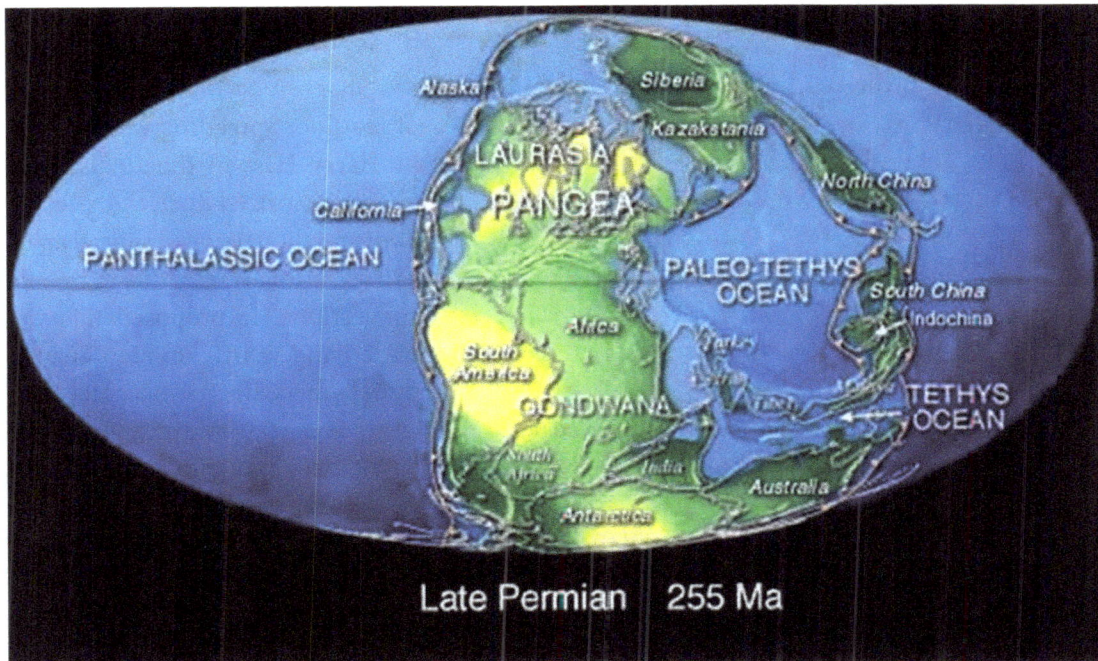

FIGURE 10 Modern reconstruction of Pangaea, ca. 255 million years ago

criticism and resistance, even though no other convincing mechanisms for the movement of the continents had been proposed except "centrifugal and tidal" forces. The latter were used by Wegener to support the explanation that the continents were moving through the Earth's crust and plowing the oceans' floors in a way similar to the way in which icebergs move. So, according to Wegener's vision, the continents are moving while the oceans remain intact. Later, much additional evidence of the movements of continents through the oceans and the continuous formation of new oceanic crust were collected. For example, it was conclusively demonstrated that the ocean is the main component in plate tectonics and that without the spreading of the oceanic floor, no plate tectonics would be possible. We now know that both continents and ocean floors form solid plates that float on the ductile asthenosphere and that the oceans play the central role as the most active zones, where new oceanic crust is constantly added through the magma generation of the underlying mantle.

In 1930, Alfred Wegener died at age fifty during an expedition to Greenland. Although his continental drift theory was not correct in many ways, his revolutionary idea of the super-continent Pangaea's break-up and his basic insights about plate tectonics are still significantly appreciated to this day. In the 1950s–1960s, the next stage in the study of the ocean floor brought startling discoveries that the key to the mechanism of plate tectonics is sea-floor spreading.

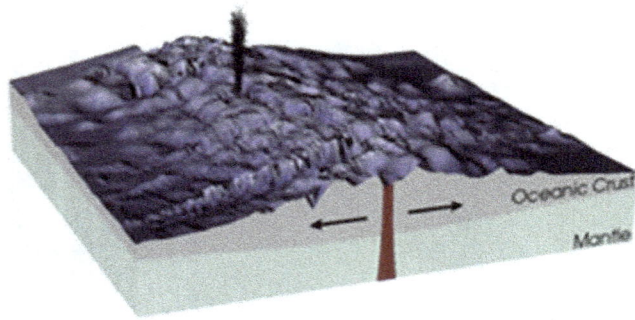

FIGURE 11 Simplified schemes of the sea floor spreading

Sea-floor spreading. The concept of sea-floor spreading was built up by Harry Hess in the 1960s. Hess (1962) theorized that the oceanic floors might have been moved apart along the ridge crests due to mantle convection. He proposed that the underlying, partly molten mantle upwells beneath the oceanic ridge crests and triggers the eruption of basaltic magma along the axis of ridges (Fig. 11). The melt is then "welded" to the existing oceanic crust, and the new portion of the oceanic crust gradually moves away from the ridge crests toward the continental margins. Such a magma-generation process is continuous, and because the mid-ocean ridges are constantly moving apart, it creates a new space along which the melt eruption repeats. The sum of these processes is called sea-floor spreading, which has operated over many millions of years and has caused formation of the 80000-km-long system of mid-ocean ridges that grew up beneath the oceanic waters.

Hess's idea of sea-floor spreading was confirmed by Vine and Matthews (1963), who discovered linear magnetic anomalies (reverse magnetic polarity) situated parallel to the axis of oceanic ridges. Geophysical observations and measurements of the magnetism in the rocks throughout the world showed that over long geological time, the Earth's north magnetic pole periodically was situated in the southern geographic hemisphere (e.g., became the south pole), and vice versa. The existence of the magnetic anomalies within the rocks of the ocean floor suggested that magnetic minerals within the rocks aligned with the Earth's magnetic field at the time when the magma cooled and solidified. Using the intensities of the magnetic field recorded in the rocks, geophysicists determined that the ocean-floor rocks contain positive and negative anomalies (Fig. 12). Based on this, they proposed that positive anomalies occur where the magnetic field is stronger, which means that these anomalies were formed when the rocks became solid, with the Earth's north magnetic field located in the northern hemisphere. In contrast, the negative magnetic intensities are weaker, and therefore, they were formed when the rock solidified with the Earth's north magnetic pole in the southern hemisphere. Vine and Matthews (1963) explained that the symmetrical repetition of "positive-negative" strips (e.g., high-intensity and low-intensity magnetism) exists because the lava that erupts on the sea floor on both sides of the ridge axis solidifies and moves away before a new portion erupts (see Figs. 11, 12). The explanation was a crucial piece of evidence for unconditional acceptance of the sea-floor spreading theory, which is still the subject of advanced studies (e.g., Müller et al., 2008).

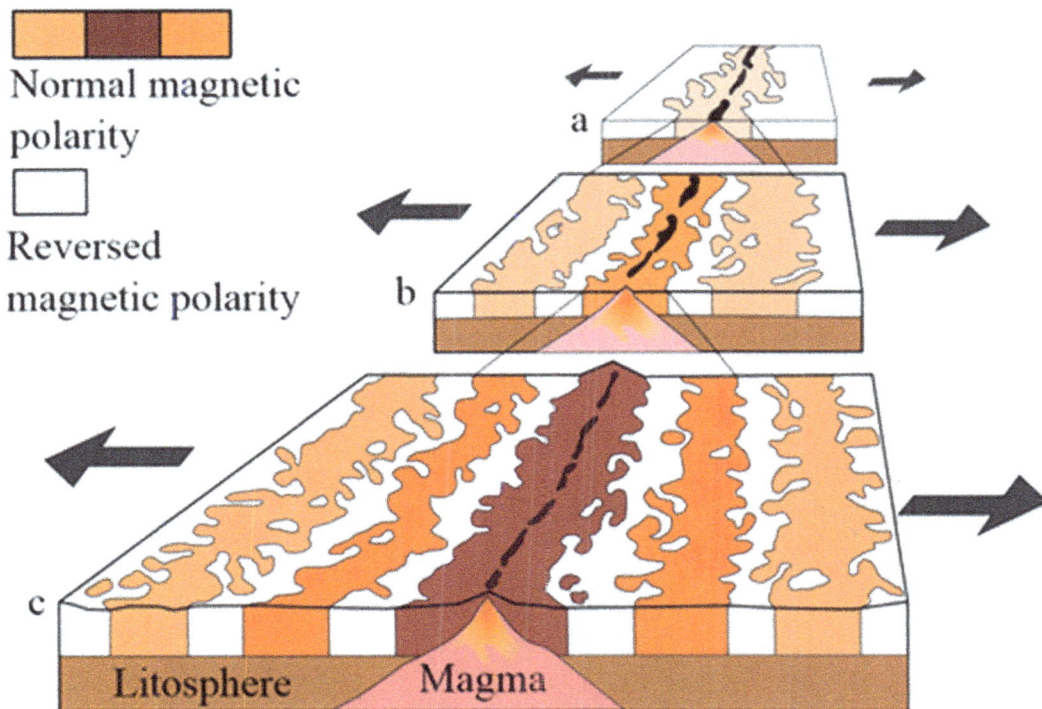

Normal magnetic polarity

Reversed magnetic polarity

Litosphere   Magma

FIGURE 12   Magnetic polarity: normal and reversed

Plate tectonics

According to plate-tectonics theory, the Earth's lithosphere is broken-up into twelve major plates that move in various directions as slowly as 5 cm per year (Fig. 13). Because of such motion, the plates sometimes collide, pull apart, or scrape against one another. The mantle convection that was created by the inhomogeneous distribution of heat inside the core-mantle regions is now seen as a driving force for plate tectonics. In a general sense, plate tectonics is a vital process of the Earth that generates and controls volcanic activities and earthquakes along the boundaries of the plates and reshapes continents and their margins by forming huge mountain chains—for example, the Himalayas in Asia and the Andes in South America.

Three types of plate boundaries are considered to be the main components of the plate-tectonics process: divergent, convergent, and transform.

Divergent boundary (plates move apart). This is a linear feature, similar to a deep fault, which separates plates, allowing them to move away from each other (Fig. 14). As the plates pull apart along the divergent boundaries, magma formed during melting of the underlying mantle rises up through this new pathway and generates chains of active volcanoes. The formation

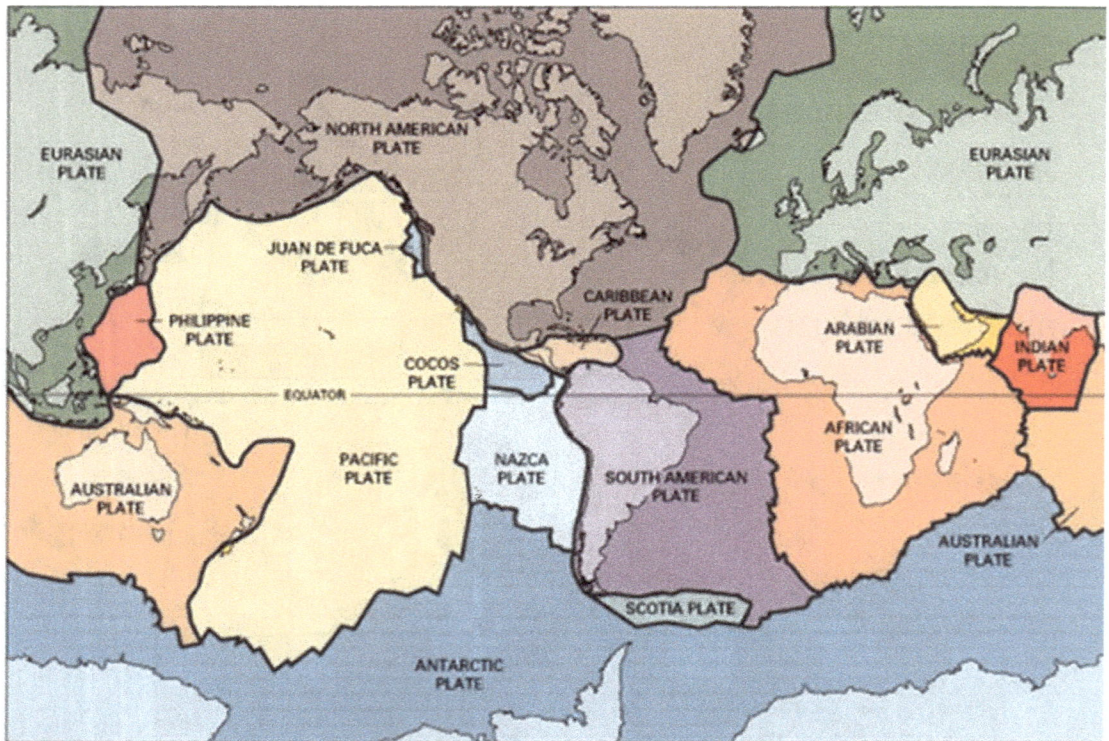

FIGURE 13 World map of the main tectonic plates

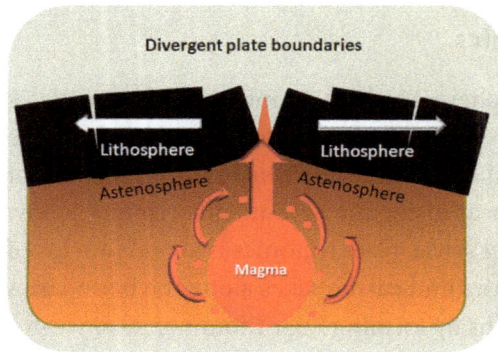

FIGURE 14 Scheme of formation of the divergent plate boundaries

of divergent boundaries is a destructive process that is caused by the mantle convection currents, which push up on the bottom of the lithospheric plate, lift it, and cause plastic rocks to flow laterally beneath the lithosphere. As a result of such lateral flow, the rigid rocks of the overlying plate are dragged and stretched thin at the crest of the mantle uplift, and finally, the plate breaks up and pulls apart. A divergent boundary formed on a continent is called a continental rift (Fig. 14), and one that is formed beneath the ocean is called an oceanic ridge. The Great Rift Valley that stretches from Syria to Mozambique in Africa is a typical example of continental rift-zone.. Oceanic ridges represent a series of mountain ranges situated beneath the oceanic waters, and they extend through

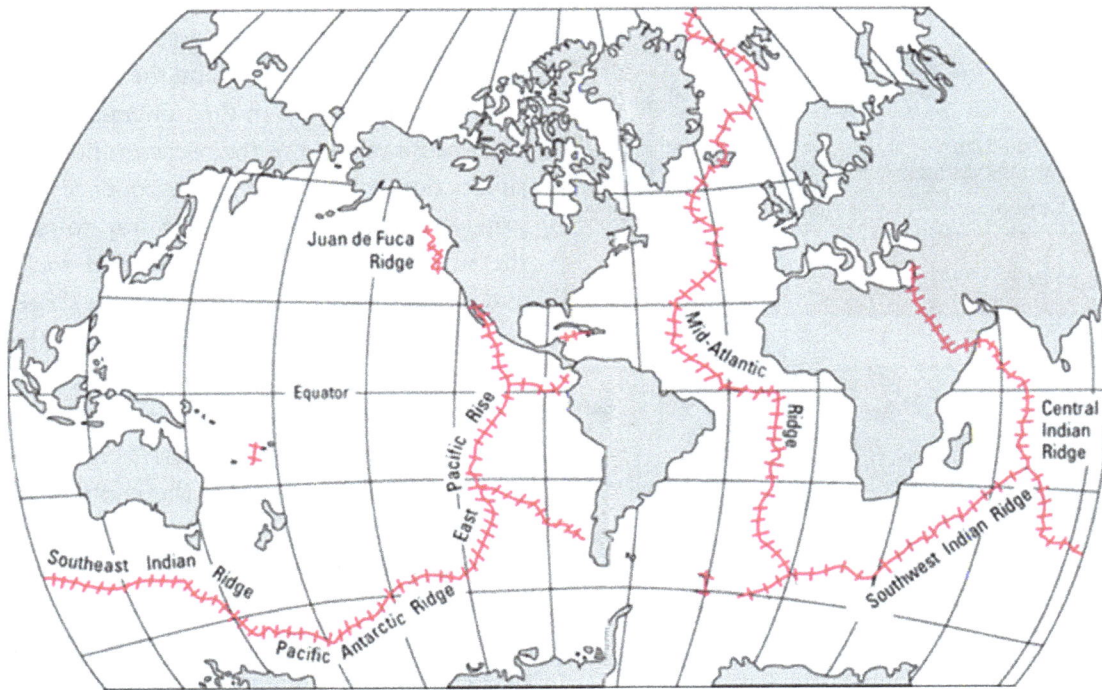

FIGURE 15 The world map of the oceanic ridges

the Atlantic, the Pacific, and the Indian oceans (Fig. 15). The formation of the oceanic ridges is associated with sea-floor spreading (see details in Fig. 11).

Convergent boundary (plates move toward each other and collide). A convergent plate boundary is formed in places where lithospheric plates are moving toward one another and eventually collide. There are three types of convergent boundaries, each defined by the types of crust (continental or oceanic) that come together: *(i) ocean-continent collision, (ii) continent-continent collision, and (iii) ocean-ocean collision.*

   (i) Ocean-continent collision. In this type of convergent boundary, a dense oceanic plate collides with a continental plate (Fig. 16a), which is less dense and therefore more buoyant. Due to such differences in densities ($d_{oceanic\ crust}$ = 2.9 g/cm^3 and $d_{continental\ crust}$ = 2.6–2.7 g/cm^3), the oceanic plate thrusts underneath because it is less buoyant than the continental lithosphere. The channel through which one plate moves beneath another is called a *subduction zone.* As a result of the oceanic plate subduction into the mantle beneath the continental lithosphere, a deep trench is formed under the oceanic waters in a place where the deep-sea plate plunges beneath the continent. Also, mountain chains grow up along the continental margin as a result of this process (Fig. 16a). The deep trench collects sediments from the continental shelf and the oceanic floor, and they may also be subducted into the mantle along with the oceanic plate.

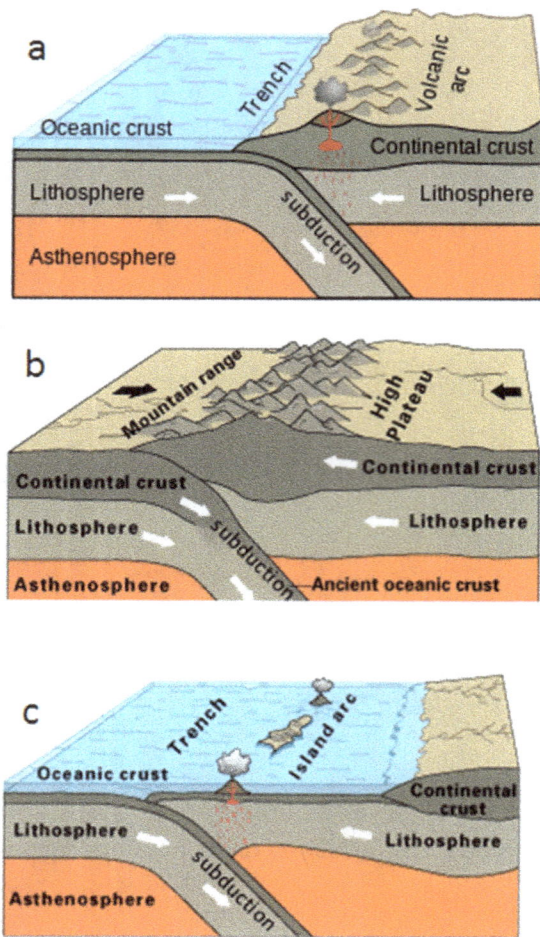

FIGURE 16 Convergent boundaries formation by (a) ocean–continent collision; (b) continent–continent collision, and (c) ocean–ocean collision

High temperature and the presence of H_2O in the rocks (porous water and the water that is chemically bonded in the minerals' structure) cause melting of the rocks in the deep subduction channel. As a consequence, the generated magmas force their way toward the surface, intruding into the solid rocks and creating volcanic arc chains on the shore.

The Andes mountain range on the western coast of South America is the best example of an ocean-continent collision, where the Nazca Plate (ocean) is continuing its subduction beneath the continental lithospheric plate of South America. Another example is the Cascade mountain range in northern California, Oregon, and Washington in the U.S., which was formed by a collision followed by the subduction of the oceanic Juan de Fuca Plate beneath the North American continent. The Mount St. Helens, a currently active volcano, is a result of the continuing subduction of the Juan de Fuca Plate. The most recent major eruption of the Mount St. Helens volcano, which occurred in 1980, was one of the most destructive volcanic events in U.S. history, leaving a barren wasteland surrounding the volcano and killing fifty-seven people. Before the eruption, several earthquakes of magnitude 4.0–4.2 were registered beneath the Mount St. Helens, and many smaller earthquakes occurred almost every hour. This and many other facts collected by geophysicists indicate that high seismic activity exists along this convergent plate boundary.

(ii) Continent-continent collision. When two continents are pushed together and collide, the oceanic crust between the continents and some fragments of the continental crust are subducted into the mantle, and the result is the formation of a mountain range (Fig. 16b). A process of mountain building is called an orogeny, and a mountain range formed during a continent-continent collision is called an orogenic belt. The best examples of huge orogenic belts are the Alps and the Atlas mountains in Europe and the Himalayas in Asia. The Alps and

the Atlas mountains were formed about 60–45 million years ago by the collision of the African and European plates, and the Himalayas were formed about 10 million years ago when India collided with the southern margin of the Asian plate. Shallow earthquakes that occur in the Himalayas suggest that this territory is still not stable, which means that some readjustments of the continental fragments inside the orogenic belt and its root zones in the underlying mantle are still going on.

When the plate-tectonics theory was accepted, it seemed to indicate that the relatively low, dense, and buoyant continental crust could not be subducted deeply into the mantle, but this assumption was proved wrong by the discovery of metamorphic rocks of continental origin containing diamond and coesite, high-pressure modifications of native carbon and mineral quartz (SiO_2), respectively. These rocks are called ultra-high-pressure metamorphic (UHPM) rocks, and they have been found on all continents. Experimental studies have shown that coesite and diamond may be formed and remain stable only at high pressures, which require subduction of continental sediments to a minimum depth of about 80–140 km (e.g., Ernst and Liou, 2008; Dobrzhinetskaya and Faryad, 2011). There are many numerical models that explain in detail such mechanisms as deep subduction of low-density crustal rocks into the deep mantle and the way they are returned to the Earth's surface (e.g., Gerya et al., 2002). However, there is still not enough knowledge to determine the volume of deeply subducted crustal materials that is returned to the shallow levels of collisional orogens and the amount of these materials that has been "amalgamated" into the deep mantle and never returned to the Earth's surface.

(iii) Ocean-ocean collision. When two oceanic plates collide, it causes the formation of an island arc, which rises as a result of the melting of the descending plate (Fig. 16c). The magma generated by the melting of the subducted oceanic crust is rich in volatile components (e.g., gases), and therefore, being less dense than its rock reservoir, it erupts on the sea floor and eventually builds an arc of volcanic islands. A deep trench and a forearc range are formed in the vicinity of the island arc (Fig. 16c). Examples of oceanic island arcs that, together with deep trenches, mark boundaries between two collided oceanic plates, are the Mariana Islands in the western Pacific Ocean and the Aleutian Islands off the coast of Alaska.

The collision of oceanic plates is not a "quiet" process, and such collisions are accompanied by powerful earthquakes on the oceanic floors along the trenchs and subduction zones. One of the recent devastating earthquakes was the Haiti earthquake, magnitude 7.0 on the Richter scale, occurred in January 2010 at the convergent boundaries between the Caribbean and North American oceanic plates (Hayes et al., 2010). During this earthquake, more than 217000 people were killed, and an additional 300000 people were injured (Bilham, 2010; Eberhard et al., 2013). Such earthquakes often cause huge oceanic waves, or tsunamis, which, when they reach land, cause incredible destruction. An example of a powerful tsunami is the one caused by the great Sumatra-Andaman earthquake, magnitude 9.1–9.3, which occurred in December 2004 in the Indian Ocean and killed more than 283000 people. The earthquake occurred at the ocean-ocean convergent boundaries between the Indo-Australian and Southeastern Eurasian plates (Lay et al.,

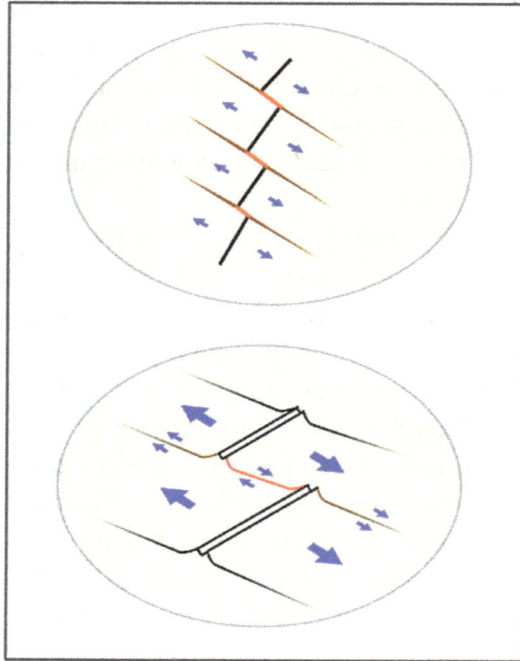

FIGURE 17 Transform boundaries

2005; Vigny et al., 2005). Another recent powerful tsunami occurred in March 2011 off the Pacific coast of Tohoku, Japan. It was caused by a magnitude 9.0 earthquake occurred along the ocean-ocean collision boundary (e.g., Koketsua et al., 2011; Sato et al., 2011). During this earthquake-tsunami disaster, 15883 were killed, 6150 were injured, and 2643 people were missing.

Transform boundaries (plates or fragments of plates that slide past each other). When one segment of a rigid lithospheric plate slides past another in response to stresses, such events form transform boundaries (Fig. 17), which are faulting systems that do not cause any destruction of the lithospheric plates as is typically observed along convergent boundaries. Transform boundaries occur mostly on the sea floor between the connecting segments of diverging mid-ocean ridges, where there is a constant upwelling of new basaltic magmas. Examples include the mid-oceanic ridge transform zones that are situated on the floor of the Atlantic Ocean between South America and Africa and the East Pacific Ridge in the southeastern Pacific Ocean. However, transform boundaries can also occur at the continental margins. For example, the San Andreas Fault in California represents a typical transform boundary that was formed 34–24 million years ago (Alwater, 1970). Earthquakes are associated with transform boundaries in both oceanic and continental margin environments.

1.2. Geological materials and processes

1.2.1. Classification of rocks

Rocks are the main building blocks of the Earth, and they consist of three main groups: (i) igneous, (ii) sedimentary, and (iii) metamorphic.

(i) **Igneous rocks** are the rocks that have crystallized from melts, and they include two groups: *intrusive* and *volcanic* rocks (Fig. 18). Intrusive rocks (they are also called plutonic rocks) are formed when magma (melt) cools and solidifies inside the Earth. The parental magma originates from the solid rocks of the continental or oceanic lithosphere as well as from the mantle. Volcanic rocks are formed by volcanic eruptions, when lava, ash, and other products of volcanic activities pour out of a volcano crater and are deposited on the Earth's surface. There are also intermediate rocks between volcanic and intrusive that form a special group—subvolcanic rocks. Igneous rocks are named based on their chemical compositions, which are usually determined in a laboratory by special chemical analysis techniques, and the main criterion for naming a rock is the bulk content of silica (SiO_2). According to the U.S. Geological Survey (USGS) classification, magmatic rocks are divided into ultramafic ($SiO_2 < 45$ wt.%), mafic ($SiO_2 < 45–52$ wt.%), and felsic ($SiO_2 > 69$ wt.%), and there are also two intermediate groups between them (Table 2).

TABLE 2: The USGS Classification of Magmatic Rocks

Type	Ultramafic < 45 wt.% SiO_2	Mafic 45–52 wt.% SiO_2	Intermediate: Mafic 52–63 wt.% SiO_2	Intermediate: Felsic 63–69 wt.% SiO_2	Felsic > 69 wt.% SiO_2
volcanic rocks	komatiite	basalt	andesite	dacite	rhyolite
subvolcanic rocks	picrite basalt, kimberlite	diabase, dolerite			aplite-pegmatite
intrusive (plutonic) rocks	lamproite, peridotite	gabbro	diorite	granodiorite	granite

The oceanic crust and the mantle of the Earth are built up with mostly ultramafic and mafic magmatic rocks, while felsic magmatic rocks compose the continental crust. Intermediate magmatic rocks occur within both continental and oceanic crusts and in island arc environments.

(ii) Sedimentary rocks. These rocks are formed by the deposition of loose particulate materials such as clay, sand, gravel, etc., followed by their lithification ("lithos" is a Greek word meaning stone), which involves sediment compaction, cementation, and even recrystallization (for example, carbonate sediments). The four most important groups of sedimentary rocks are: (a) terrigeneous, (b) chemical-biochemical, (c) organic, and (d) others (Table 3).

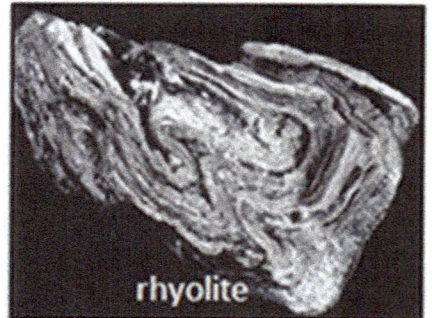

FIGURE 18 The main group of the igneous rocks (intrusive and volcanic)

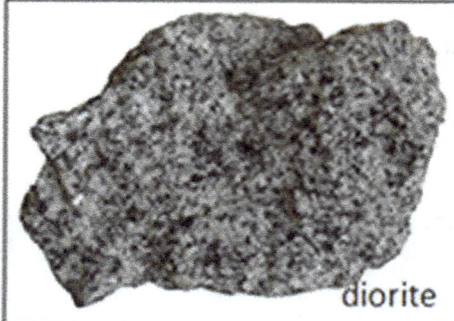

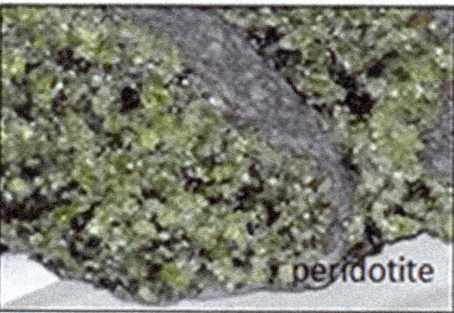

TABLE 3: Classification of Major Types of Sedimentary Rocks

Terrigeneous (detrital or clastic)	Chemical and Biochemical	Organic	Others
Conglomerate, Breccia	Evaporate (rock salt, rock gypsum, travertine)	Coal	Oolitic hematite, Banded iron formations
Sandstone, Greywacke	Carbonate rocks (limestone, dolostone)	Lignite, Bituminous shale	
Siltstone, Claystone	Siliceous (chert, diatomite)		
Shale	Phosphate		

(a) *Terrigeneous* sedimentary rocks are also called detrital or clastic, and they are formed by weathering (disintegration of small pieces or particles) of any pre-existing rocks that have been transported to any depositional basin. They are characterized by a broken or fragmental (clastic) texture consisting of clasts (e.g., larger pieces of gravel), matrix (fine-grained sedimentary particles that surround the clasts), and cement that keeps all of the fragments together. The chemical composition of the cement varies, although in many cases, cements consist of carbonate minerals, silica, and iron oxides. Terrigeneous rocks include conglomerates and breccia, sandstone, greywacke, siltstone, claystone and shales, and some of them are shown in Fig. 19.

(b) *Chemical and biochemical* sedimentary rocks form within a basin from chemical and biological components dissolved in sea water. In reality, it is difficult to separate the chemical and biochemical processes because during sedimentation, the chemical components such as silica and carbonate are intermixed with those that assisted a biological process such as shells growth.

Evaporate. The typical type of chemical sedimentary rocks is the evaporate, which consists of a solid material formed after the evaporation of sea water. The evaporate (Table 3) includes rock salt (NaCl and KCl or their mixture), gypsum ($CaSO_4 \cdot 2H_2O$), and travertine, which consists of $CaCO_3$ deposited around hot springs and inside caves. Although travertine has a chemical composition similar to that of the regular carbonate sediments such as limestone, it is considered to be a separate rock with its own name. Travertine and carbonate sediments are distinguished based on their geological occurrences (Fig. 20).

Carbonate sediments are usually represented by limestone ($CaCO_3$) and dolostone [(Ca,Mg)CO_3] or their mixtures. These rocks are easy to distinguish in the field by testing them with drops of hydrochloric (HCl) acid—the carbonate-rich rocks will fizz due to the following reaction:

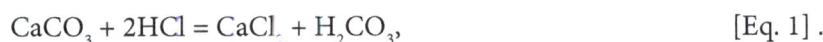

$$CaCO_3 + 2HCl = CaCl_2 + H_2CO_3, \hspace{3cm} [Eq. 1] .$$

FIGURE 19 Terrigenous sedimentary rocks

Many carbonate rocks are rich in fossils, and in such cases, they are called fossiliferrous limestone or dolostone.

Siliceous rocks are dominated by silica (SiO_2), and two major groups, diatomite and chert, are recognized. Diatomite was formed from silica-secreting micro-organisms such as diatoms living in sea water (e.g., radiolarians or sponges), whereas chert consists mostly of microcrystalline quartz formed by chemical reactions of SiO_2 in solution.

Phosphate rocks (phosphorite or phosphatic shales) were formed by precipitation of calcium phosphates (a wide group of minerals containing Ca^{2+} and PO_4^{3-}) derived mainly from micro-organisms.

FIGURE 20 Sedimentary rocks

Organic rocks are represented principally by coal, which consists of organic material that was formed primarily from ancient plant debris accumulated in oxygen-deficient environments such as swamps. In the first stage of coal formation, organic matter is transformed into peat, which eventually becomes lignite (brown coal). The lignite is transformed into soft bituminous coal, which finally becomes hard anthracite coal.

Other rocks include some exotic rock formations such as oolitic hematite and banded iron formations, which are rich in iron and silica.

Sedimentary rocks are formed in a wide variety of environments on the Earth's surface and in sea water as well as in transitional areas between sea and land. Specifically, these environments include rivers (fluvial deposits), alluvial fans, lakes (lacustrine environments), swamps, deserts (aeolian environments), marine environments (continental shelf, deep-sea fans, continental slopes, abyssal plains, and reefs), and transitional environments (beach and barrier islands, deltas, estuaries, and lagoons).

(iii) **Metamorphic rocks** (metamorphism is a Greek word meaning "meta" [change] and "morph" [form]). Rocks of this type are formed from any pre-existing (i.e., parental) rocks (igneous, sedimentary, and even other metamorphic rocks) if they are buried deep in the Earth. There, they are subjected to a solid state recrystallization under the varying temperature and pressure conditions and become metamorphic rocks. Recrystallization is a process of changing the size and chemical compositions of rocks that form new minerals due to chemical reactions. As a result, the common textural characteristics and appearance of the rocks may also be significantly transformed. Metamorphic rocks are associated with different geological settings and plate-tectonic regimes. They occur within shields and cratons (the ancient, stable parts of the continental lithosphere) and at convergent boundaries—ocean-continent, ocean-ocean, and continent-continent collisions. In active plate-tectonic settings, the subduction zone is a place where rocks are subjected to high-pressure and high-temperature metamorphism. Then, due to tectonic exhumation, the metamorphic rocks are eventually returned to the Earth's surface, where they are amalgamated with other rocks in the collision orogens.

The metamorphism may completely obliterate all textural and mineralogical characteristics of the parental rocks, and because of that, they have their own specific names. For example, metamorphosed granite is called gneiss, or granite-gneiss, to emphasize that this gneiss originated from granite. Also, if some specific mineral is present, its name is added as a prefix (e.g., garnet gneiss); (Fig. 21). Shale, after metamorphism, becomes a schist, and because of its special chemical and mineralogical composition, it can be called garnet schist or garnet-mica schist. Furthermore, if basalt (a volcanic rock) is metamorphosed, it is called amphibolite, blue schist, or eclogite (Fig. 21). The final name of the metamorphosed basalt depends on the temperature and pressure conditions at which a parental basalt was recrystallized and on the details of its primary geochemistry. Metamorphosed limestone is called marble, and metamorphosed chert is called, for example, quartzite because it consists mostly of mineral quartz. Unfortunately, in many other cases, there are no specific rules for the naming of metamorphic rocks, and the existing nomenclature approach should be used to avoid misunderstanding (http://pubs.usgs.gov/of/2004/1451/sltt/appendixB/appendixB.pdf).

1.2.2. Rock cycle

The rock cycle is the term that allows us to bring together rock types and the continuous processes of their formation, erosion, destruction, and eventual recycling (melting and/or full recrystallization). The process is nonlinear, and there are no sequences that must be followed or specific timing and duration that the rocks will remain in one form or another. The three types of solid rock—igneous, sedimentary, and metamorphic—may be transformed into one another

FIGURE 21 Metamorphic rocks

by different actions of plate tectonics and surface processes (Fig. 22). For example, subducted sediments will be recrystallized into metamorphic rocks during their residence in deep subduction zones, where temperatures and pressures are high enough to completely obliterate their primary sedimentary origin. These metamorphic rocks will be returned back to the Earth's surface, where they will be further eroded, disintegrated by weathering, and eventually transformed

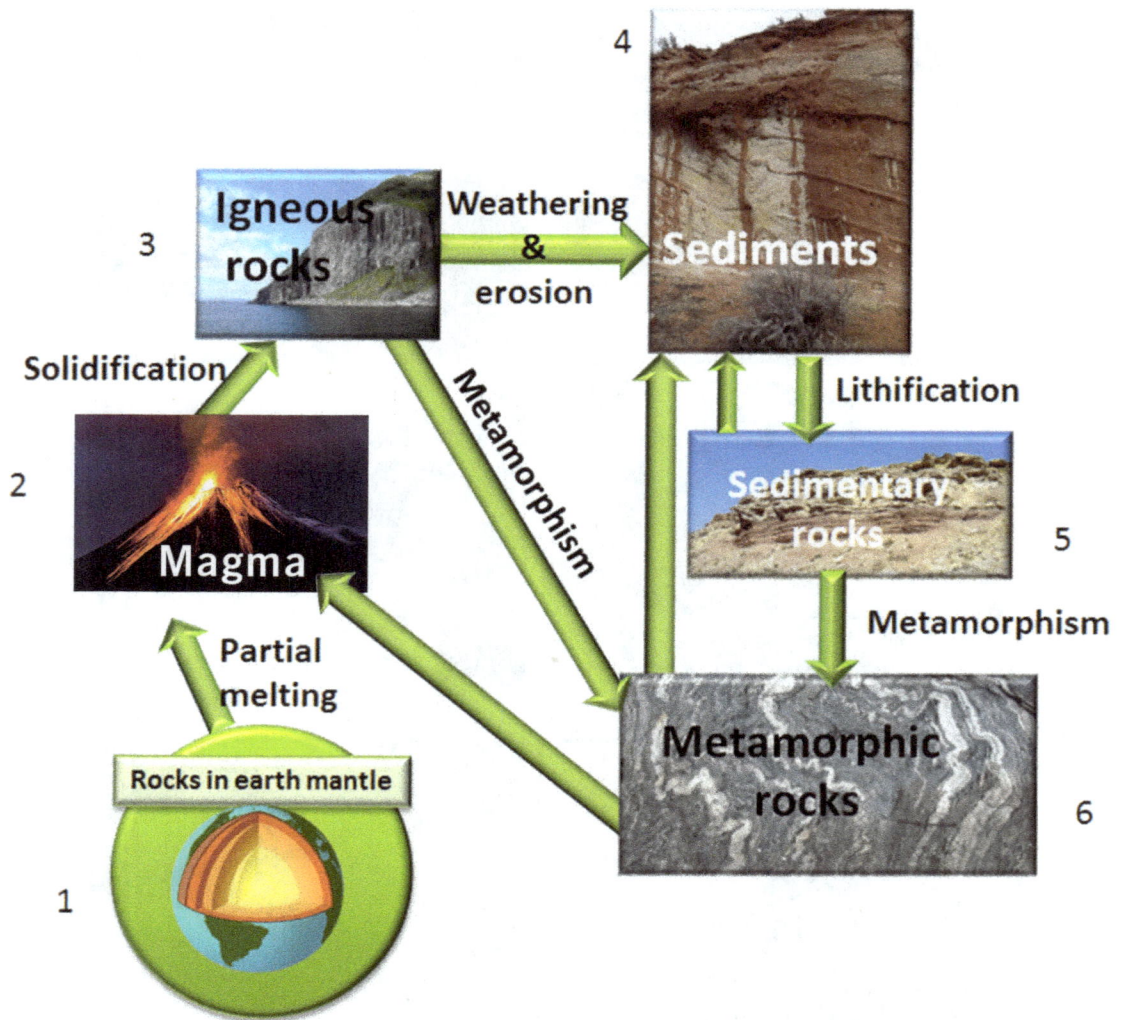

FIGURE 22   Rock cycle

again into sedimentary rocks. Both sedimentary and metamorphic rocks in the situation of ocean-continent collision, for example, will be molten and returned back to the surface through volcanic eruptions to be solidified and eventually converted again into the sedimentary rocks (Fig. 22). The rock-cycling diagram is an important visualization of the Earth as a continuous system that consists of many interacting geological processes that integrate themselves to form a complex whole. Behind this simple rock-cycle diagram (Fig. 22), geologists see interactions among Earth's water, air, land, and deep-Earth processes, led by plate tectonics and internal heat energy that cause rocks to change from one type to another.

1.2.3. Weathering processes

Weathering is a process of disintegration of the solid rocks exposed on the Earth's surface that is followed by transportation of the loosened particles by water, wind, and ice. There are two types of rock weathering: mechanical and chemical.

Mechanical weathering leads to the disintegration of the rocks into smaller and smaller fragments and particles, mostly by frost action (Bland and Rolls, 1998). However, there are several physical processes that can cause mechanical weathering: (i) frozen water action, (ii) contraction and expansion caused by day-night thermal gradients, (iii) exfoliation, and (iv) living organism and plant actions.

(i) Frozen water action. Surface and atmospheric water trickles down into the abundant microcracks and pores in rocks. If there are sharp changes in the day-night temperatures, such as movement back and forth over the freezing point, the water will freeze in the night, causing a change in the volume by 10% or more. This is because the water expands when it freezes and occupies a greater volume than when it is in liquid form. Due to repetition of this process, the rocks will eventually be broken into small, angular fragments. Such weathering is very dynamic in high mountain deserts, where day-night temperature changes are significant.

(ii) Expansion and contraction. This mechanism includes expansion and contraction of the rocks caused by thermal heating and cooling. It is closely associated with frost action, though no moisture is required, only heating and cooling.

(iii) Exfoliation. This form of weathering leads to the formation of curved plates of rock that eventually are stripped from the host-rock body. As a result, exfoliation domes, dome-like hills, or rounded boulders will appear within massif rocks, mostly of granitic composition. The mechanism of exfoliation is still not completely understood, but it has been suggested that it is probably due to "unloading," or drastically reduced pressure at the Earth's surface during very slow uplifting of the granitic domes. Granitic massifs in Joshua Tree and Yosemite national parks in California and the batholith in the Quiet City of Rocks near Oakley, Idaho, represent exceptional examples of exfoliation weathering.

(iv) Actions of plant roots and living organisms. Plant roots and burrowing animals also contribute to rock cracking and disintegration. Lichens and mosses grow on rocks, wedging their tiny roots into pores and crevices, and larger trees and shrubs may grow in the cracks of boulders. Ants, earthworms, rabbits, woodchucks, and other animals dig holes in the soil, providing pathways for air and water to reach the bedrock and weather it. Finally, bacteria also contribute to the weathering of rocks and minerals (Uroz et al., 2009). All of these things increase the total bedrock surface area that is exposed to weathering processes.

Chemical weathering. This is a process of weakening and subsequent disintegration of rocks by chemical reactions. Chemical weathering is a combination of the action of rainwater, oxygen, carbon dioxide, and the acids produced by plant and animal decay, and the rate of chemical weathering is accelerated in warm and humid climates (Kump et al., 2000). Chemical weathering is controlled by three main reactions—oxidation, hydrolysis, and carbonation (carbonic acid activity)—which may take place together or in different combinations and sequences.

Oxidation. The oxygen available in the air is dissolved in water and "attacks" exposed rocks that contain metallic elements abundant in silicate minerals. Iron, a commonly known element, produces red or rust coloring in rocks if it is oxidized. This reaction is represented by Eq. 2:

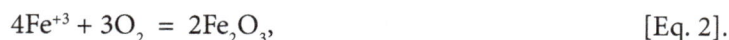

$$4Fe^{+3} + 3O_2 = 2Fe_2O_3, \qquad [Eq.\ 2].$$

Hydrolysis. Chemical reactions of this type are governed by water, and the typical example is hydrolysis of silicate minerals into clays, which are the final products of the reaction of the mineral orthoclase ($KAlSi_3O_8$) with water. Orthoclase is a common mineral in granitic rocks that, during chemical weathering by a hydrolysis reaction, is broken down and produces kaolinite (a mineral of the clay group), silicic acid, and potassium, as represented in Eq. 3:

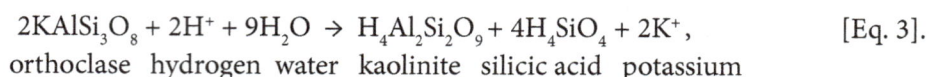

$$2KAlSi_3O_8 + 2H^+ + 9H_2O \rightarrow H_4Al_2Si_2O_9 + 4H_4SiO_4 + 2K^+, \qquad [Eq.\ 3].$$
orthoclase hydrogen water kaolinite silicic acid potassium

Carbonation. The carbonation reaction takes place when carbonic acid reacts with rock-forming minerals. Carbonic acid dissolves or breaks down minerals in rocks, making them weak and less resistant to any other type of weathering or erosion. Carbonic acid is formed by the reaction between, for example, rain water and carbon dioxide emitted into the atmosphere by industrial activity. This reaction is represented by Eq. 4:

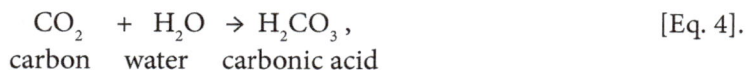

$$CO_2 + H_2O \rightarrow H_2CO_3, \qquad [Eq.\ 4].$$
carbon water carbonic acid

Carbonic acid reacts with the mineral calcite ($CaCO_3$) by breaking down its structure, producing elemental calcium and a new portion of bicarbonate, as shown in Eq. 5:

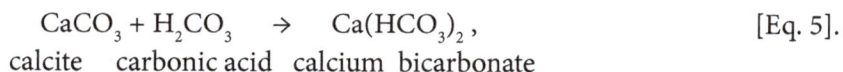

$$CaCO_3 + H_2CO_3 \rightarrow Ca(HCO_3)_2, \qquad [Eq.\ 5].$$
calcite carbonic acid calcium bicarbonate

1.2.4. Geological hazards

The Earth is dynamic, and its physical and chemical processes continually act on or beneath its surface. These processes include core-mantle activity, mantle plumes and convection, radioactive decay, lithospheric plate movements, magmas generation, and the interactions between the geosphere, the atmosphere, and the hydrosphere. These processes produce geological hazards of different scales and different amounts of destructive power.

There are two main categories of geological hazards: *visible hazards* and *silent hazards*. Visible geological hazards such as earthquakes, tsunamis, volcano eruptions, landslides, flooding, karst phenomena (sinkholes), dust storms, and others related to this group act instantly and can cause tremendous loss of life, destruction of property, loss of vital services, social and economic disruptions, and environmental damages. Silent geological hazards include geochemical contaminations of land and water reservoirs with arsenic, mercury, radon, asbestos, and other toxic chemical agents and particulate matters. They mostly affect environments, bio-ecosystems and human health, and although they are invisible at the beginning, they can be as harmful and destructive as visible geological hazards.

Although the greatest portion of the global human populations resides in areas of active geological processes and intensively uses the Earth's resources, there is a lack of widespread, multidisciplinary education that would help us to understand the nature of the geological and geochemical processes giving rise to disasters. Because it is difficult to predict when natural disasters will take place and how long they will last, the reduction of such uncertainty through education becomes an important issue. Widespread knowledge of the basic geological processes and properties of the Earth's materials, in combination with existing technologies, could assist rescue agencies and civil defense organizations to more quickly obtain understanding of the scale of damage caused by geological hazards and how to cope with them.

References

Alfè, D., M. J. Gillan, and G. D. Price. 1999. The melting curve of iron at the pressures of the Earth's core from *ab-initio* calculations. *Nature* 401:462–464.

Alwater, T. 1970. Implications of plate tectonics for the Cenozoic tectonic evolution of western North America. *Geological Society of America Bulletin* 8:3513–3536.

Bland, W., and D. Rolls. 1998. *Weathering: An introduction to the scientific principles.* New York: Arnold.

Bilham, R. 2010. Lessons from the Haiti earthquake. *Nature* 463:878–879.

Boehler, R. 2000. High-pressure experiments and the phase diagram of lower mantle and core materials. *Review of Geophysics* 38:221–245.

Carlson, R. L., and G. S. Raskin. 1984. Density of the ocean crust. *Nature* 311:555–558.

Dasgupta, R., A. Mallik, A. Tsuno, A. C. Whithers, G. Hirth, and M. M. Hirschman. 2013. Carbon-dioxide-rich silicate melt in the Earth's upper mantle. *Nature* 493:211–215.

Dobrzhinetskaya, L., and S. W. Faryad. 2011. Frontiers of ultrahigh-pressure metamorphism: View from field and laboratory. In: L. F. Dobrzhinetskaya, S. W. Faryad, S. Wallis, and S. Cuthbert (Eds.), *Ultrahigh-pressure metamorphism: 25 years after the discovery of coesite and diamond.* London: Elsevier, pp. 1–39.

Duffy, T. S. 2008. Mineralogy at the extremes. *Nature* 451:269–270.

Eberhard, M. O., S. Baldridge, J. Marshall, W. Mooney, and G. J. Rix. 2010. The MW 7.0 Haiti earthquake of January 12, 2010. USGS/EERI advance reconnaissance team report. U.S.Geological Survey open-file report 2010–1048. http://pubs.usgs.gov/of/2010/1048/.

Ernst, W. G., and J. G. Liou. 2008. High- and ultrahigh-pressure metamorphism—past results, future prospects. *American Mineralogist* 93:1771–1786.

Fowler, C. M. R. 2005. *The solid Earth: An introduction to global geophysics*. Cambridge, U. K.: Cambridge University Press.

Gerya, T. V., B. Stöckhert, and A. L. Perchuk. 2002. Exhumation of high-pressure metamorphic rocks in a subduction channel: A numerical simulation. *Tectonics* 21:1–19.

Gillan, M. J., G. A. De Wijs, G. Kresse, L. Vočadlo, D. Dobson, D. Alfè, and G. D. Price. 1998. The viscosity of liquid iron at the physical conditions of the Earth's core. *Nature* 392:805.

Hayes, G. P., R. W. Briggs, A. Sladen, E. J. Fielding, C. Prentice, K. Hudnut, P. Mann, F. W. Taylor, A. J. Crone, R. Gold, T. Ito, and M. Simons. 2010. Complex rupture during the 12 January 2010 Haiti earthquake. *Nature Geoscience* 3:800–805.

Hess, H. 1962. History of ocean basins. In: *Petrologic studies, a volume in Honor of A. F. Buddington*. New York: Geological Society of America, pp. 599–620.

Hofmann, A. W. 1988. Chemical differentiation of the Earth: The relationship between mantle, continental crust, and oceanic crust. *Earth and Planetary Science Letters* 90:297–314.

Jackson, I. 2000. *The Earth's mantle—composition, structure, and evolution*. Cambridge, U. K.: Cambridge University Press.

Koketsu, K., U. Yokota, N. Nishimura, Y. Yagi, S. Miyazaki, K. Satake, Y. Fujii, H. Miyake, S. Sakai, Y. Yamanaka, and T. Okada. 2011. Aunified source model for the 2011 Tohoku earthquake. *Earth and Planetary Science Letters* 310:480–487.

Kump, L. R., S. L. Brantley, and M. A. Arthur. 2000. Chemical weathering, atmospheric CO_2, and climate. *Annual Review of Earth Planetary Science* 28:611–667.

Lay, T., H. Kanamori, C. Ammon, M. Nettles, S. Ward, R. Aster, S. Beck, S. Bilek, M. R. Brudzinski, R. Butler, H. R. DeShon, G. Ekström, K. Satake, and S. Sipkin. 2005. The great Sumatra-Andaman earthquake of 26 December 2004. *Science* 308:1127–1133.

Lean, J. 2005. Living with a variable sun. *Physics Today* 58:32–38.

Müller, R. D., M. Sdrolias, C. Gaina, and W. R. Roest. 2008. Age, spreading rates and spreading symmetry of the world's ocean crust. *Geochemistry, Geophysics, Geosystems* 9: Q04006, doi:10.1029/2007GC001743.

North American geologic-map data model - steering committee science language for composite-genesis materials, v. 1.0. *Science Language Technical Team, Composite-Genesis Subgroup* 12/18/04. http://pubs.usgs.gov/of/2004/1451/sltt/appendixB/appendixB.pdf.

Rudnick, R. L., and D. M. Fountain. 1995. Nature and composition of the continental crust: A lower crustal perspective. *Review in Geophysics* 33:267–309.

Sato, M., T. Ishikawa, N. Ujihara, S. Yoshida, M. Fujita, M. Mochizuki, and A. Asada. 2011. Displacement above the hypocenter of the 2011 Tohoku-Oki earthquake. *Science* 332:1395.

Uroz, S., C. Calvaruso, M-P Turpault, and P. Frey-Klett. 2009. Mineral weathering by bacteria: Ecology, actors and mechanisms. *Trends in Microbiology* 17:378–87.

Vigny, C., W. J. F. Simons, S. Abu, R. Bamphenyu, C. Satirapod, N. Choosakul, C. Subarya, A. Socquet, K. Omar, H. Z. Abidin, and B. A. C. Ambrosius. 2005. Insight into the 2004 Sumatra–Andaman earthquake from GPS measurements in Southeast Asia. *Nature* 436:201–206.

Vine, F. J., and D. I-I. Matthews. 1963. Magnetic anomalies over oceanic ridges. *Nature* 199:947.

Wegener, A. 1924. *The origin of continents and oceans*. York, U. K.: Methuen & Co.

Image Credits

Figure 1: Adapted from: Copyright © Lsmpascal (CC BY-SA 3.0) at http://commons.wikimedia.org/wiki/File:Telluric_planets_size_comparison.jpg.

Figure 4: United States Geological Survey, "World geological provinces," http://commons.wikimedia.org/wiki/File:World_geologic_provinces.jpg. Copyright in the Public Domain.

Figure 6: Copyright © National Geophysical Data Center, National Oceanic and Atmospheric Administration, U.S. Department of Commerce (CC BY-SA 3.0) at http://commons.wikimedia.org/wiki/File:Age_of_oceanic_lithosphere.jpg.

Figure 7: Anasofiapaixao, "earth-cutaway-schematic-english," http://en.wikipedia.org/wiki/File:Earth-cutaway-schematic-english.svg. Copyright in the Public Domain.

Figure 10: Adapted from: C. R. Scotese, "Modern reconstruction of Pangaea, ca. 255 million years ago," Adapted from: http://geomaps.wr.usgs.gov/parks/pltec/sc255ma.html.

Figure 11: NASA/GSFC/Robert Simmon, "Ridge render," http://commons.wikimedia.org/wiki/File:Ridge_render.jpg. Copyright in the Public Domain.

Figure 12: Chmee2, "Oceanic.Stripe.Magnetic.Anomalies.Scheme," http://commons.wikimedia.org/wiki/File:Oceanic.Stripe.Magnetic.Anomalies.Scheme.svg. Copyright in the Public Domain.

Figure 13: "Tectonic plates," http://commons.wikimedia.org/wiki/File:Tectonic_plates.svg. Copyright in the Public Domain.

Figure 15: United States Geological Survey, "The Mid-ocean Ridge," http://pubs.usgs.gov/gip/dynamic/baseball.html. Copyright in the Public Domain.

Figure 16: United States Geological Survey, "Convergent Boundary Formation," http://pubs.usgs.gov/gip/dynamic/understanding.html. Copyright in the Public Domain.

Figure 18a: National Park Service, U.S. Department of the Interior, "Granodiorite hand specimen," http://www.nps.gov/goga/learn/education/granite-and-granodiorite-faq.htm. Copyright in the Public Domain.

Figure 18b: Copyright © Siim Sepp (CC BY-SA 3.0) at http://commons.wikimedia.org/wiki/File:Diorite.jpg.

Figure 18c: Mark A. Wilson, "GabbroRockCreek1," http://commons.wikimedia.org/wiki/File:GabbroRockCreek1.jpg. Copyright in the Public Domain.

Figure 18d: Copyright © Dinshaw Dadachanji (CC BY-SA 3.0) at http://www.newworldencyclopedia.org/entry/File:Peridot_in_basalt.jpg.

Figure 18e: Copyright © Michael C. Rygel (CC BY-SA 3.0) at http://commons.wikimedia.org/wiki/File:Flow_banded_rhyolite.JPG.

Figure 18f: "andesite," http://volcanoes.usgs.gov/Imgs/Jpg/Lassen/And-Brokeoff_large.jpg. Copyright in the Public Domain.

Figure 18g: Copyright © Zureks (CC BY-SA 3.0) at http://commons.wikimedia.org/wiki/File:Igneous_rock_Santoroni_Greece.jpg.

Figure 18h: GeoRanger, "KomatiiteCanada 682By512," http://commons.wikimedia.org/wiki/File:KomatiiteCanada_682By512.jpg. Copyright in the Public Domain.

Figure 19a: Jstuby, "Lehigh conglom," http://commons.wikimedia.org/wiki/File:Lehigh_conglom.jpg. Copyright in the Public Domain.

Figure 19b: United States Geological Survey, "Aquia Creek Sandstone," http://gallery.usgs.gov/images/01_08_2013/x16Fw32VUp_01_08_2013/large/AquiaCreekSandstone.JPG. Copyright in the Public Domain.

Figure 19c: Mark A. Wilson, "DebrisFlowDepositRestingSpringsPass," http://commons.wikimedia.org/wiki/File:DebrisFlowDepositRestingSpringsPass.JPG. Copyright in the Public Domain.

Figure 19d: Copyright © Pollinator (CC BY-SA 3.0) at http://en.wikipedia.org/wiki/File:Shale_8040.jpg.

Figure 20a: Copyright © Tasma3197 (CC BY-SA 3.0) at http://commons.wikimedia.org/wiki/File:Mammoth_Hot_Springs_Travertine_Terrace.JPG.

Figure 20b: Copyright © GFDL (CC BY-SA 3.0) at http://commons.wikimedia.org/wiki/File:Alabaster.jpg.

Figure 20c: United States Geological Survey and the Mineral Information Institute, "ChertUSGOVjpg," http://commons.wikimedia.org/wiki/File:ChertUSGOVjpg.jpg. Copyright in the Public Domain.

Figure 20d: Mark A. Wilson, "CoombefieldBlocks," http://commons.wikimedia.org/wiki/File:CoombefieldBlocks.jpg. Copyright in the Public Domain.

Figure 20e: Copyright © Mizu_Basyo (CC BY-SA 3.0) at http://upload.wikimedia.org/wikipedia/commons/f/f5/Mongolian_rock_salt.jpg.

Figure 20f: United States Geological Survey, "Figure 6 - Pliocene Panaca Formation in southeastern Nevada," http://minerals.usgs.gov/news/newsletter/v2n1/4indmins.html. Copyright in the Public Domain.

Figure 21a: United State Geological Survey, "Skagit-gneiss-Cascades," http://en.wikipedia.org/wiki/Rock_(geology)#mediaviewer/File:Skagit-gneiss-Cascades.jpg. Copyright in the Public Domain.

Review Questions

1. Name and describe the size of the terrestrial group planets.

2. What is the age of the Earth?

3. There are six eras that characterize geological time and the evolution of the Earth starting from the beginning:

   a. Hadean era
   b. Archean era
   c. Proterozoic era
   d. Paleozoic era
   e. Mesozoic era
   f. Cenozoic era

   Briefly describe the main geological events that distinguish these eras from one another. Which era was the most long-lasting? Which one was the shortest?

4. What are the most common elements making up the Earth?

5. What are the three major layers of the Earth?

6. What are the two types of crust? How do they differ physically and compositionally? What is the average density of the oceanic crust? What is the average density of the continental crust? Why is the oceanic crust denser than the continental crust?

7. What is the lithospheric plate? How many lithospheric plates are on the Earth? What are the differences between the continental and oceanic lithospheres?

8. What is the mantle of the Earth, and what are its two subdivisions? What is the thickness of the mantle? What is the average chemical composition of the mantle? What is the average density of the mantle? Is Earth's mantle

hot? If yes, what is the temperature at a depth of 620–660 km? What is the temperature at a depth of 2700 km?

9. What are the main characteristics of the asthenosphere, and what is its role in plate tectonics?

10. What are the composition and physical properties of the core, including both the inner and outer core?

11. What is the theory of plate tectonics? Who formulated the primary concept of plate tectonics? Who formulated the final concept of plate tectonics, and how does the final concept differ from the primary one?

12. What are continental drift and sea-floor spreading?

13. How many lithospheric plates are there on the Earth's surface?

14. What are the three types of plate boundaries, and how are they different?

15. How do plate boundaries create mountains and volcanoes?

16. There are three types of plate-convergent boundaries: ocean-continent collision, continent-continent collision, and ocean-ocean collision. What are their main characteristics?

17. What is an orogenic belt or orogen? Provide the most typical examples of orogenic belts.

18. What is a mid-ocean ridge?

19. Is the collision of oceanic plates a "quiet" process? How are collisions of oceanic plates related to geological hazards?

20. What is the main driving force behind plate tectonics?

21. What are the three main types of rock?

22. The main criterion chosen to name magmatic rocks is the bulk content of silica (SiO_2). What are the names of the magmatic rocks which, according to the USGS classification, contain:

i.   < 45 wt.% SiO_2
ii. 45–52 wt.% SiO_2
iii. 52–63 wt.% SiO_2
iv. 63–69 wt.% SiO_2
v.   > 69 wt.% SiO_2

Where does each occur within the Earth?

23. What are the names, characteristics, and occurrences of the major types of sedimentary rocks? What are examples of each that you have seen in lecture and discussion sessions?

24. What are metamorphic rocks? Where do they occur?

25. What is the rock cycle? How are rocks recycled and converted into new rocks?

26. What are the processes of rock weathering? Review the following types of weathering:

i.   mechanical (frozen water action; expansion and contraction; exfoliation; actions of plants and living organisms);
ii. chemical (oxidation, hydrolysis, carbonation).

27. Provide examples of the main reactions that govern chemical weathering, and explain how they fit into the rock cycle.

28. Describe geological hazards, both visible and silent, and tell why we should study them.

Quizzes (see answers on page 307)

1. Crust is neither destroyed nor formed along
   a. Oceanic-continental boundaries
   b. Transform boundaries
   c. Convergent boundaries
   d. Oceanic-oceanic boundaries

2. When two plates move toward each other, the following will occur:
   a. Transform boundary
   b. Magnetic reversal
   c. Rift valley
   d. Convergent boundary
   e. Divergent boundary
   f. Mid-ocean ridge

3. Which of these did Wegener use to support his theory of continental drift?
   a. Similar rocks and similar fossils on different continents
   b. Convection currents in the asthenosphere and sea-floor spreading
   c. Fossils from ancient organisms and convection currents in the asthenosphere
   d. Sea-floor spreading and similar rocks on different continents

4. Early studies of the ocean floor helped to develop the theory of plate tectonics because they showed that:
   a. Oceanic crust is younger near a deep-ocean trench
   b. Oceanic crust is oldest near the center of a mid-ocean ridge
   c. Oceanic crust is nearly the same age as continental crust
   d. The age of oceanic crust increases with its distance from a mid-ocean ridge

5. Magmatic rocks are called ultramafic because they contain:
   a. > 45 wt.% CaO
   b. ~42–50 wt.% SiO_2
   c. <45 wt.% SiO_2
   d. >65 wt.% SiO_2

6. Which era characterizes the earliest stage of the Earth's development?
   a. Proterozoic
   b. Cenozoic
   c. Hadean
   d. Archean
   e. Mesozoic
   f. None of the above

Minerals Around Us

Chapter 2

Earth's materials—from smaller to larger scale 2.1.

Human life is surrounded by the Earth's materials that have originated from rocks and minerals. Without these materials, there would be no buildings in which to live and work, no telephones or computers with which to communicate, and no cars, planes, or trains to provide transportation. To extend the significance of the Earth's materials, it is also worth saying that if there were no rocks and minerals, there would be no food, because soil is a mixture of disintegrated minerals and organic matter. Without minerals, there would not be toothpaste, pencils, silverware, makeup, jewelry, paintings, sculptures, and so many other things that we use or enjoy in our daily lives. We live in an era of intensive use of the Earth's materials, from smaller scales such as for our personal lives to larger scales such as fuel and energy production and technological progress. Interactions of humans with mineral resources have both beneficial and harmful effects, and thus, the Earth's materials should be considered as a fundamental component of public health and the environment. Many epidemic diseases and health conditions of populations can be addressed to utilization of rocks and minerals. Natural geological processes such as earthquakes, volcanic eruptions, and weathering disintegrate rocks and minerals and create particulate materials, such as tiny fibers and toxic gases, that are well known to be hazardous materials. The materials, in addition to causing harm to local populations, may be transported through the atmosphere over long distances and cause intercontinental pollutions of land, oceans, and air. Anyone involved with or concerned about the long-term effects of minerals on public health and the environment should start from gaining an understanding of the fundamental principles of mineralogy.

2.2. Principles of mineralogy

2.2.1. Chemical elements

If you analyze the periodic table of chemical elements (Fig. 23), you will see that currently, there are a total of 118 different elements. About ninety of them are found inside terrestrial rocks and minerals, while the remaining twenty-eight elements have been synthesized in laboratories and nuclear accelerators.

Since the Earth's major layers (the continental and oceanic crusts and the upper and lower mantle) consist of rocks, and because rocks, in turn, consist of different minerals, the ninety chemical elements should be considered to be the building blocks of our planet (Fig. 24). Amazingly, the different combinations of only ninety chemical elements were able to create the diversity of terrestrial mineralogy, which now includes more than 3800 mineral species. Chapter 1 has already presented the most abundant elements of the continental and oceanic crusts (Fig. 5) and the mantle (Fig. 8). Similar elements compose extraterrestrial bodies and planets and cosmic space, and some of them are essential for biological life.

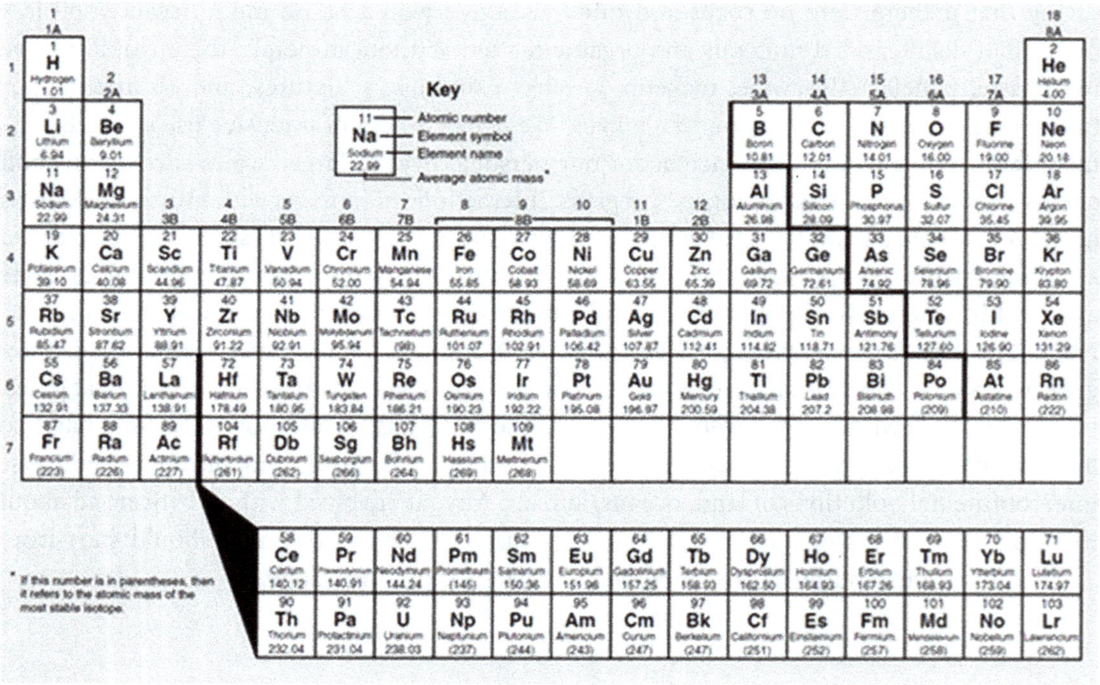

FIGURE 23 Periodic Table of the Elements

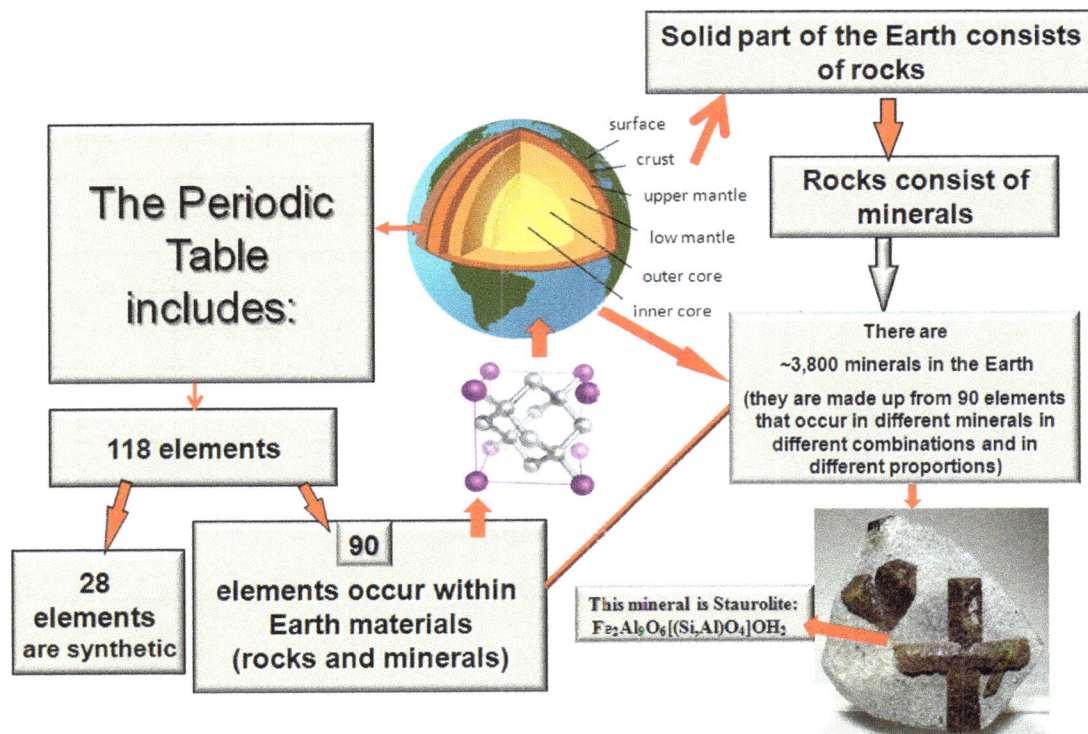

FIGURE 24   Elements, minerals, and rocks are building blocks of the Earth

It is interesting that 99% of the mass of the human body is built up by only six elements (in at.%): oxygen (65), carbon (18.5), hydrogen (9.5), nitrogen (3.2), calcium (1.5), and phosphorous (1.0); (Fig. 25). The rest includes five other essential elements that occur in smaller concentrations (in at.%): potassium (0.4), sulfur (0.3), sodium (0.2), chlorine (0.2), and magnesium (0.1). Trace elements that occur in extremely small quantities (percent per million [ppm], ‰) include boron, cobalt, chromium, copper, fluoride, iodine, iron, manganese, molybdenum, selenium, silicon, tin, vanadium, and zinc (Fig. 25).

2.2.2. Mineral definition

According to the Mineralogical Society of America (MSA), a mineral substance is defined as a naturally occurring, homogeneous solid, inorganically formed, with a well-defined chemical composition (or range of compositions) and an ordered atomic arrangement (structure) that has been formed by geological processes, either on the Earth or in extraterrestrial bodies. To understand this definition, the five most important characteristics of minerals are discussed below. Indeed, they represent the basic knowledge that is required for understanding minerals.

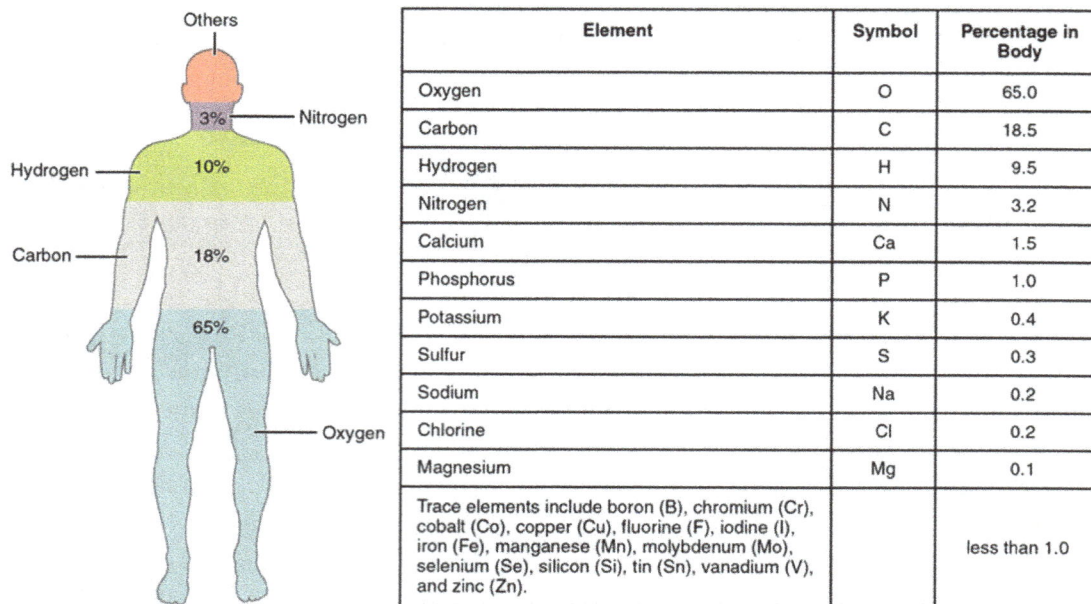

Element	Symbol	Percentage in Body
Oxygen	O	65.0
Carbon	C	18.5
Hydrogen	H	9.5
Nitrogen	N	3.2
Calcium	Ca	1.5
Phosphorus	P	1.0
Potassium	K	0.4
Sulfur	S	0.3
Sodium	Na	0.2
Chlorine	Cl	0.2
Magnesium	Mg	0.1
Trace elements include boron (B), chromium (Cr), cobalt (Co), copper (Cu), fluorine (F), iodine (I), iron (Fe), manganese (Mn), molybdenum (Mo), selenium (Se), silicon (Si), tin (Sn), vanadium (V), and zinc (Zn).		less than 1.0

FIGURE 25 The main elements that compose the human body

2.2.2.1. Naturally occurring

This means that a mineral must be formed by any natural (usually geological) process inside the Earth or on the Earth's surface. This part of the definition is correct because minerals compose rocks, and the Earth is built up from the rocks, except for the outer core, which is a melt. Extraterrestrial bodies and planets also consist of minerals, and there is no controversy about calling the processes of mineral formation in these extraterrestrial bodies and planets geological because "rocky" meteorites, asteroids, and planets consist of the same minerals that exist on the Earth. However, the kinetics of chemical reactions, pressures, and temperatures of their formation may be different from those on Earth.

The words "naturally occurring," nevertheless, are widely debated with respect to the definition of minerals. Should we call a mineral such a material as a kidney stone in a human body? Kidney stones in general are minerals of the phosphate group, which includes apatite $[Ca_5(PO_4)_3(OH,F,Cl)]$ and whitelockite $[Ca_9(Mg,Fe)(PO_4)_6(PO_3OH)]$. The process of kidney-stones formation is natural, but it is a biological process rather than a geological process, and biological processes are organic by their nature. It is better to call them "kidney stone minerals."

Another question related to the definition of minerals as "naturally occurring" is how we should classify gemstones or any other mineral-like industrial compounds synthesized in a laboratory. In most cases, minerals created in the laboratory have almost the same physical and chemical properties as their counterparts found in nature. For example, a natural mineral ruby (Al_2O_3) of gem quality has a red color, and its synthetic counterpart synthesized in the laboratory has the same color and the same physical and chemical properties. A synthetic diamond often looks the same as a natural diamond, even though they have been formed in different environments—the first was synthesized by humans in a laboratory, and the second was created naturally by a geological process. Thus, by definition, rubies and diamonds synthesized in the laboratory are not minerals, so it is correct to call them "synthetic rubies" and "synthetic diamonds" to distinguish them from rubies and diamonds that occur within natural rocks. The same principles should be applied to any other minerals and their synthetic counterparts.

2.2.2.2. Homogeneous solid

Crystalline material, to be classified as a mineral, must be solid and homogeneous. Within ~3800 known minerals, there are two exceptions—mercury (Hg), which belongs to the group of native elements, and water ice (H_2O-ice), which belongs to the group of hydroxides. Mercury is the only metal that occurs in natural rocks in the form of a liquid, whereas H_2O-ice exists in a solid state at the temperature of 0°C and below. Acceptance of mercury as a mineral has more historical roots than connections to the scientific definition because mercury was one of the important alchemical substances in ancient times. Also, interestingly enough, many people do not think of H_2O-ice as a mineral. However, in reality, H_2O-ice has been officially approved as a mineral by the Commission on New Minerals, Nomenclature, and Classification (CNMNC) of the International Mineralogical Association (IMA), which was founded in 1958. Commission members, representatives of thirty-four countries including the U.S., vote separately on each new mineral and its name. After approval, the discoverers of the new mineral are expected to publish a full description of the mineral in a scientific journal within two years or risk forfeiting the Commission's approval. Detailed descriptions of all known minerals may be found in the *Handbook of Mineralogy* (http://www.handbookofmineralogy.org/search.html?p=all).

Minerals exhibit diverse physical properties that are used for their identification (e.g., Shumann, 1993; Levin, 2006). The physical properties include the following.

Crystal structure. This is the order of atom arrangements that determine a crystal's form and symmetry (see Section 2.1.1.5).

Hardness. The hardness of minerals is compared on the Mohs hardness scale, a ten-point scale running from the softest (1) mineral, talc, to the hardest (10) mineral, diamond (Fig. 26).

Mohs hardness scale

1. — talc
2. — gypsum (human fingernails have a hardness of 2.5)

Mineral	Mohs Hardness	Image
Talc	1	
Gypsum	2	
Calcite	3	
Fluorite .	4	
Apatite	5	
Feldspar	6	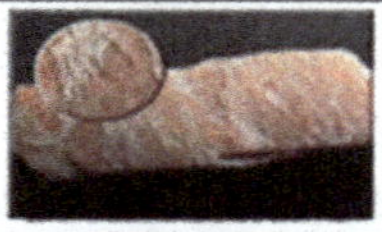
Quartz	7	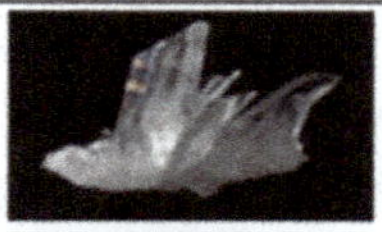
Topaz	8	
Corundum	9	
Diamond	10	

FIGURE 26   Mohs' hardness scale of minerals

3. — calcite (a copper penny also has a hardness of 3)
4. — fluorite
5. — apatite (penknives and glass plates have a hardness of 5.5)
6. — orthoclase
7. — quartz (window glass)
8. — topaz
9. — corundum
10. — diamond

Luster. This is the appearance of a mineral in reflected light.

Color. This is one of the most obvious properties of minerals, though color is not always a reliable tool for minerals identification. There are only a few minerals that have specific characteristic colors—e.g., malachite (green), azurite (intense blue), sulfur (intense yellow). In addition, many minerals can have the same color—i.e., chromium diopside can be as green as malachite, and a white color is typical for gypsum, quartz, halite, and calcite. Finally, one mineral may have many different colors; for example, tourmaline can be black to black-bluish, brown to brown-yellowish, as well as blue, green, red, yellow, pink, green-pink, and colorless.

Streak. This is the color of a mineral left after scratching on an unglazed ceramic plate. The streak color can be different from the color of the hand specimen.

Cleavage. This is the tendency of a mineral to split along certain preferred planar surfaces that correspond to the weakest atomic bonding (Fig. 27).

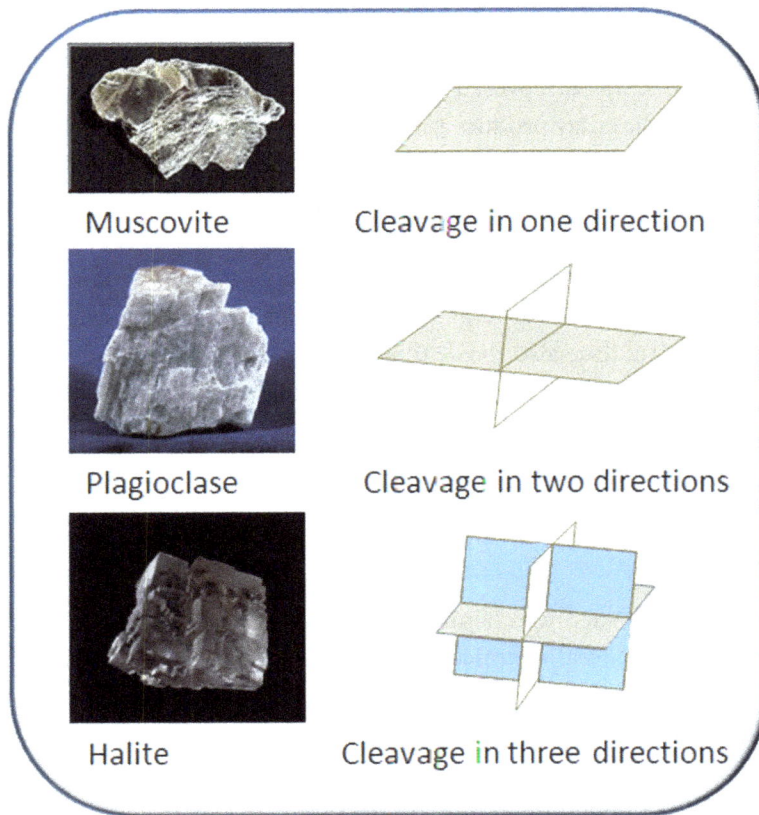

FIGURE 27 Cleavage in minerals

Fracture. This is how a mineral breaks against its natural cleavage planes.

Specific gravity. The density of a mineral compared with water. By "hefting" hand samples, minerals can be put into groupings of low, medium, and high densities. Most rock-forming minerals have a specific gravity of ~2.7 g/cm³, most metallic minerals ~5 g/cm³, and gold 19.3 g/cm³.

Others. Any other remarkable or unique properties, such as a specific odor, "opalescence," magnetism, double refraction, fluorescence, acid test, etc.

2.2.2.3. Inorganically formed

This means that a mineral must not be formed by any other process than inorganic (e.g., biological processes are defined as organic). It should be noted that the inorganic origin definition is not absolutely precise because modern chemistry deals with laboratory synthesis of many organic materials without any biological organisms being involved. The question is whether an organic material always belongs to a biological substance. In addition, a new complication emerged recently when it was discovered that some living organisms (e.g., humans, mollusks, and some forms of bacteria) may produce minerals.

Another example is related to human bones, which consist of the mineral apatite intimately mixed with a biological substance, collagen. Recent studies have shown that the collagen governs the nucleation, growth, and orientation of bone apatite (Wang et al., 2012), and it is impossible to separate or distinguish whether the mineral or organic (biological) substances play the leading role in such processes. An emerging new research direction considering the evolution of minerals over geological time also shows that, indeed, many mineral species were created by only biological processes (Hazen, 2010; Hazen et al., 2013). Therefore, the part of the definition of a mineral as "inorganically formed" awaits a new formulation in the future. Meanwhile, the "inorganic" characteristic may be considered to mean that the mineral does not have C-C double bonds and does not consist of only hydrogen (H).

2.2.2.4. A well-defined chemical composition (or range of compositions)

A mineral must have a "well-defined chemical composition," which means that the composition can be represented by a general chemical formula. For example, mineral quartz, which is a common component of the Earth's continental crust, consists of silicon and oxygen, and its chemical formula is SiO_2. The mineral olivine, the major component of the Earth's mantle, consists of magnesium, iron, silicon, and oxygen, and its general chemical formula is $(Mg,Fe)_2SiO_4$. However, this mineral does not have fixed amounts of Mg and Fe, and their abundance depends on the specific chemical environment in which the olivine was crystallized. If olivine crystallized from Mg-rich silicate melt, it is called forsterite, and its chemical formula is Mg_2SiO_4. If olivine crystallized from Fe-rich silicate melt, its name is fayalite, and its chemical formula is Fe_2SiO_4. Also, it is not difficult to imagine that olivine with a chemical formula $(Mg,Fe)_2SiO_4$ is a mixture of forsterite and fayalite because, in addition to the silica group (SiO_4), which is common to all types of olivine, it contains both Mg (a component of forsterite) and Fe (a component of fayalite). However, such a mixture is not a mechanical one, but rather it is called a *solid solution*. Stated differently, some minerals have a well-defined range of chemical compositions—e.g., solid solution. The principles of writing mineral formulas are slightly different than those for writing formulas for chemical compounds, but in this short review, this subject is omitted.

2.2.2.5. An ordered atomic arrangement (structure)

Minerals are distinguished from any solid, amorphous, or liquid substance by a specifically ordered arrangement of atoms, which is a periodic array of the atoms, called a crystalline structure or a crystalline lattice. To understand how a crystalline lattice is built up, you should consider atoms as being spheres of different radii and imagine a silica tetrahedron (Fig. 28) that consists of one low-radii silicon atom surrounded by four large-radii oxygen atoms. In such a combination, negatively charged anions of oxygen (O^{-2}) bond to positively charged cations of

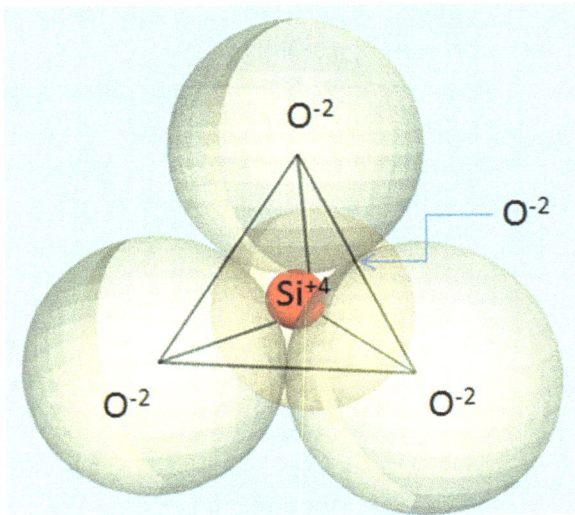

FIGURE 28   Silica tetrahedron—the fundamental building block of the silicate minerals

silicon (Si^{+4}), creating so-called four-fold coordination. In minerals of the silicate group, the silica tetrahedron creates repeatable units linked together, with silica tertrahedra attached to one another by sharing oxygen atoms, while other metallic atoms occupy spaces between the oxygen atoms. Because silicon atoms can share oxygen atoms, there are a variety of ways to build silicate structures, and this gives rise to a set of basic groups of silicate minerals: framework silicates (feldspar and quartz—major minerals of the continental crust), isosilicates (olivine—the main mineral of the Earth's mantle), and chain silicates (pyroxene

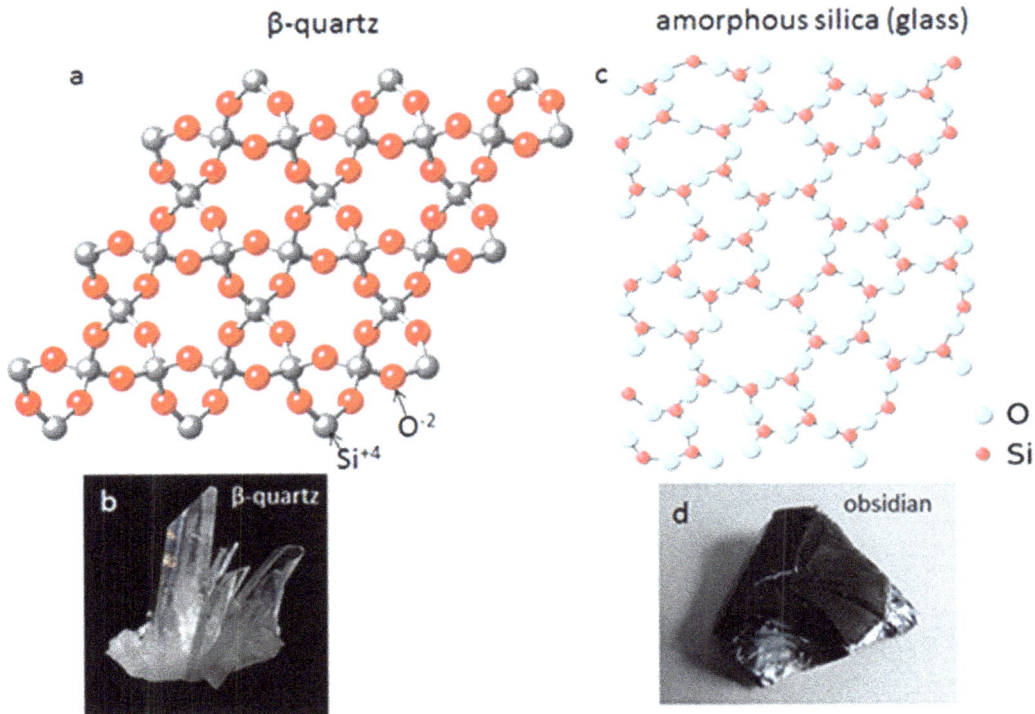

FIGURE 29   Structure of β-quartz ($SiO2$) and amorphous $SiO2$ (glass)

and amphibole groups—mainly mantle and oceanic crust). The geometries of the resulting atomic frameworks are reflected in certain physical properties of minerals.

Amorphous materials such as volcanic glass or glass of windows, dish glass, and other types of glasses do not have ordered atomic arrangements in the same way that minerals do. They have silica tetrahedra, but they are arranged only on a local scale without systematic repetition and without numerous combinations of the connections between individual silica tetrahedra (Fig. 29).

2.3. Minerals in human life, industry and science

Since ancient times, minerals have been the basic raw materials used by humans in their everyday lives, and eventually, they became irreplaceable for economic and technological developments.

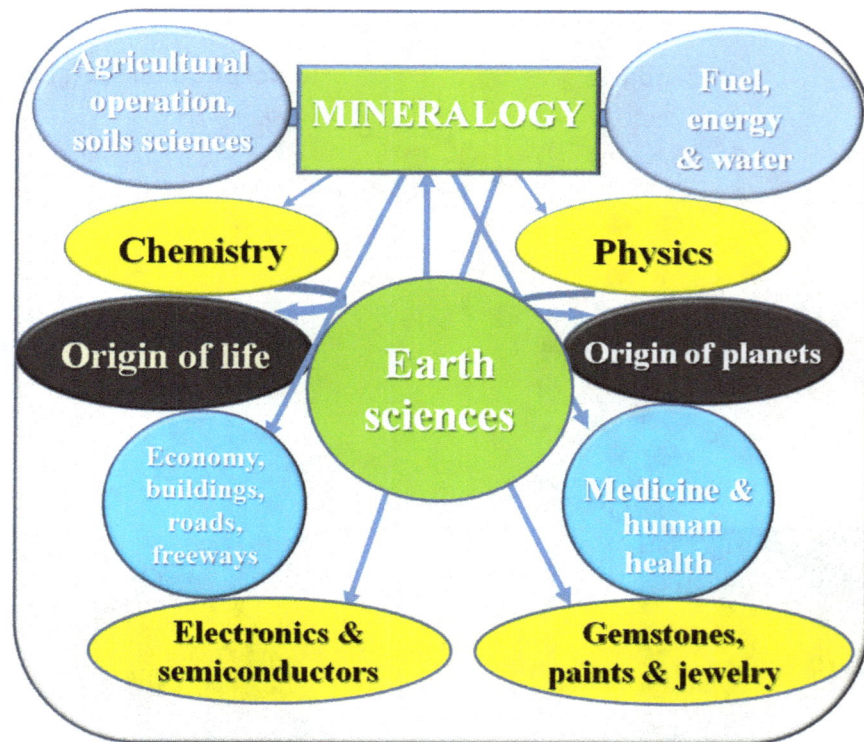

FIGURE 30   Mineralogy and its connection to other scientific disciplines and human activities

In this course, you will learn that minerals and mineralogy, a science which studies minerals, are very important components of other disciplines. The schematic in Fig. 30 shows that knowledge about rocks and the minerals themselves are significant parts of many basic disciplines and sciences as well as social and economic development.

The disciplines include the agricultural and soil sciences; the food and cosmetics industries; domestic materials fabrication (such as dishes, cutlery, cookers, appliances, sinks, bathtubs, etc.); basic and other sciences, including chemistry, physics, and the Earth sciences; planetary sciences, the life sciences, and medicine; environmental sciences, fuel and water resources development; economic and infrastructure development; high technologies, electronics, and the semiconductor industry; and gemstones and the jewelry industry, among many others. The examples below illustrate the most important applications of minerals, their uses in daily life, and their interactions with other sciences and different activities of societies.

2.3.1. The agricultural sector and soil sciences

The success of the agricultural sector is based on many factors, and one of the most important is soil productivity. Soil consists almost completely (95–99%) of tiny fragments of weathered rocks and minerals mixed with 1–5% of organic matter and products of the life activities of micro-organisms. At times, however, soil exhaustion driven by a desire to increase food production may lead to starvation followed by economic and societal collapses (Abraham, 2002). Therefore, the global demands for food will require more and more application of fertilizers on agricultural fields in order to sustain crop yields, and the most available source of many fertilizers is minerals and rocks.

Minerals–fertilizer. Minerals used for production of fertilizers are usually rich in Ca, S, K, and P. Some minerals used for fertilizers occur in sedimentary rocks such as phosphates, which form marine deposits or originate from weathered magmatic rocks. The latter could also be a primary source for minerals of the apatite group [$Ca_5(PO_4)_3(OH,F,Cl)$]. The deposits of evaporate are rich in minerals such as sylvite (KCl), carnalite ($KMgCl_3 \cdot 6H_2O$), gypsum ($CaSO_4 \cdot 2H_2O$) and halite (NaCl). They are formed in isolated brine basins and shallow shelf environments, and their formation requires a hot and dry climate, which facilitates the evaporation followed by precipitation of sylvite, halite, carnalite, and gypsum. Some of them—e.g., gypsum deposits—are also found in hot springs, where they form from volcanic vapors.

Phosphate rocks. These are the basic material for the production of phosphate-bearing fertilizers. They represent a vital component for remediation of soil productivity, and therefore, they are essential for continuing food production. No substitutes for phosphorus are known in agriculture. The demand for phosphate fertilizers remains very high because of the need for food supplies to feed an increasing world population and because of increasing soils exhaustion

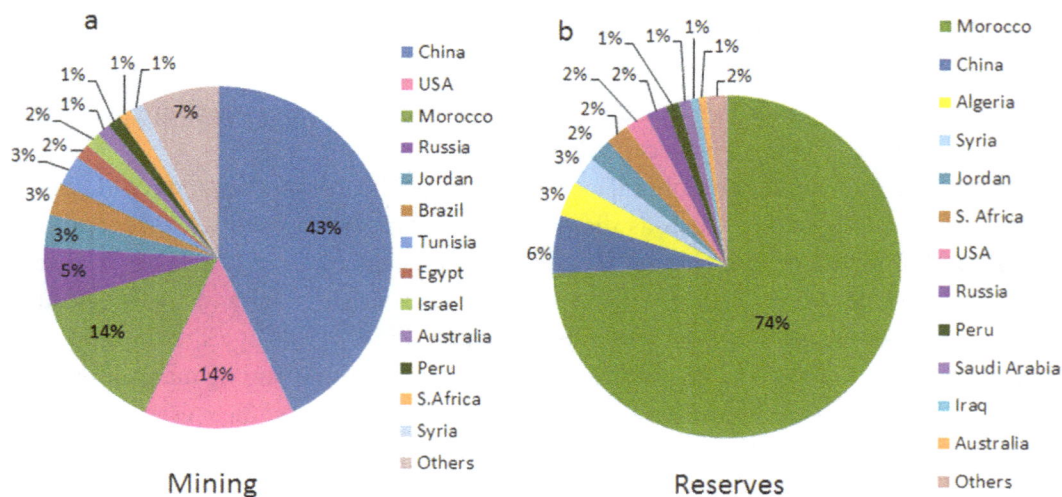

a

■	China
■	USA
■	Morocco
■	Russia
■	Jordan
■	Brazil
■	Tunisia
■	Egypt
■	Israel
■	Australia
■	Peru
■	S.Africa
■	Syria
▩	Others

Mining

b

■	Morocco
■	China
■	Algeria
■	Syria
■	Jordan
■	S. Africa
■	USA
■	Russia
■	Peru
■	Saudi Arabia
▩	Iraq
■	Australia
▩	Others

Reserves

FIGURE 31   World mine (a) production and (b) reserves of phosphate rocks in 2012; data are adopted from the USGS data base http://minerals.usgs.gov/minerals/pubs/commodity/phosphate_rock/mcs-2013-phosp.pdf

caused by humans' agricultural activities. More than 80% of the world's phosphate mining and raw-material productions come from China, Morocco, Russia, Tunisia, Jordan, and the United States, with China being the world's largest producer, accounting for about 43% of the global capacity (Fig. 31a). However, the reserves of phosphate rocks in China comprise only 6% of the world's reserves (compare diagrams a and b in Fig. 31). The other 20% of world phosphate mining is contributed by Brazil, Israel, Egypt, Australia, Syria, Peru, South Africa, and other countries. Phosphate ores mined by Russia, Brazil, and South Africa originate from intrusive rocks, which also yield rare earth elements (REE). Therefore, they are used to extract not only mineral apatite $[Ca_5(PO_4)_3(OH,F,Cl)]$ for fertilizers but also REE that are widely used for the modern electronics industry. The largest reserves (~74%) of phosphate sedimentary rocks in the world are in Morocco, although this country mines only 14% of the world's phosphates (Fig. 31). Large phosphate deposits exist on the continental shelves and on seamounts in the Atlantic and Pacific oceans, and these will probably be mined in the future when new technologies become available. According to the USGS, there are a total of more than 300 billion tons of phosphate rock in the world.

Potash—a source of the element potassium (K). Potash is a commercial name for sylvite (KCl) and carnalite $(KMgCl_3 \cdot 6H_2O)$ ores. According to the USGS commodity database of 2013, estimated world resources of potash are about 250 billion tons. About 97% of large-scale world evaporate

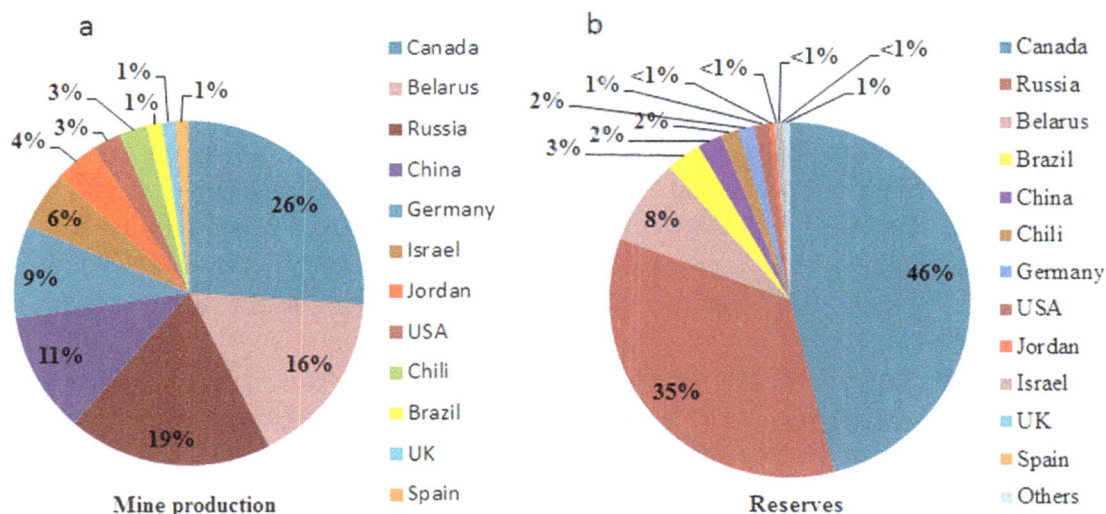

FIGURE 32 (a) Potash world mine production and (b) reserves in 2012 (the USGS database (http://minerals.usgs.gov/minerals/pubs/commodity/potash)

mining and commercial production of potash are located in Canada, Russia, Belarus, China, Germany, Israel, the United States, Jordan, Brazil, Spain, and the United Kingdom (Fig. 32a). Other deposits of commercial-value evaporates are known in Poland, the Netherlands, Denmark, and Thailand. The world reserves of potash are distributed as follows: Canada (46%), Russia (35%), Belarus (8%), Brazil (3%), China (2%), Chile (2%), and Germany (2%), with other countries accounting for 1% or less (Fig. 32b). Potash is one of the critical fertilizers for agricultural sectors because it is responsible for the formation of protein in crops.

Gypsum—$CaSO_4 \cdot 2H_2O$. Gypsum is widely used as a fertilizer as well because it is an efficient source of calcium (Ca) and sulfur (S.) Plants and agricultural crops need calcium for growth, and sulfur contributes to the formation of protein.

2.3.2. Fossil fuel, energy, and water resources

Human societies cannot be developed without electricity and water because both are essential for the sustainable economic and technological functioning of modern countries. Electricity is produced by turbines that are operated by steam, which is produced by heating water through the burning of fossilized fuels (coal, natural gas, and oil). The turbines give power to a generator, which includes a large magnet spinning inside copper windings, and electrons flowing from the copper create electricity. Even from such a short description, you can see how many materials are needed to create electricity, and all of them primarily come from natural geological resources that are modified by industrial processes and technologies. According to data of the U.S. Energy Information Administration (EIA), about 68% of all electricity in the U.S. is produced using fossil fuels, which include coal (37%), natural gas (30%), and oil (1%). Uranium ores are used to produce about 19% of the electricity with nuclear generators, and hydropower provides 7%. The other 5% of the electricity is produced from so-called renewable sources of energy such as biomass (1.42%), wind (4.46%), geothermal (0.41%), solar (0.11%), and other gases (<1%); (see http://www.eia.gov).

Another component required for electricity production is water. In addition, water is an essential ingredient for the biological functioning of humans, and therefore, energy, fossil fuels, and water resources represent a unique example of sustainable connections between geological resources, minerals, and human life. According to data from the United Nations Environment Program (UNEP), the total volume of water on the Earth (including permanent ice and snow and underground water) is ~1.4 bln.km³ (see http://www.unwater.org). However, only ~2.5% (e.g., ~35 mln.km³) of the total volume consists of freshwater which includes ground water, rivers, and lakes, whereas the other ~97.5% of Earth's water is saltwater contained in oceans and salty lakes. For a detailed breakdown of global water of different types, see Fig. 33. The growing industrial, mining, and agricultural consumptions of freshwater coupled with the world's growing population have a strong negative impact on the sustainability of local ecosystems and require consistency in parallel development of environmental protection policies and natural resources management.

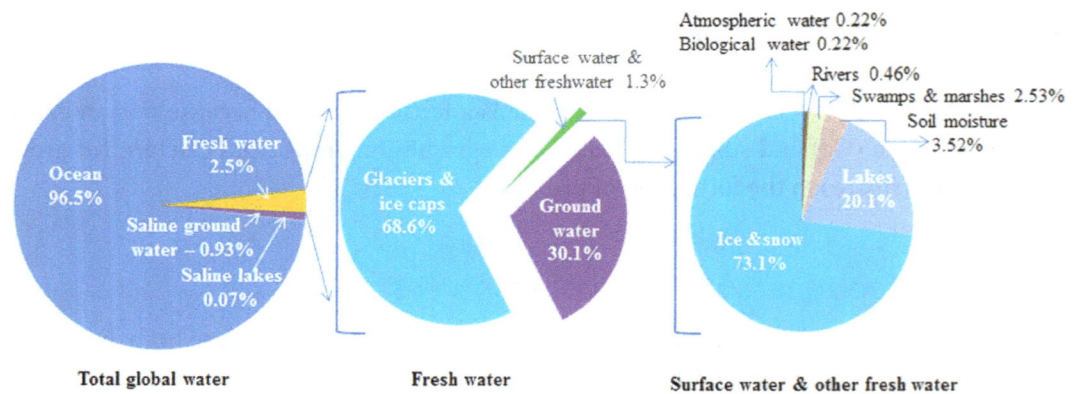

FIGURE 33 Global water distribution (data adapted from Shiklomanov, 1993)

2.3.3. Environmental sciences

The environmental sciences are based on the biological, chemical, or geological-mineralogical aspects of the environments in which people live. Environmental protection of lands, water reservoirs, and the atmosphere from contamination produced by industrial and mining activities has both direct and indirect links to mineralogy. The most visible and sometimes dramatic impacts of mining and ore-extraction processes on environments include:

(i) Land disturbances. If such disturbances are accompanied by removal or destruction of trees, shrubs and other vegetation, and excavation and filling of soil and dirt, the soil erosions will be facilitated by wind and runoff water.

(ii) Chemical contaminations. Agricultural fields and underground and surface water supplies can become contaminated by toxic chemicals released from mining and ore-extraction activities. Dissolution of heavy and toxic metals from the ores mined both underground and close to the Earth's surface can be caused by runoff and underground water, which often leads to toxic leakage into drinking water reservoirs. The most dangerous pollutants originate from sulfide ore mining and from the chemical extraction of copper, nickel, zinc, and other metals that release sulfuric acid onto the land.

(iii) Spreading dust from active on-surface mines and mining tails stored in abandoned mining areas. There is a huge dust problem generated by coarse particles like silt and sand, especially around open pits, quarries, and landfills. All of these by-products create hazardous environments that are dangerous to public health; cause destruction of land; chemically contaminate rivers, lakes, and ponds, which may cause the annihilation of fish and wildlife habitats; and destroy fragile links between local ecosystems and biodiversity. Small- to medium-size mining companies are often closed due to bankruptcy, and therefore, they cannot afford to pay for clean-up of the contaminated areas. In many countries, by law, mining permissions include environmental and rehabilitation codes that require mine operators to minimize the impact on the local ecosystems starting at the exploration stage, continuing through the mining process, and carrying on after a mine has been closed. As a result, when the mining is over, the area of interest is supposed to be returned to its original state. Many companies follow precautions during their active mining periods by constantly checking the purity of water reservoirs and replanting the seeds around a site. This keeps the native species around a site healthy and unaffected by the mining process, thus maintaining plant life as a part of the wider ecosystems and protecting local biodiversity.

Modern environmental sciences emphasize understanding mineralogical resources not only as a positive economic factor, but also because of their possible destructive impacts. International organizations such as the World Bank, the United Nations, consulting and engineering firms, and planning boards and government agencies constantly seek solutions for the many hazardous impacts of mining on human health, and they try to resolve the negative issues caused by mining and distribution of mineralogical resources.

2.3.4. Economy, politics, and conflict minerals

Mineral resources are nonrenewable, and they disproportionally occur in the Earth according to specific geological situations that are necessary for their formation. Nevertheless, historically, countries rich with mineral resources are not necessarily rich and technologically well-developed. Many countries in Africa and South America, for example, though they are rich in mineral resources, have extremely high levels of poverty and low rates of economic growth in their non-resource sectors and infrastructures.

The absence of a positive correlation between mineral resources and economic growth is in many cases a result of unstable governments and ineffective controls on the mining, processing, and distribution of minerals. In such countries, because the mineral resources are the main endowment and revenue source, local political conflicts and civil wars very often take place. For example, a civil war in the Democratic Republic of Congo (DRC), Africa, has lasted for more than a decade as different armed groups of people fight for controls over the vast deposits of diamond, wolframite [tungsten–$(Fe,Mn)WO_4$], cassiterite (tin–SnO_2), and other industrial minerals. Over six million people have been killed in this war since 1996 (Nest, 2012). In addition, armed conflicts for coltan [a generic name for the minerals columbite–$(Fe,Mn)(Nb,Ta)_2O_6$ and tantalite–(Fe,Mn) $(Ta,Nb)_2O_6$)] are continuing now along the eastern border between the DRC, Rwanda, and Uganda (Fig. 34).

These armed conflicts have been blamed for the commission of violent crimes against local populations of the DRC, including murder, rape, and the forced labor of children. Due to that, coltan is considered one of the *conflict minerals*, which are mined under conditions of armed conflicts and human rights abuses. In addition to coltan, three other minerals—wolframite, cassiterite, and gold—originating in the DRC are also referred to as conflict minerals (Fig. 35).

Why is coltan so important? Coltan ores are used for extraction of the elements niobium (Nb) and tantalum (Ta). Due to their strong heat resistance, both of them can hold a high electrical charge, making them vital elements for the manufacturing of electrical capacitors. The latter are the microelectronic elements controlling current flow inside the miniature circuit boards used in cell phones, laptop computers, and many other electronic devices. Because the demand for Nb-Ta ores by the high-technology industry is constantly increasing,

FIGURE 34 (a) Columbite-tantalite and (b) tantalite; (c) the raw ore of both minerals, also known as Coltan, originated from the DRC, Africa, and is considered a conflict mineral

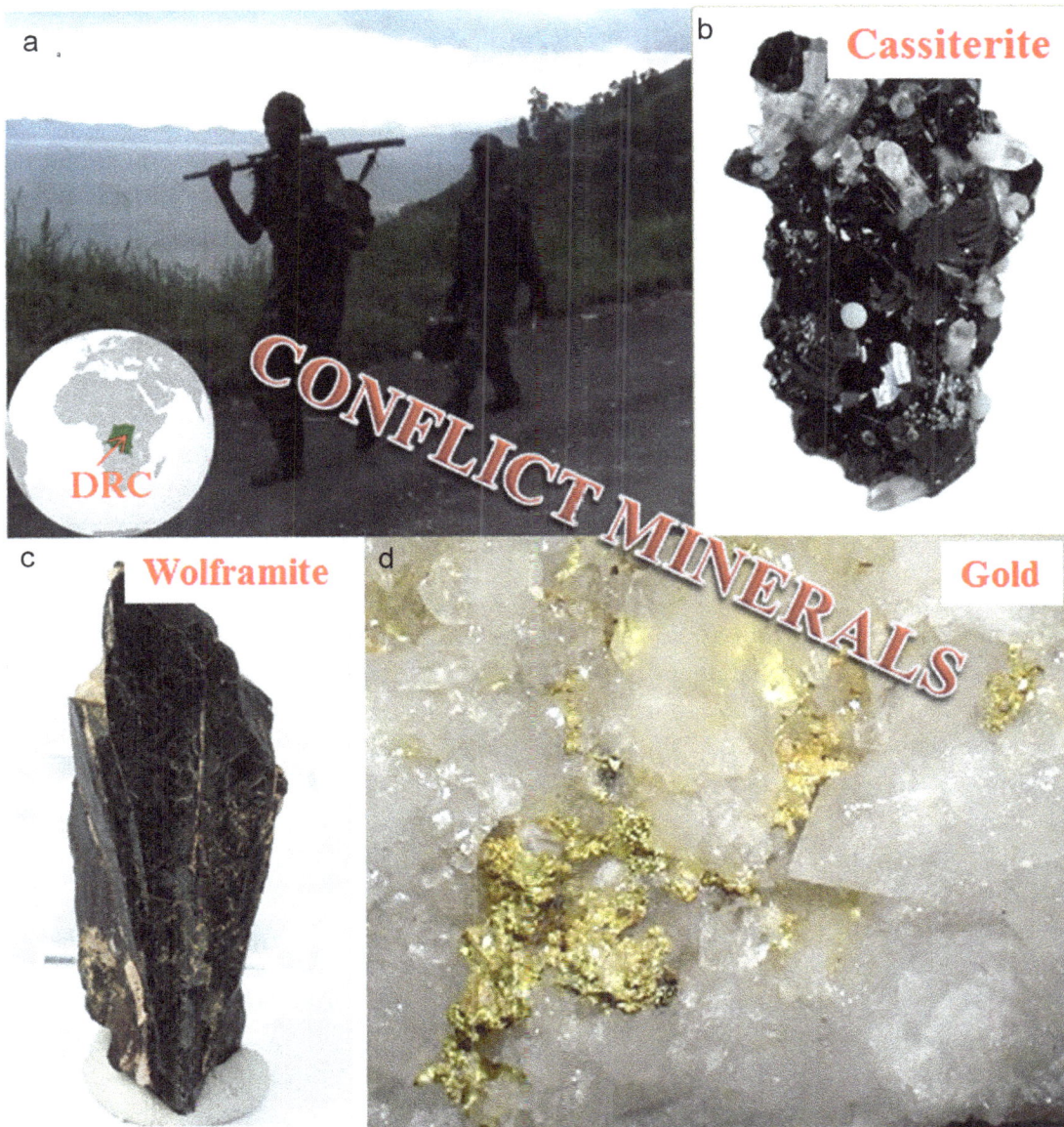

FIGURE 35 (b) Cassiterite; (c) wolframite; and (d) gold mined in the DRC are also included in the list of conflict minerals

the price for the Nb-Ta raw materials also continues to rise. For example, the tantalite ore price has risen from ~US$70–75 per kilogram (kg) during the period of 2005–2011 to ~US$200–280 per kg during the period from 2011 to 2013, and there were also spikes of up to US$400–500 per kg in 2001–2002. Though the DRC possesses 64% of the world's reserves of coltan, its population is racked by poverty and war due to smuggling, the corruption of the government and administration at all levels, military conflicts between rival political groups and tribes, and illegal minerals exploitation.

The international organizations are trying to take some actions to increase the responsibility of manufacturers for governance over conflict minerals used in the electronics industry. One example is the U.S. Dodd-Frank Act of 2010, Section 1502, according to which all U.S. manufacturers of electronics, medical devices, tools, and other products that may use conflict minerals are responsible for examination of their supply chains to determine whether conflict minerals are in fact being used in their products. Currently, other countries such as the European Union (EU) and Canada are also taking actions to ensure that their companies are not using conflict minerals in their supply chains. In addition to establishing numerous actions for ending the wars over minerals by international efforts, the DRC and some other African countries need to strengthen their own strategic management of mineral resources. Legal institutions capable of aligning mineral resources mining and extraction, and legal trading with their long-term goals of national developments must be strengthen.

In the modern world, in which resources, economies, politics, cultures, and the individual lives are constantly interacting one with another, international efforts to regulate these complex relationships will eventually have ramifications. This tendency gives hope that in the future, the problems related to conflict minerals from the DRC will be significantly diminished, as is currently happening with regard to conflict diamonds since the implementation of the Kimberley Process Certification Scheme (KPCS). The KPCS was endorsed by the United Nations General Assembly (UNGA) and the United Nations Security Council (UNSC) in 2003. The fact that diamonds were supporting military conflicts came to international attention during the extremely brutal civil wars in Sierra Leone, Africa, in the 1990s. The KPCS includes eighty member countries, and its regulations are focused first on the prevention of the trade in diamonds from funding armed conflicts against legitimate governments, and second on alleviating the poverty and physical and human rights abuses perpetrated on local populations. The KCPS allows raw diamonds to be exported and imported only within countries that are members of the Kimberley Process. This provides confidence that eventually, no conflict diamond will be able to enter the legitimate diamond supply chain or be used for illegitimate purposes.

2.3.5. Industrial minerals, materials science, and engineering

"Industrial minerals" is a generic name for a huge family of beneficial raw materials, including minerals, ores, and rocks, that are used for the construction or manufacturing of buildings, railways, airports, freeways, local roads, water-filtering media, toxic chemical spill absorbents, abrasives, automobiles, sports supplies, tools, cement preparation, ceramics, and glass, to mention just a few applications. Many industrial minerals or their components are included as ingredients in many of the cosmetic, painting, pigment, home decoration, gardening, and other products used in people's everyday lives. According to the Industrial Mineral Association of

North America (IMA-NA), each American citizen uses ~24 tons of industrial minerals per year through direct and indirect consumptions and services.

Examples of industrial minerals include quartz (used for the production of glass); feldspars and clay minerals (for cement and ceramics); sand and gravel (for buildings, roads, and freeways); zeolites and diatomites (used as filtering compounds); granite, gabbro, and quartzite (polished slabs are used for kitchen countertops and walls decoration); borates, gypsum, and sulfates (used as agricultural additives); calcite and dolomite, which are rock-forming minerals of limestone, marble, and dolostone (used for industrial construction, building decorations, countertops, and agricultural additives); clay and micas (for cosmetics); rutile (TiO_2; used for the preparation of pigments for painting materials); corundum (Al_2O_3) and non-jewelry-quality diamonds (used as abrasives); as well as many others. Industrial minerals are not included in the metal commodities, but their mining companies, merchandise producers, and manufacturers need to be responsible for quality in order to succeed in a highly competitive market boosted by rapidly advancing technologies. Industrial minerals are usually low-cost materials, because they are extracted using respectively non-expensive open-pit and quarrying operations depending upon their scales and locations.

There are also many other industrial minerals, which due to their specific atomic structures and physical properties, are of the focus of materials science and engineering. One of the fascinating examples is a group of minerals that in mineralogy is called *native elements*, which includes metals (gold group: copper, gold, silver; platinum group: palladium, platinum, osmium; iron group: iron and iron-nickel alloys), semimetals (arsenic, bismuth, antimony), and non-metals (graphite, diamond, and sulfur); (see Table 4).

TABLE 4:   Minerals of Native-Elements Group

Metals	Semimetals	Non-metals
Gold group:	Arsenic (As)	Graphite (C)
Copper (Cu)	Bismuth (Bi)	Diamond (C)
Gold (Au)	Antimony (Sb)	Sulfur (S)
Silver (Ag)		
Platinum group:		
Palladium (Pd)		
Platinum (Pt)		
Osmium (Os, Pt, Ir)		
Iron group		
(mostly in meteorites)		
Iron (Fe)		
Kamacite (Fe,Ni)		
Taenite (Ni,Fe)		

The structure of metals is characterized by strong metallic bonds, which give them excellent electrical conductivity. Uses of such metals as copper, gold, iron, and silver go back to prehistorical periods of human existence. In the modern world, many of them are used in industrial engineering—e.g., copper is required for electrical wiring. Other unique and important minerals of the native-elements group are diamond and graphite (Fig. 36), which have myriad industrial, nanotechnological, and societal applications. Materials science, which focuses on both the fundamental properties and the technological applications of materials, carries on intensive studies of metallic alloys, metalloids, and nonmetallic native elements. This knowledge promotes further modification of the materials synthesized

FIGURE 36 (a) Diamond and (c) graphite are carbon polymorphs that have dissimilar structure: (b) cubic for diamond and (d) hexagonal layers for graphite

in the laboratory and helps in the creation of new materials using as analogies the atomic structures that have been learned from natural minerals. For example, many of these modifications are related to changing the electrical conductivity of diamond by doping it with boron. Diamond has the highest cubic symmetry (Fig. 36a,b), with each carbon atom being attached to four neighboring carbon atoms at the same distance, making diamond structure very strong. This is why diamond has a higher density than any other Earth- or man-made minerals and materials. Diamond as a hardest material is successfully used in saw blades and other supper-hard cutting tools and drills of different scales, and as a perfect abrasive material. Due to its atomic configuration, diamond is a great thermo-conductor, it can tolerate extreme pressures and extreme temperatures, and can remain neutral within reactive chemical environments. When diamond is doped with controlled-level impurities, it can be used for many electronic devices. For example, if diamond is doped with boron, it becomes a semiconductor, which opens a new avenue for many innovative technological applications (Kraft, 2007). The boron-doped polycrystalline diamond synthesized by methods of chemical vapor deposition (CVD) has already been experimentally tested for possible applications in radiation detectors, sensors, electrodes, and electrical heating elements (Sussmann, 2009). In addition to technological uses, diamond of gemstone quality, which is mined from specific volcanic rocks (kimberlite) and alluvial placers, is an object of admiration and a well-marketed symbol of love and faith. Everyone is familiar with the advertisement created by the De Beers company, "A diamond is forever," which makes it a fundamental part of wedding ceremonies around the world.

Another important industrial mineral from the group of carbon is graphite, which is used for countless applications in materials science, industry, high technology, and electronics (Fig. 36c,d). The regular uses of graphite are in the extraction and purification of metal in the electrolysis process and in fuel cells, as well as for manufacturing lithium-ion batteries, which are components of computers, laptops, cell phones, and many other portable electronic devices that are in great demands. Because graphite, like diamond, is chemically inert, it is used as a filtering material during the production and separation of many reactive and highly corrosive chemicals. Due to its high lubricating property, graphite is used in heavy machinery to reduce the friction between moving parts and in many other mechanical applications such as sliding and/or guide bearings, piston rings, steam joint rings, vacuum pumps, and compressors. The aircraft and aerospace industries use graphite to manufacture engine cases, blast tubes, rocket nozzles, edge components, thermal insulators, etc. Graphite is also used in the nuclear power plant industry, and its high-density modifications are used in many other technologies to produce composite materials of high strength. The leading producers of raw graphite are Brazil, Canada, China, India, and South Korea. China alone produces about 75–80% of the raw graphite in the world.

A novel material that is extracted from graphite is graphene, a single, one-atom-thick layer of carbon detached from flakes of graphite (Fig. 37a). Graphene is 200 times stronger than steel, has high conductivity, and has amazing flexibility without breaking. The graphene layer was isolated in 2004 from graphite-mineral due to the simple method of exfoliation by two physicists from the University of Manchester (U.K.), A. K. Geim and K. S. Novoselov, who were awarded a Nobel Prize in 2010. Graphene has thousands of potential innovative applications, including touch-screen computers, mobile phones, ink preparation, transistors, and efficient battery electrodes, to note just a few (see, e.g., Geim and Novoselov, 2007; Mertens, 2013). The real commercial viability of graphene's applications is still under consideration, but even though its marketing uncertainties are not completely resolved, graphene is often called "the world's next wonder material".

Two other forms of very specific atomic arrangements of carbon that are also considered to be at the leading edge of electronics are fullerenes, which include C_{60}, and nanotubes (Fig. 37b–d). Nanotubes (Fig. 37c) are represented by a tube-shaped material consisting of single atoms of carbon connected to each other in a honeycomb arrangement. To better understand its structure, you should imagine that you have rolled one graphene layer into a tube and that the tube has a hole with a diameter of about one nanometer (nm). The average size of a human hair is 1–10 micrometers (μm), and 1 μm = 1000 nm, so a carbon nanotube appears to be 1000–10000 times smaller than the diameter of a human hair. The carbon nanotube was discovered in 1991 by the Japanese physicist S. Iijima in the soot of arc discharge (Iijima, 1991). In the historical description of nanotube discoveries, it is mentioned that nanotube-structured carbon was observed earlier, in the 1950s, by the Russian scientists L. Radushkevich and V. Lukyanovich and again in the 1970s by the Japanese scientist M. Endo. However, carbon nanotubes triggered a worldwide interest only after publication by S. Iijima in 1991, which was followed by countless experiments and the establishment of the specific field of this new discipline of nanotechnology. You can find more information in O'Connel (2006) and online at: http://nanogloss.com/nanotubes.

The carbon-nanotube material is very tiny and light, but it is also very strong. Its commercial applications include electronics, optics (computer chips, sensors, compact batteries to power automobiles, and highly efficient power lines), and series of nano-biotechnological and medicinal applications because carbon is compatible with the human body.

Fullerene C_{60} was discovered in 1985 by a group of researchers led by R. Smalley and R. Curl, both from Rice University, Houston, Texas, USA, and H. Kroto from the University of Sussex, Brighton, U.K., all of whom were awarded the Nobel Prize in 1996. They vaporized graphite at very high temperatures using a high-energy laser light beam, and with mass spectrometry, they identified a strong peak corresponded to a molecule that consists of 60 atoms of carbon (Kroto et al., 1985). This molecule has an unusual structural arrangement of carbon atoms reminiscent of a soccer ball (Fig. 37b), so C_{60} is also called a "buckyball." Fullerene C_{60}, as well as other derivatives of nanocarbon, has considerable potential for development of innovative applications in electronics and nanotechnologies, including medicinal applications. The latter are based on the configuration of the C_{60} molecule, which is like an empty mini-cage, and therefore can

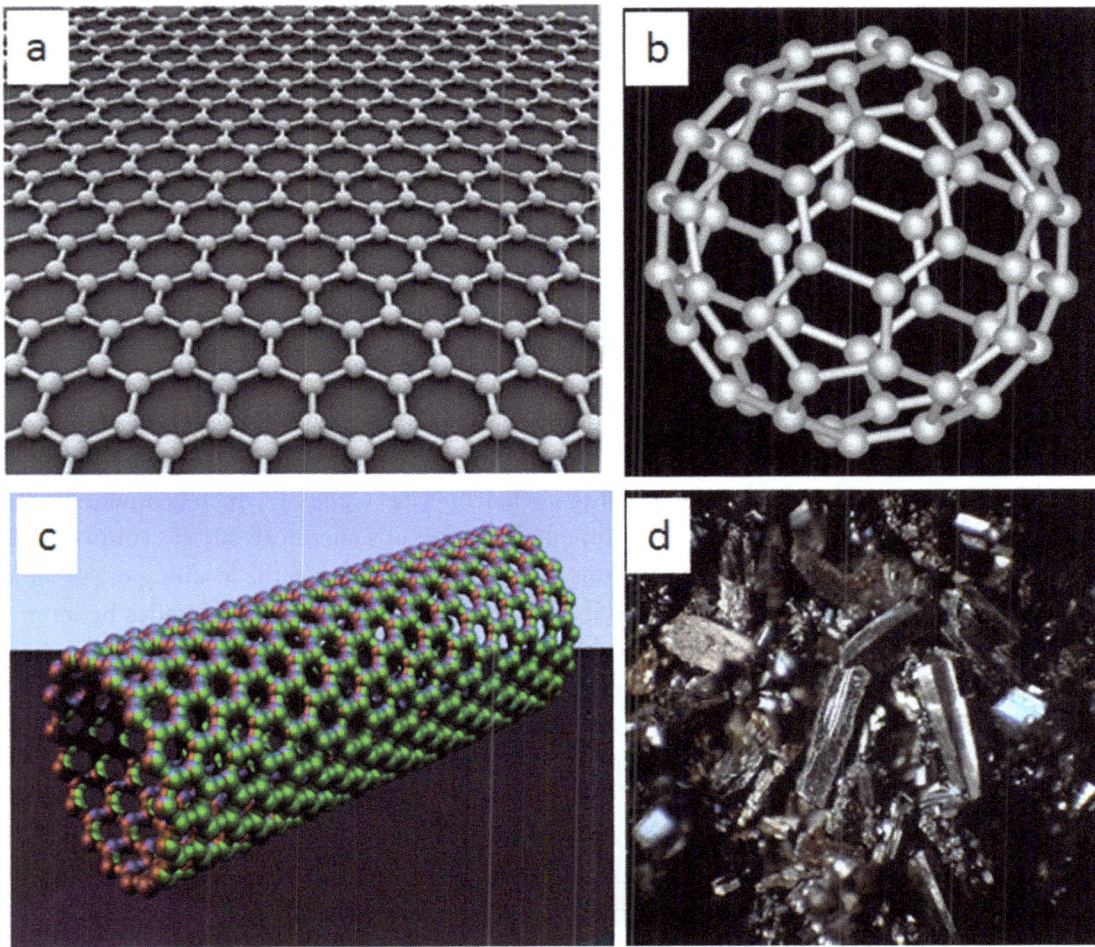

FIGURE 37 Structure of carbon nanomaterials: (a) graphene; (b) fullerene C60; (c) fullerene nanotube; (d) C60 fullerene in crystalline form

be used as a "container" to hold drugs that eventually can be used to target human cells in a controlled manner.

Experimental studies have shown that fullerene C_{60} filled with drugs can be used in therapy for diseases such as multiple sclerosis, HIV infection, some forms of cancer, radiation exposure, allergic and infectious diseases, and general inflammation. Because of the high lubrication properties of the C_{60} spherical molecules, this material is also considered as a potential future component of artificial joints that would replace damaged hips and knees after orthopedic surgeries. However, despite such an immense field of possible applications, there is still concern about the toxicity of fullerenes (C_{60} and other nanocarbon materials), and it is not yet clear if this amazing material will ever be used in medicine in the future (Kepley, 2012).

Another excellent example of minerals engineering is synthetically created yttrium-aluminum-garnet (YAG), which became the main component of lasers that are used to produce high-intensity light. The YAG laser is used for manufacturing and engraving steels and for the production of plastics, metallic alloys, and composite materials, and it is also widely used in laser surgery on eyes and in cosmetic surgery and dentistry.

2.3.6. Minerals in the biotechnological and medical applications

Example of biotechnological applications. The separation of metals from ores and their surrounding rocks is usually carried out by hydrometallurgical methods that involve additives of toxic chemicals. For example, a heap leaching of chalcopyrite ($CuFeS_2$), which is mined for the extraction of copper, involves many environmentally unfriendly chemicals such as sulfuric acid, nitric acid, chloride, and ammoniac solutions. Microbial assistance in the leaching of sulphide minerals is a new and fast-developing discipline called *bioleaching*, which provides better environmental protection of geographical regions where mining and metallic ore treatments and metals extraction take place. For example, the bacteria families *Acidithiobacillus ferrooxidans* and *Acidithiobacillus caldus* act as sulfur and iron oxidizers (Watling, 2006), and many studies are now focused on understanding of the mechanisms by which bacteria provide electrochemical interaction with mineral surfaces and how they contribute to sulfide dissolution processes. Observations, at least, suggest that enzymes—the main product of microbial metabolism—act as a source for some kinds of "acids" that directly dissolve the metal. However, the details of this mechanism are still waiting to be discovered.

Minerals in medical applications. For several decades, pure titanium (Ti), titanium alloy ($TiAl_6V_4$), and stainless steel have been used for orthopedic implants. The corrosion resistance of such implants is strengthened by coating their surfaces with a very thin film of Cr_2O_3, which provides a protective layer. The Cr_2O_3-coated surface is the ultimate interface with tissue after implantation, which minimizes cell toxicity and inflammation (Balazic et al., 2007).

Plaster casts that are used for wrapping broken arms or legs are made from gypsum ($CaSO_4 \cdot 2H_2O$), and medical thermometers contain liquid mercury (Hg), which is a mineral of the native-elements group. A thin powder of $BaSO_4$ (mineral barite) mixed with water is used for X-ray examination of the intestinal tract and stomach, and the low solubility of barium sulfate minimizes the toxic effects that might occur with other barium compounds.

There are new, emerging disciplines of biominerals and biomineralization that are directly related to the medical geology field (Dove et al., 2005). "Biomineral" is a specific term used to describe a mineral produced by organisms, and the biominerals, indeed, represent composite materials that consist of both inorganic (mineral) and organic (biological) matter. They are beyond the scope of this book, but readers are encouraged to seek more information in the *Biomineralization Sourcebook: Characterization of Biominerals and Biomimetic Materials* (DiMasi and Gower, 2014).

Hazardous minerals 2.4.

Hazardous minerals are those that directly or indirectly (through their byproducts of mining, treatment, and manufacturing) have negative effects on the health of humans and other organisms and contaminate environments. Most of ~3800 known minerals are safe to handle, but about ten to forty of them (researchers disagree on the exact number) have to be handled with caution, and dozens of them may seriously affect human health by causing occupational and even epidemic diseases. The full list of hazardous minerals is not completely established yet, and if you were to conduct a literature survey, you would find that different authors suggest different numbers of toxic/hazardous minerals. On the one hand, this is because there is usually a lag of twenty to forty years between exposure and the onset of symptoms of hazardous-mineral-triggered diseases, so very often, there simply is not enough available statistical data to determine the primary sources of illnesses. On the other hand, health problems arising from geological materials and processes are less recognized by physicians, whose training is different from that of occupational medicine specialists, policymakers, and local administrative authorities. In other words, the public health problems connected with minerals and their mining, treatment, and fabrication are more common than most people would imagine. The International Union of Geological Sciences (IUGS) has estimated that the health of more than three billion people worldwide may be seriously affected by hazardous geological processes and the Earth materials.

The ten most dangerous minerals listed by G. Brown (Stanford University) in his mineralogy course (for details, see http://emsi.stanford.edu/10DangerousMineral.html) include minerals of the "asbestos" group [crocidolite–$Na_2(Fe^{2+},Mg)_3Fe^{3+}2Si_8O_{22}(OH)_2$ and chrysotile–$Mg_3Si_2O_5(OH)_4$], the zeolite group [erionite–$(NaK_2MgCa_{1.5})[Al_8Si_{28}]O_{72}\cdot28H_2O)$], silicates and oxides (K-feldspar–$KAlSi_3O_8$, phenacite–$BeSiO_4$, and quartz–SiO_2), sulfides (cinnabar–HgS, galena–PbS, pyrite–FeS_2), fluorite (CaF_2), and hydroxylapatite [$Ca_5(PO_4)_3(OH)$]. However, this list is minimal and can be extended by adding other minerals that are not less dangerous than those listed by Brown—e.g., fibrous amphiboles, arsenopyrites, and minerals of the uranium group, which are characterized by high radioactivity, among others (Table 5).

TABLE 5   List of Most Dangerous Minerals

Mineral Name and Chemical Formula	Generic Name	Description
*Crocidolite (amphibole of riebeckite group): $Na_2(Fe^{2+},Mg)_3Fe^{3+}_2Si_8O_{22}(OH)_2$	Crosidolite-asbestos	*Known as blue asbestos. Causes lung diseases, including lung and mesothelial cancer.
Amosite (amphibole of cummingtonite-grunerite solid solution series): $(Mg_{2.1},Fe_{4.9})Si_8O_{22}(OH)_2 - Fe_7Si_8O_{22}(OH)_2$	Amosite-asbestos	Known as brown asbestos. The real name of this mineral is grunerite; its name amosite originates from a locality in Africa named Amosa, where this mineral is mined. According to the American Cancer Society, exposure to amosite asbestos creates a higher risk of cancer in comparison with other types of asbestos.
Tremolite (amphibole of tremolite-actinolite solid solution series):$Ca_2Mg_5Si_8O_{22}(OH)_2$, associates with talc–$Mg_3Si_4O_{10}(OH)_2$ and vermiculite– $(Mg,Ca,K,Fe^{2+})_3(Si,AL,Fe^{3+})_4O_{10}(OH)_2O_4 \cdot H_2O$	Tremolite-asbestos	It is not mined on its own. It is fibrous, and it is found in large amounts, often intergrown with talc and vermiculite. Miners and millers are at higher risk for developing lung cancer and other respiratory conditions.
Actinolite (amphibole of tremolite-actinolite solid solution series): $Ca_2(Mg,Fe^{2+})_5Si_8O_{22}(OH)_2$, associates with vermiculite	Actinolite-asbestos	Actinolite develops needle-like, radially fibrous crystal shapes. It is intergrown often with vermiculite. Inhaling of actinolite particles causes a high risk of developing lung cancer and other respiratory conditions.
Anthophyllite	Anthophylite-asbestos	It is a rare abestiform mineral that does not have a long history of commercial use. The risk of developing mesothelioma is much less than it is from exposure to other types of asbestos.
*Chrysotile (serpentine group): $Mg_3Si_2O_5(OH)_4$	Chrysotile-asbestos	*Known as white asbestos. Chronic exposure can cause a progressive lung disease.
*Hydroxylapatite (phosphate group): $Ca_5(PO_4)_3(OH)$		*Forms deposits in human heart valves and arteries.
*Erionite (zeolite group): $(NaK_2MgCa_{1.5})(Al_8Si_{28})O_{72} \cdot 28H_2O$		*A fibrous zeolite (sometimes referred to as a molecular sieve). Known to induce malignant mesotheliomas in humans.
*Phenacite (beryllium ortosilicate): $BeSiO_4$		*Beryllium (Be)-containing dust is highly poisonous.
*K-Feldspar (tectosilicates group): $KAlSi_3O_8$ *This is one of the most common minerals in the Earth's continental crust, comprising about 22% by volume of crustal rocks.		*Contains small quantities of radioactive uranium, a major source of lead (Pb). Radioactive uranium decays to form radon gas, which has been implicated by the EPA as potentially responsible for about one-third of all lung cancer deaths in the U.S. among nonsmokers.
*Pyrite (sulfides group): FeS_2		*Main source of acid mine waters associated with sulfide mine tailings. Arsenic (As)-containing pyrite. Oxidation of pyrite, which is catalyzed by certain types of bacteria, releases trace quantities of toxic metals such as As.

Mineral Name and Chemical Formula	Generic Name	Description
*Galena (sulfides group): PbS		*The main source of lead. When released into the environment in various forms, it can cause developmental and nervous-system disorders in fetuses and children. Implicated in cardiovascular disease in adults.
*Cinnabar (sulfides group): HgS		*The main source of mercury. Causes developmental and nervous-system disorders in fetuses and children.
*Fluorite: CaF_2		*The major fluorine-containing mineral. Too much fluorine in human diets can cause a very severe bone disorder, resulting in an irreversible disease referred to as skeletal fluorosis.
*Quartz (oxides group): SiO_2	Crystals of quartz are not toxic.	*Quartz in fine particulate form has been known since biblical times to cause respiratory effects (silicosis or silicotuberculosis).
Arsenopyrite (sulfide group): AsFeS		Heated or altered arsenopyrite produces lethally toxic, corrosive and carcinogenic vapors. Arsenic is harmful to the nervous system, and it has been associated with certain types of skin cancer. Some studies also show a possible link with lung, bladder, liver, colon, and kidney cancers.
Uraninite (oxide group): $(U,Th)O_2$		Uranium minerals and their decay products and associated trace elements may create environmental hazards, and they are dangerous for human health due to their radioactivity. Uranium minerals should be kept in a sealed container, out of the light, and should be handled as little as possible.

Minerals marked with * and their explanations are adapted from Brown's list (http://emsi.stanford.edu/10DangerousMineral.html), while others (without a mark) are added using the USGS database on asbestos minerals (http://minerals.usgs.gov/minerals/pubs).

In many publications that are not related to the mineralogical sciences, one may read that minerals of the serpentine group—for example, chrysotile—are called "asbestos," and another mineral that belongs to the riebeckite-amphibole group, crocidolite, is also called "asbestos." *Asbestos is a generic term for a group of silicate minerals that have fibrous texture and that have been widely used in commercial products.* The EPA and many other health-safety organizations and related institutions use the term "asbestos" to characterize at least six minerals (Table 5) that have a fibrous texture that produces small, sharp particles when it is mechanically disintegrated. These particles, which have a "needle"-like shape and may also be called asbestos fibers, have been shown to cause certain diseases (asbestosis, lung cancer, and mesothelioma—a special type of lung cancer). These diseases were diagnosed mostly in that group of people who used to work in asbestos-minerals mining or ore processing, sorting, and even delivery, where they had been exposed to the toxic asbestos particles. The symptoms of asbestos-related diseases usually become well pronounced about twenty to forty years after inhalation, and the result of their development can be fatal. These and other issues will be addressed in detail later (see Chapter 7).

Another group of dangerous minerals mentioned in Table 5 are sulfides, which are a main source of metals. Acid mine waters associated with sulfide mine tailings and rain water interacting with tailings release extremely toxic pollutants such as arsenic, mercury, and others, which can be serious threats to public health and can cause a long-term destruction of biodiversity and environments. Other minerals such as the oxides, ortosilicates, and tectosilicates mentioned in Table 5 should also be taken into account as possible health threats, though some of them are less dangerous than the asbestiform and the sulfide minerals. In general, one should avoid breathing the dust released from the mechanical disintegration of any minerals, whether they are from industrial or mining operations, or from crushing minerals from your personal collection with a knife or a hammer. After handling minerals during your laboratory studies, trying to determine their hardness, luster, specific gravity, and other physical properties, you should thoroughly wash your hands. Also, if your tongue usually tells you exactly how much salt to add to your soup, you should never apply this technique to any minerals because you may be badly contaminated by invisible toxic particles. Finally, minerals should not be accessible to small children.

References

Abrahams, P. W. 2002. Soils: Their implications to human health. *The Science of the Total Environment* 291:1–32.

Balazic, M., J. Kopac, M. J. Jackson, W. Ahmed. 2007. Review: Titanium and titanium alloy applications in medicine. *International Journal of Nano and Biomaterials* 1:3–34.

DiMasi, E., and L. B Gower. 2014. *Biomineralization Sourcebook: Characterization of Biominerals and Biomimetic Materials*. Florence, KY: CRC Press, p. 420.

Dove, P. M., J. Yoreo, and S. Weiner. 2005. Biomineralization. *Reviews in Mineralogy and Geochemistry* 54:331.

Geim, A. K., and K. S. Novoselov. 2007. The rise of graphene. *Nature Materials* 6:183–191.

Hazen, R. M. 2010. Evolution of minerals. *Scientific American* 302:58–65.

Hazen, R. M., R. T. Downs, L. Kah, and D. Sverjensky. 2013. Carbon mineral evolution. *Reviews in Mineralogy and Geochemistry* 75:79–107.

Iijima S. 1991. Helical microtubules of graphitic carbon. *Nature* 354:56–58.

Kepley, C. 2012. Fullerenes in medicine: Will it ever occur? *Journal of Nanomedicine and Nanotechnologies* 3:e111.

Kraft, A. 2007. Doped diamond: A compact review on a new, versatile electrode material. *International Journal of Electrochemical Science* 2:355.

Kroto, H. W., J. R. Heath, S. C. O'Brien, R. F. Curl, and R. E. Smalley. 1985. C$_{60}$: buckminster fullerene. *Nature* 318:162–163.

Levin, H. 2006. *The Earth through time.* 8th ed. Hoboken, NJ: John Wiley & Sons, Inc., p. 560.

Mertens, R. 2013. *The graphene handbook.* Ron Mertens Publisher, p. 111. http://www.graphene-info.com/handbook.

Nest, M. 2012. *Coltan.* Cambridge and Malden, MA: Polity Press, p. 220.

O'Connell, M. (Ed.). 2006. *Carbon nanotubes: Properties and applications.* Florence, KY: Taylor & Francis Group, LLC, p. 313.

Rosing, M. T. 2008. On the evolution of minerals. *Nature* 456:457–458.

Shiklomanov, I. A. 1993. World fresh water resources. In: P. H. Gleick (Ed.), *Water in crisis: A guide to the world's fresh water resources.* New York: Oxford University Press, pp. 1–24.

Shumann, W. 1993. *Rocks, minerals and gemstones.* New York: HarperCollins Publisher, p. 381.

Sussmann, R. S. 2009. *CVD diamond for electronic devices and sensors.* West Sussex, U.K.: John Wiley & Sons Ltd, p. 165.

Wang, Y., T. Azaïs, M. Robin, A. Vallee, C. Catania, P. Legriel, G. Pehau-Arnaudet, F. Babonneau, M-M. Giraud-Guille, and N. Nassif. 2012. The predominant role of collagen in the nucleation, growth, structure and orientation of bone apatite. *Nature Materials* 11:724–733.

Watling, H. R. 2006. The bioleaching of sulfide minerals with emphasis on copper sulfides: A review. *Hydrometallurgy,* 84:81–108.

Web resources

Anthony, J. W., R. A. Bideaux, K. W. Bladh, and M. C. Nichols (Eds.), Handbook of mineralogy. Chantilly, VA: Mineralogical Society of America; access online http://www.handbookofmineralogy.org.

http://emsi.stanford.edu/10DangerousMineral.html: Ten dangerous minerals.

http://www.eia.gov: The U.S. Energy Information Administration.

http://nanogloss.com/nanotubes: Introduction to Nanotechnologies.

http://www.unwater.org: UN Water.

Image Credits

Figure 23: Adapted from: "Chemistry Reference Sheet," http://www.cde.ca.gov/ta/tg/sr/documents/chemtoe.pdf.

Figure 24a: Adapted from: Copyright © Mats Halldin (CC BY-SA 3.0) at http://commons.wikimedia.org/wiki/File:Jordens_inre-numbers.svg.

Figure 24b: Copyright © Rob Lavinsky (CC BY-SA 3.0) at http://commons.wikimedia.org/wiki/File:Staurolite-26463.jpg.

Figure 24c: Yassine Mrabet, "Carbon lattice diamond," http://commons.wikimedia.org/wiki/File:Carbon_lattice_diamond.png. Copyright in the Public Domain.

Figure 25: Copyright © OpenStax College (CC by 3.0) at http://en.wikipedia.org/wiki/File:201_Elements_of_the_Human_Body-01.jpg.

Figure 26a: United States Geological Survey and the Mineral Information Institute, "Talc block," http://commons.wikimedia.org/wiki/File:Talc_block.jpg. Copyright in the Public Domain.

Figure 26b: Copyright © JJ Harrison (CC BY-SA 2.5) at http://commons.wikimedia.org/wiki/File:Gypsum_var._selenite_from_Andamooka_Ranges_-_Lake_Torrens_area,_South_Australia.jpg.

Figure 26c: Copyright © Rob Lavinsky (CC BY-SA 3.0) at http://commons.wikimedia.org/wiki/File:Calcite-Laumontite-27308.jpg.

Figure 26d: Copyright © Eurico Zimbres and Tom Epaminondas (CC BY-SA 2.0 Brazil) at http://commons.wikimedia.org/wiki/File:Fluorita_green.jpeg.

Figure 26e: OG59, "Apatite crystals," http://commons.wikimedia.org/wiki/File:Apatite_crystals.jpg. Copyright in the Public Domain.

Figure 26f: United States Geological Survey, "Potassium feldspar," http://geomaps.wr.usgs.gov/parks/rxmin/mineral.html. Copyright in the Public Domain.

Figure 26g: Copyright © JJ Harrison (CC BY-SA 2.5) at http://commons.wikimedia.org/wiki/File:Quartz,_Tibet.jpg.

Figure 26h: Copyright © Rob Lavinsky (CC BY-SA 3.0) at http://commons.wikimedia.org/wiki/File:Topaz-tuc1016b.jpg.

Figure 26i: Copyright © Rob Lavinsky (CC BY-SA 3.0) at http://commons.wikimedia.org/wiki/File:Corundum-dtn14b.jpg.

Figure 26j: United States Geological Survey and the Mineral Information Institute, "Rough Diamond," http://commons.wikimedia.org/wiki/File:Rough_diamond.jpg. Copyright in the Public Domain.

Figure 27a: United States Geological Survey, "Mica," http://geomaps.wr.usgs.gov/parks/rxmin/mineral.html. Copyright in the Public Domain.

Figure 27b: Jurema Oliveira, "PlagioclaseFeldsparUSGOV," http://commons.wikimedia.org/wiki/File:PlagioclaseFeldsparUSGOV.jpg. Copyright in the Public Domain.

Figure 27c: Copyright © Didier Descouens (CC BY-SA 4.0 International) at http://commons.wikimedia.org/wiki/File:Selpologne.jpg.

Figure 28: Adapted from: "Silica tetrahedron," http://www.esri.com/news/arcuser/0807/graphics/nongeo_1-lg.jpg.

Figure 29a: Ben Mills, "Beta-quartz-CM-2D-balls," http://commons.wikimedia.org/wiki/File:Beta-quartz-CM-2D-balls.png. Copyright in the Public Domain.

Review Questions

1. How are elements, rocks, and minerals related to one another?

2. What is the definition of a mineral?

3. Are kidney stones minerals? What is their chemical composition, and to what minerals can they be compared?

4. Do diamonds and rubies made in a laboratory have the same composition and crystalline structure as their counterparts from the rocks? Are they minerals? What are the chemical compositions of ruby and diamond?

5. Why is glass having a composition of SiO_2 not a mineral, but quartz having a similar composition of SiO_2 is?

6. How many mineral species are known on the Earth?

7. Do all minerals have to be solid?

8. Is H_2O-ice a mineral? If yes, explain why.

9. What is the name of the organization that accepts new minerals?

10. List the main physical properties of minerals.

11. What is the Mohs hardness scale? What minerals have hardness 10 and 1?

12. Assume that you have an unknown mineral and need to determine its hardness using a mineral hardness kit. You also know that your unknown mineral is softer than fluorite and harder than gypsum. What is the hardness of your unknown mineral? How can you check it if you do not have a mineral hardness kit?

13. What is a cleavage of a mineral?

14. What is the specific gravity of a mineral?
15. What is the basic component of the mineral structure?

16. What is the difference between a solid amorphous material and a solid crystalline material?

17. Does mineralogy have connections to other sciences, industry, politics, and the economy?

18. What important elements/compounds in fertilizer are often provided by minerals?

19. What minerals provide calcium, magnesium, phosphorous and potassium for fertilizers?? List three to four of them and provide their chemical formulas.

20. Why is apatite important, and why is it mined?

21. What is potash? What minerals are mined to supply potash for the agricultural industry?

22. Where is the mineral gypsum used besides the agricultural sector?

23. Which three natural Earth materials are called fossilized fuel?

24. What minerals and native elements are important sources of metal?

25. What is the environmental downside of extracting metals from metal-bearing minerals?

26. What percentage of electricity does the U.S. produce from coal? From natural gas? From oil? From uranium ore in nuclear generators? By hydropower?

27. Are metal-bearing minerals in short supply?

28. What percentage of electricity does the U.S. produce from renewable sources of energy? What renewable sources of energy do you know? Do they have a future, and what will be necessary to increase their production?
29. How large is the total volume of water on the Earth?

30. How large is the volume of freshwater on the Earth?

31. Does the growth of population affect the sustainability of water resources on the Earth?

32. What does environmental science study?

33. What is the connection between industrial mining of mineral resources and environmental science?

34. What is the most visible negative impact that the mining industry has on the environment?

35. Are mineral resources and fossilized fuels renewable?

36. What countries are the major suppliers of conflict mineral resources?

37. Does a positive correlation always exist between mineral resources and economic growth in developing countries? Do political conflicts between countries involve mineral resources?

38. What does the term "conflict mineral" mean?

39. What is coltan, and what is it good for? Why was coltan classified as a conflict mineral?

40. Name three other conflict minerals that are found in the Democratic Republic of Congo (DRC).

41. How large are the resources of coltan in the DRC?

42. What does the U.S. Dodd-Frank Act of 2010 state? Can minerals be important sources of energy?

43. What does the Kimberley Process Certification Scheme regulate? When and by what organizations was the KPCS endorsed?

44. What minerals are called industrial minerals?

45. Where are industrial minerals used? What materials are mined for energy-generating technologies?

46. What does the term "native element" mean in mineralogy? What minerals belong to the native-elements group? Why are they important for industrial engineering and technologies?

47. Crystalline carbon includes diamond and graphite. How are diamond and graphite different? How are their crystalline structures different?

48. What are industrial diamonds and industrial graphite used for?

49. Graphene, nanotubes, and carbon-60 (buckyball) are novel carbon-nanoscale materials. How are they different from one another?

50. What is the implication of carbon-nanomaterial for modern technologies? Are they promising for medical applications?

51. How are the laser industry and the mineral garnet (YAG) related?

52. What are biominerals and biomineralization?

53. What technique is more environmentally friendly: hydrometallurgical leaching of metals from ores with strong chemicals such as HCl, H_2SO_4, and other strong acids, or leaching with the involvement of specific bacteria?

54. What minerals or their compounds are used for biomedical applications? Why are carbon nanoparticles such as C_{60} suitable for medical applications? Is there a downside to their use?

55. Name six minerals of the asbestos group and explain why they are dangerous to human health.

56. List the most important precautions that should be taken during and after you work with minerals in the field or in laboratories.

Quizzes (see answers on page 307)

1. Glass is not a mineral because...
   a. It is made from different elements than minerals are.
   b. It is not crystalline (e.g., there is no repetitively organized molecular structure).
   c. It is never naturally occurring (it only exists as a synthetic product).
   d. It doesn't have a fixed chemical composition.

2. Minerals are used in society...
   a. As fertilizers.
   b. As essential nutrition for the human body.
   c. In electronic devices.
   d. In personal hygiene products (e.g., toothpaste).
   e. All of the above.

3. The U.S. production of electricity consists of...
   a. Solar energy (90%) and domestic oil and gas (10%).
   b. Renewable energy sources (2.5%).
   c. Fossil fuels (coal, gas, and oil–97.5%).
   d. Nuclear generator energy (40%) and fossil fuels (60%).
   e. None of the above.

4. Conflict minerals...
   a. Are mined in the U.S. without permission from a local administration.
   b. Include all minerals of the native-elements group because many of them are noble metals important for electronics.
   c. Are mined in conditions of armed conflict and human rights abuses and may be used to support civil wars and conflicts between neighboring countries in Africa.
   d. All of the above.

5. Asbestos minerals are dangerous for human health because...
    a. They contain traces of mercury, which is a toxic element.
    b. They are extremely flammable at room temperature.
    c. They are radioactive.
    d. They consist of tiny, needle-like fibers that, if you inhale them, become stagnated in the lungs for a long time, often causing cancer.

6. Which one of the listed carbon materials is graphene?
    a. It consists of carbon atoms that are situated at equal distances from one another.
    b. It consists of a single layer of carbon atoms that are packed in a honey-comb-like configuration.
    c. Each molecule consists of 60 carbon atoms arranged in the form of a buckyball.
    d. It consists of a single linear chain of carbon atoms.
    e. Its carbon atoms are organized as tubing with a hole in the middle.

Medical Geology

Chapter 3

What is medical geology? 3.1.

Medical geology is a multidisciplinary field of science that focuses on understanding the relationships among geological materials/processes, biological environments, and human health. Public health problems resulting from the production of metals have been identified in many parts of the world since the industrial era began. The sciences of toxicology and occupational medicine emerged in response to the health problems caused by the industrialization and the development of new technologies and materials. The clinical data on public health also showed that there are huge negative consequences of excess exposure to chemical elements and particulates that are spread in natural environments due to both geological hazards and extensive mining and ore processing. Chemical pollutions and physical destructions of the natural environment and ecosystems, particularly when both geological processes and products of human activities are involved, deserve more attention from scientists, politicians, and society.

In response to such demands, a working group formed in 1996 began to study different aspects of medical geology under the auspices of the IUGS. The tasks of the working group included dissemination of knowledge on hazardous geological processes and Earth's materials that would make people aware of the invisible harmful effects of prolonged exposure on human health. To build up and strengthen interest in medical geology, the working group invited scientists from both, the developed and the developing countries, emphasizing training, exchange of information, and planning of field and laboratory studies. The Commission on Geological Sciences for Environmental Planning (COGEOENVIRONMENT) of the IUGS has established medical geology as an independent discipline and has defined it as "a science dealing with the influence of ordinary environmental factors on the geographical distribution of health problems in man and animals" (Selinus et al., 2007). In 2004, the International Medical Geology Association (IMGA) was founded to promote awareness among scientists, medical specialists, politicians, business communities and the general public on geological hazards and the increasing usage of hazardous Earth's materials for technology. The IMGA expressed concerns about the limited cooperation between geoscientists, public health workers, the business community, and policymakers to solve a range of complex environmental health problems. Nevertheless, the efforts of the IUGS and the IMGA in promoting medical geology have already had significant results. The Hemispheric Conferences on Medical Geology that were held in Puerto Rico (2005), Brazil (2007), Uruguay (2009), and Italy (2010) were followed by a number of other international conferences and sessions at International Geological Congresses and at the International Symposium on Environment and Health in China (2014). Also, many contributions relating to the definition and description of the tasks ahead, the directions of research, and special themes of medical geology have been published (e.g., Finkelman et al., 2005; Skinner and Berger, 2001; Selinus and Frank, 2000; Selinus and Alloway 2005; Dissanayake and Chandrajith, 2009; Randive, 2012). These efforts demonstrate how medical geology is developing as a strong multidisciplinary framework that unites geology and the health sciences—disciplines that until now have existed apart from one another (Fig. 38). Further merging of the geological and health sciences supported by modern technologies and analytical instruments looks very promising, and such an interconnection will definitely unravel some of the interesting aspects of human interaction with the Earth's environments.

3.1.1. Historical roots that connect geological materials to human health

The roots of medical geology extended to the ancient "pre-scientific" time, when no molecular biology or atomic crystallography, mineral chemistry, physics of condensed matter, or medicine based on fundamental scientific achievements existed. In those times, it was not just curiosity, but mostly empirical investigations growing out of the necessity to heal wounds and survive made humans explore natural rocks, minerals, and other Earth's materials, including herbs, algae, insects, birds, fishes and animals. It is conceivable that the interactions among human health, minerals, and the environment have been appreciated throughout all of human history. Rocks and minerals contain the majority of naturally occurring chemical elements and compose all of the continental plates, which are the fundamental fragments of our planet that, in turn, provide cycling of many elements, including those that support sustainable life. In the classical period (~500–300 BCE), ancient Greek philosophers and physicians understood that health and some minerals were causally related and that environmental factors affected geographical distributions of some human diseases. For example, Aristotle (~300 BCE) noted

FIGURE 38 Medical geology and its connections

lead poisoning in sick miners who worked in metallic ore-extraction mines. Hippocrates (~477 BCE) described arsenic minerals such as orpiment (As_2S_3) and realgar (As_4S_4) to be "corrosive, causing burning of the skin, with severe pain." As far back as 2500 years ago, he also formulated wider relationships among the health of the population, geological resources, and the cleanness of the atmosphere, saying: "…if you want to learn about the health of a population, look at the air they breathe, the water they drink, and the places where they live" (Hippocrates, ~477 BCE).

In our day, all of the mentioned interactions are underlying factors of modern medical geology. Historical evidence of geomedical approaches can also be found in Marco Polo's descriptions of his traveling to China. An Italian merchant, Marco Polo embarked on an adventurous journey from Venice, Italy, to China, which took him about twenty-four years. In some of his travel notes, which were later translated into many languages, he described: "a poisonous plant...which if eaten by horses has the effect of causing the hoofs of the animal to drop off" (see English version: Polo and Piza, 2004). What Marco Polo found in the thirteenth century is now believed to be plants that are capable of accumulating selenium (Se), a toxic element of which very high concentrations can be found in some specific rocks and soils (Selinus, 2004). From Polo's description of the symptoms that caused the fatal illness of their horses and the geographic location of that area, Selinus (2004) hypothesized that this could have happened in the Hubei Province of China, where the selenium concentration in natural rocks and soils is extremely high. Thus, Marco Polo's travel notes may be considered to be elements of an ancient medical geology that was formulated only in many centuries (Selinus and Alloway, 2005).

Another example of the "pre-scientific" utilization of minerals and rocks for healing is clay minerals (rock clay), the use of which dates back thousands of years. In the second century CE, Claudius Galenus, the famous Roman surgeon and physician (known also as Galen of Pergamon), noticed that clay was used for healing sick or injured animals. Using these observations, he developed a special clay therapy that successfully healed wounds in human skin. Biblical references mention the healing powers of clay, and many ancient Greek, Egyptian, and some Islamic historical records tell us that clays were used both internally and externally for prevention of gastrointestinal illnesses, physical healing of injuries, sustaining life, and promoting general health. The benefit of clay was also appreciated by the Native American Indians, the natives of Mexico and South America, and civilizations that preceded the Aztecs. They believed in the spiritual significance of clay and used clay as food for body purification and the healing of wounds, as well as for ceremonial events and trading between tribes. In modern times, clay minerals are used for some pharmaceutical and cosmetic applications.

The mechanism of the healing properties of clay is poorly understood, and it is under investigation in many research laboratories around the world (see review by Carratero, 2002). At least, by now it is becoming clear that in the antibacterial actions, the clay particles do not physically penetrate the bacterial cells (e.g., Williams et al., 2008). Instead, as Williams et al. (2008) have explained, the special pH (8.8–7.8) of the water in contact with the clay probably provides specific conditions of chemical oxidation that are adverse to the growth of pathogenic bacteria. The unit pH is a measure of how acidic or alkaline water is. The total pH range of acidity-alkalinity is measured from 0 to 14. Pure water has a pH = 7 (neutral), acid solutions have a pH < 7, and alkaline solutions have a pH > 7. Williams et al.'s research concluded that the clays act like self-contained chemical mini-laboratories where clay somehow buffers the aqueous chemistry, and the latter is the most important process that destroys pathogenic bacteria. Further research for more detailed understanding of the mechanism that governs clay's healing processes is currently in progress.

Special mention should be made of the traditional Chinese medicine (TCM) and Ayurvedic medicine (IAM) in India, both of which, in addition to using numerous herbs, use many different minerals. The medicinal value of more than 350 minerals was collected in China over at least 2000 years, and approximately sixty of them are still used today (Yu et al., 1995). However, the TCM has no scientific foundation, and there are no systematic statistical observations with regard to either the side effects or the benefits of these minerals in terms of human health. The TCM and the IAM were established through long experience of trials and mistakes, and they continue to be official medicines in modern China and India, where their traditional medicine coexists with the Western medicine.

In the West, modern medicine does not accept many TCM and IAM drugs for treatments of maladies, mainly because of the unknown mechanisms involved and the uncontrolled side effects. This makes sense, because minerals have complex chemistries and include a great diversity of trace elements including those known as toxic. In addition, there are no two minerals that have a similar diversity of microelements (trace elements) with the same value of their concentrations. This is because all of these parameters strongly depend on the geological and geochemical conditions in which the mineral of interest was formed. For example, mineral pyrite (FeS_2) will have the same number of main elements (Fe and S) required by its chemical valence balance, but the concentrations of the trace elements and their speciation will be different for pyrite that has originated from volcanic rocks, pyrite that is hosted by hydrothermal veins, and pyrite from metamorphic rocks. Traditional medicines use toxic mineral drugs such as cinnabar (HgS), minium (Pb_4O_3), realgar (As_4S_4), orpiment (As_2S_3), and calomel (Hg_2Cl_2) (see, e.g., Yu et al., 1995; Lui et al., 2008). They use very small doses of these mineral drugs, which they believe are not harmful, but there is no scientific definition of how "small" a small dose is, but only practitioners' experience. An expression of Paracelsus (1493–1541), who was a creator of toxicology, that "…the right dose differentiates a poison and a remedy" clearly points out how crucial it is to choose a non-harmful dosage. Put differently, a negative effect can originate from either increases or decreases in the concentrations of various trace elements that are available in many naturally occurring biological and geological substances. The concentration and diversity of the trace elements were usually unknown in that time because they simply could not be measured due to the absence of any analytical methods implied to minerals. Therefore, it is difficult to imagine now that the "pre-scientific" practitioners were able to justify without analytical methods the therapeutic effects of the mineral drugs that they used for healing. Apparently, in traditional medicine, there is a gap between practical observations of the effects of treatment with mineral drugs, choices of non-harmful dosages, and understanding of the mechanisms of mineral drugs' interactions with biological systems.

From the point of view of the modern geosciences, the best example to demonstrate this gap is arsenic-bearing mineral drugs. In ancient times, since Aristotle (384–322 BCE), these mineral drugs were used for treatment of skin diseases, for bald head scab disinfectant, and for psoriasis, inflammation, detoxication, diarrhea, abdominal pain, and possibly even leukemia (Liu et al., 2008). The minerals of the arsenic group—orpiment and realgar—mostly occur within hydrothermal veins, which form specific mineralogical ore assemblages that may also include native Pb, Hg, and other metals and metalloids and their derivates. The tiny, several-nanometer-to-micrometer size inclusions of native Hg (a metal liquid) may be hosted by realgar and/or orpiment, as we now know from studies of the geochemistry of polymetallic ore deposits. The liquid Hg is extremely toxic, and if practitioners could not sort out how high the possible Hg concentration might be in a bulk sample of any realgar or orpiment, they would not be able to balance the benefits and risks. Although some descriptions of successful treatment by traditional medicine are known, they all need analytical verification of the trace element species and their concentrations by modern methods and instruments, and a full scientific justification is required for their consumption.

Several decades ago, arsenic evoked concern within world health organizations because traditional medicines were becoming more popular as alternatives to and supplements for the Western medicine (e.g., Kumar et al., 2006). Academic studies of the time were focused first on analyzing the risk of adverse effects and the potential benefits of arsenic minerals using data available in traditional medicine. Western medical knowledge is still a work in progress, and it includes intensive studies of the influence of arsenic and other toxic minerals and elements on human health. For example, in 1942, the U.S. implemented an interim standard for arsenic in drinking water to be not more than 0.05 ppm (or 50 µg/L). In 1995, the World Health Organization (WHO) recommended a drinking water standard of 0.01 ppm (or 10 µg/L), and the U.S. Environmental Protection Agency (EPA) also lowered its arsenic-dose standard in drinking water to not more than 0.01 ppm (or 10 µg/L) in 2001, with the ruling delayed to 2002 (e.g., Hughes et al., 2011). However, many studies have also revealed a beneficial effect of arsenic in the treatment of some forms of cancer (Antman, 2001), and in 2001, the U.S. Food and Drug Administration (FDA) approved arsenic trioxide for leukemia chemotherapy (e.g., Soinget et al., 2001; Lallemand-Breitenbach and de Thé, 2013). Direct treatment of some forms of leukemia with As-trioxide without chemotherapy is a parallel action of modern medicine (e.g., Au et al., 2003; Ghavamzadeh et al., 2006).

Other toxic minerals mentioned in this chapter—cinnabar (HgS), minium (Pb_4O_3), and calomel (Hg_2Cl_2)—also have connections to both "pre-scientific" medicine and cases of poisoning. Cinnabar, a mercury sulfide ore, was widely used from the fourth to the first century BCE in Greece for extraction of mercury by separating it from sulfur in the presence of copper and vinegar (Lins and Oddy, 1975). After melting cinnabar ore and copper together, a by-product, copper amalgam, was subjected to heating and condensation to release pure mercury that was used for the gilding of silver and gold. The silver-mercury and gold-mercury amalgams were spread over the base of any other metal surfaces and heated to evaporate the mercury. At the end of this stage of gilding, a thin film of gold or silver remained adhered to the surface of a metal. In ancient Greece, it was considered prestigious to make gold statues, leaves, and wreaths, with the latter serving as awards for victors in both athletic competitions and wars. From excavations by modern archeologists, we know that the gilding of decorative wood for palaces and the gilding of ceramics had also been undertaken for centuries, not only in Greece, but also in Egypt, China, and many other civilizations. These historical facts provide indications for modern toxicologists that during the gilding procedures, the vapors of mercury, which have no color and no odor, were directly inhaled by many workers involved in this process, which could have caused damage to the brain, kidneys, and lungs, and even death. Later, during the Bronze Age, the mercury-based gilding technique was replaced by leaded tin bronze with a switch from forging to casting. Though lead alloys or lead oxides—e.g., the mineral minium (Pb_4O_3)—are also toxic, the available evidence supports the assumption that the potential for toxic exposure might have been deduced (Harper, 1987).

With regard to the toxicity of calomel (Hg_2Cl_2), it worth referring to historical facts and speculations surrounding death of a famous historic figure, Napoleon Bonaparte (1769–1821), who was an emperor of France from 1804 to 1814. After Napoleon's army was defeated in 1815 at the famous Battle of Waterloo, the British government exiled Napoleon to Saint Helena, a small island in the South Atlantic, where his health began to deteriorate, with symptoms of either stomach ulcers or cancer, followed by his death in 1821. Research conducted on the authenticated hairs of Napoleon by the Harwell Nuclear Research Laboratory, London, in 1960 and by the Toxicology Crime Laboratory of the U.S. Federal Bureau of Investigation (FBI) in 1995 revealed toxic levels of arsenic. Weider and Fournier (1999) hypothesized that Napoleon was poisoned by a cathartic orange-flavored drink called *orgeat*, which consisted of calomel (Hg_2Cl_2) flavored with the oil of bitter almonds. These substances together formed mercury cyanide, which is lethal. The authors also believe that Napoleon was poisoned with arsenic which over time has weakened him, making the associated debility appear to be a natural illness and thus allaying any suspicions prior to the final lethal phase.

3.1.2. Crossing boundaries of different disciplines

Although minerals and the products of their decomposition during heating and other technological processes were known for centuries, the direct relationships and mechanisms of the interactions of chemical elements and compounds derived from minerals with human populations was not established until four or five decades ago. It was not possible for such a long time because there were difficulties to integrate the knowledge of different experts, each viewing the problem from their distinctive disciplinary perspectives. For a multidisciplinary team that includes a variety of experts from different scientific fields, it is necessary to have a common vision in understanding of the problems to be solved because usually each specialist may consider his or her contribution as only the field of his or her discipline.

Interdisciplinary study is a complex process which includes answering a question, testing working hypotheses, modifying too broad tasks to be dealt with adequately by a single discipline, and integrating existing insights to construct a more comprehensive understanding. Interdisciplinary courses, such as "Medical Geology," "Environmental Sciences and Urbanization," or this course, "Minerals and Human Health," as well as many others similar to them, are now in demand to prepare a new generation of geologists, biologists, environmental scientists, occupational medicine specialists, and physicians to work together. Archeologists have also noted links between human health and geological/environmental factors. Modern element analysis of bone materials is an excellent approach for studying the diet and nutritional status of prehistoric humans and animals. Furthermore, fossils not only provide the geological record of evolution and timing, but also remind us of the crucial role that mineralized tissues play in the biology of modern organisms.

One more important aspect for progress in interdisciplinary integration is the availability of sophisticated analytical instruments and powerful computers. They can help scientists to collect multidisciplinary data and gain a better understanding of how the geochemical and mineralogical environments affect the health and well-being of human populations.

3.1.3. Medical geology and environmental management issues

Medical geology is a complex discipline in which subjects may proceed by many two-way streets, and therefore, it is important to emphasize the most vital directions and management issues. Recent international studies supported by the World Bank, UNEP, the British Geological Survey (BGS), IUGS, and the U.S. Geological Survey (USGS) have focused on such topics as:

i. The search for correlations between cancer and high concentrations of arsenic in drinking water in Bangladesh;
ii. Mercury and arsenic contamination of land and ground water related to gold mining in some regions of Africa and Asia;
iii. Dental and skeletal deformities caused by high concentrations of fluorides in water in Africa, Asia, and Central Europe;
iv. High concentrations of uranium in drinking water in Jordan;
v. Heart disease triggered by insufficient content of selenium in China;
vi. Goiter induced by iodine deficiency in Sri Lanka and China.

In a general sense, the current and future directions of medical geology (Fig. 39) include the following problems:

i. Local and intercontinental dust storms;
ii. Toxic metals and metalloids, mercury and arsenic emissions from coal combustion, the accumulation of these toxic substances in the environment, and the impacts of mercury and lead mobilizations in regions where artisanal gold mining is conducted;
iii. Human exposure to fibrous minerals such as those in the asbestos group and zeolites from mining and ore processing, from the construction industries, and from natural outcrops and tailings;
iv. Biogeochemical cycling and the potential geomedical relevance of iodine, radon, and selenium;

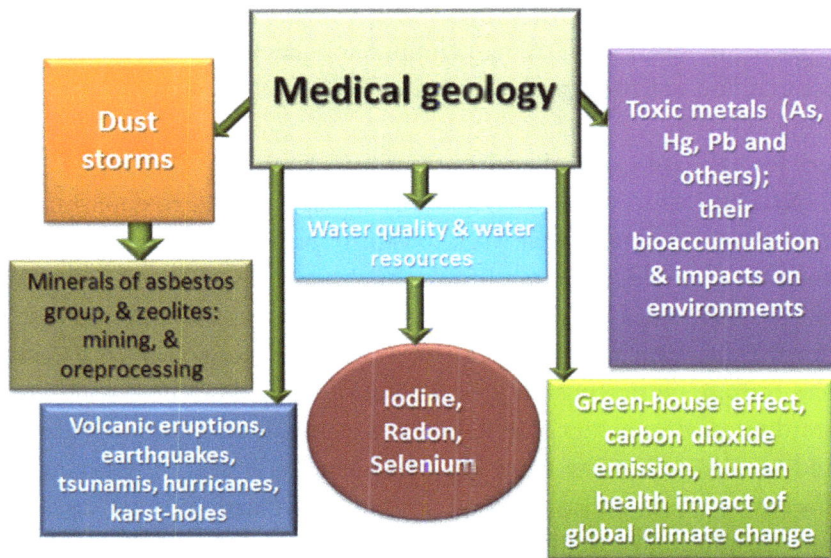

FIGURE 39 Medical geology: current and future studies

v.   Water resources and water quality;

vi.   The residual health impacts of geologic processes such as volcanic emissions, earth-quakes, tsunamis, hurricanes, karst-holes (sinkholes), etc.;

vii. The greenhouse effect, carbon dioxide emissions, and the health impacts of global climate change.

More than three billion people in the world are afflicted by environmental health issues, but some of these problems could be avoided, prevented, or mitigated if knowledge about geological processes and Earth's materials were taken into consideration by planners, manufacturers, policymakers, and government organizations. In addition, extensive interdisciplinary researches are needed to provide new insights on the interactions between the Earth's processes and industrial activities. This knowledge should be used to estimate the ratios between natural hazard input, which may not be manageable, and human activity-induced hazards that can be minimized. The general public and the news media very often have no deep understanding of the complicated interactions among geological conditions, atmospheric processes, and biodiversity, and they take into account only anthropogenic factors (i.e., the influences of human beings on nature), which are easier to observe during the scale of human lives. However, the medical geology and neighboring fields of the basic sciences should pursue studies that will help to find quantitative units of measurement that would show the extent of the impacts of industrial-anthropogenic activities on global changes in the geological and atmospheric processes of the Earth.

In the U.S., there are several government and public agencies and programs that direct and coordinate research and activities related to the environment and the health of populations. These include the following.

The Centers for Disease Control and Prevention—National Center for Environmental Health (CDC–NCEH) (http://www.cdc.gov/nceh/) is responsible for establishing and coordinating different national programs that work toward improving the health of the population. The center promotes knowledge about a healthy environment to prevent and avoid illnesses, disabilities, and premature deaths from non-infectious, non-occupational factors. The NCEH houses a National Environmental Public Health Tracking Network, which is one of the best Internet sources (http://ephtracking.cdc.gov) for information about diseases that are possibly caused by hazardous environments. The network is a unique one because it brings together data that would usually be kept by many separate agencies. It also provides maps related to diseases and the environment, and appropriate scientific information. States, cities, universities, and professional organizations use this information to help them in making crucial decisions related to environmental health resources that protect people and save lives.

The Agency for Toxic Substances and Disease Registry (ATSDR) also works under the auspices of the CDC, and it bears responsibility for addressing threats from toxic substance exposures (http://www.atsdr.cdc.gov). The ATSDR database refers to more than 250 toxic substances, including such extremely dangerous ones as arsenic, cadmium, asbestos, radon, mercury, lead, and others. The ATSDR's task is to identify places in the U.S. where people might be exposed to hazardous substances in the environment, to determine the severity of hazards, and to make recommendations about what people should do in order to be safe. The ATSDR works closely with communities, environmental groups, state and tribal governments, and other federal agencies.

The Environmental Protection Agency (EPA) monitors the quality of the environment, creates standards, and engages experts from related scientific fields and from a diversity of other disciplines, including economics, law, public policy, communications, and information sciences, in order to provide sustainability for the environment and human health (www.epa.gov). The EPA's goals include ensuring air quality, taking actions on climate change, protecting safe and sustainable water resources, developing and applying innovative technologies for water management and large-scale water reuse, and assuring the safety of chemicals and ecosystems so as to promote the health of the human populations.

References

Antman, K. H. 2001. Introduction: The history of arsenic trioxide in cancer therapy. *Oncologist* 6:1–2.

Au, W. Y., C. R. Kumana, M. Kou, R. Mak, G. C. F. Chan, C. W. Lam, and Y. L. Kwong. 2003. Oral arsenic trioxide in the treatment of relapsed acute promyelocytic leukemia. *Blood* 102:407–408.

Carretero, M. I. 2002. Clay minerals and their beneficial effects upon human health. A review. *Applied Clay Science* 21:155–163.

Dissanayake, C. B., and R. Chandrajith. 2009. *Introduction to medical geology*. Erlangen Earth Conference Series. Berlin, Heidelberg: Springer, p 320.

Ghavamzadeh, A., K. Alimoghaddam, S. H. Ghaffari, S. Rostami, M. Jahani, R. Hosseini, A. Mossavi, E. Baybordi, A. Khodabadeh, M. Iravani, B. Bahar, Y. Mortazavi, M. Totonchi, and N. Aghdami. 2006. Treatment of acute promyelocytic leukemia with arsenic trioxide without ATRA and/or chemotherapy. *Annals of Oncology* 17:131–134.

Finkelman, R. B., J. A. Centeno, and O. Selinus. 2005. The emerging medical and geological association. *Transactions of the American Clinical and Climatological Association.* 116:155–165.

Harper, M. 1987. Possible toxic metal exposure of prehistoric bronze workers. *British Journal of Industrial Medicine* 44:652–656.

Hughes, M. F., B. D. Beck, Y. Chen, A. S. Lewis, and D. J. Thomas. 2011. Arsenic exposure and toxicology: A historical perspective. *Toxicological Sciences* 123:305–332.

Kumar, A., A. G. Nair, A. V. Reddy, and A. N. Garg. 2006. Bhasmas: Unique ayurvedic metallic-herbal preparations, chemical characterization. *Biological Trace Elements Research* 109:.231–254.

Lallemand-Breitenbach, V., and H. de Thé. 2013. Retinoic acid plus arsenic trioxide, the ultimate panacea for acute promyelocytic leukemia? *Blood* 122:2008–2010.

Lins, P. A., and W. A. Oddy. 1975. The origins of mercury gilding. *Journal of Archaeological Science* 2:365–373.

Liu, J., Y. Lu, Q. Wu, R. A. Goyer, and M. P. Waalkes. 2008. Mineral arsenicals in traditional medicines: Orpiment, realgar, and arsenolite. *Perspectives in Pharmacology* 326:363–368.

Polo, M., and R. Pisa. 2004. *The Travels of Marco Polo*. Produced by C. Franks, R. Connal, J. Williams, and PG Distributed Proofreaders. [EBook #10636]. http://www.gutenberg.org.

Randive, K. R. 2012. *Elements of geochemistry, geochemical exploration and medical geology*. Singapore: Research Publishing Service.

Selinus, O., and A. Frank. 2000. Medical geology. In Moller, L. (Ed.). *Environmental Medicine*. Stockholm: Joint Industrial Safety Council.

Selinus, O. 2004. Medical geology: An emerging specialty. *Terrae* 1:8–15.

Selinus, O., and B. J. Alloway. 2005. *Essentials of medical geology: Impacts of the natural environment on public health*. Amsterdam: Academic Press.

Selinus, O., R. B. Finkelman, and J. A. Centeno. 2007. The medical geology revolution—The evolution of an IUGS initiative. *Episodes* 30:1–5.

Skinner, H. C., and A. R. Berger. 2003. *Geology and health:closing the gap.* Oxford, U.K.: Oxford University Press.

Soignet, S. L., S. R. Frankel, D. Douer, M. S. Tallman, H. Kantarjian, E. et al. 2001. United States multicenter study of arsenic trioxide in relapsed acute promyelocytic leukemia. *Journal of Clinical Oncology* 19:3852–3860.

Weider, B., and J. H. Fournier. 1999. Activation analyses of authenticated hairs of Napoleon Bonaparte confirm arsenic poisoning. *American Journal of Forensic Medical Pathology* 20:378–382.

Williams, L. B., S. E. Hayde, F. R. Geise, Jr., and D. D. Eberl. 2008. Chemical and mineralogical characteristics of French green clays used for healing. *Clays and Clay Mineralogy* 56:437–452.

Yu, W., H. D. Foster, and T. Zhang. 1995. Discovering Chinese mineral drugs. *Journal of Orthomolecular Medicine* 10:31–59.

Web Resources

http://www.cdc.gov/nceh
http://ephtracking.cdc.gov
http://www.atsdr.cdc.gov
www.epa.gov

Image Credits

Review Questions

1. What is the definition of medical geology?

2. What does medical geology study?

3. What has been the role of the International Union of Geosciences (IUGS) in the establishment of the medical geology discipline?

4. When and under what auspices were the first working groups of medical geology formed?

5. Why do we need such a discipline as medical geology?

6. Will developing countries benefit from introducing medical geology into their educational curriculums?

7. Should we be concerned about the cleanliness of the air we breathe and the water we drink? If yes, explain why.

8. Describe and analyze some prehistoric roots of medical geology.

9. Why did prehistoric people start to use minerals to treat some illnesses? How did they collect knowledge about the positive and adverse effects of minerals?

10. Did "pre-scientific" physicians know about the long-term adverse effects of their herb and mineral-drug medicines?

11. What kind of traditional medicine is still in official use along with Western (scientific) medicine?

12. Is consumption of traditional medicine mineral drugs dangerous? If so, explain why.

13. Do you know any historical examples of mineral drugs being used for the assassination of politicians or public figures?

14. What kinds of knowledge related to modern medical geology did we gain from Marco Polo's notes about his traveling to China in the thirteenth century?

15. What kinds of difficulties arise in interdisciplinary studies?

16. Medical geology as a discipline was formulated only recently, though both positive and adverse effects of minerals on human health have been known for centuries. Why did it take so long for medical geology to be recognized?

17. Medical geology crosses the boundaries of what disciplines and human activities?

18. Do you think that the medical geology discipline has a future? Will it be modified, and if so, in what directions?

19. What is the role of the U.S. Environmental Protection Agency (EPA)? Is this a private or a government organization?

20. If you do a search on the Internet through the National Environmental Public Health Tracking Network, what kinds of special information will you be able to find there?

21. What kinds of recommendations would you expect to hear from the Agency for Toxic Substances and Disease Registry? What kinds of questions related to medical geology would you address to this agency?

Quizzes (see answers on page 307)

1. Medical geology is a discipline that studies…
   a. Public health.
   b. Environmental and geological sciences.
   c. Toxic metals released from mining operations.
   d. Products of coal combustions.
   e. All of the above.
   f. Only a, b, and c.

2. Consumption of traditional medicine mineral drugs may be dangerous because…
   a. Natural minerals that are used for drugs may contain toxic trace elements for which the mechanism of interaction with biological cells in most cases is not scientifically verified.
   b. Though positive and adverse effects of traditional medicine drugs are known (as many believe), this is only empirical (i.e., observational) experience, not a scientific data.
   c. Both of the above.
   d. Only b is correct.
   e. None of the above.

3. Marco Polo, a famous Italian merchant who traveled to China in the thirteenth century, described in his diary that his horses were poisoned to death by eating some plant which, as we know now, contained selenium (Se), a toxic element. To what geological process would you attribute that plant enrichment with Se?
   a. A local earthquake that released an underground gas rich in Se, which was deposited on the leaves of the plants and contaminated them.
   b. Weathering of the Se-rich rocks followed by the formation of Se-rich soil.
   c. Se-rich rains that contaminated the plants.
   d. All of the above.

Geological Hazards—Mineral Dust, Aerosols, and Dust Storms

Chapter 4

Earth's atmosphere composition and air pollution: Background

Humans and any biological organisms cannot live without oxygen because it is needed to support basic metabolic reactions. It is also well known that the proper functioning of humans depends on the ratio of oxygen and CO_2 that we inhale; if one inhales air that contains too much CO_2, the consequences can be lethal. The Earth's atmosphere consists of 20.946% oxygen (O_2), 78.084% nitrogen (N_2), 0.9340% argon (Ar), 0.037680% carbon dioxide (CO_2), and other gases such as hydrogen (H_2), neon (Ne), helium (He), krypton (Kr), and methane (CH_4); (see Fig. 40).

In some non-scientific sources, one may find information that *the concentration of oxygen varies with elevation* and that at high altitudes, travelers become sick *because of a deficiency of oxygen*. Though high-mountain sickness is a real phenomenon, the concentration of oxygen in the Earth's atmosphere is constant around the globe. However, the air in the high mountains becomes less compressed and is therefore "thinner" if compared with air at or near sea level, where most humans live. The expression "thinner air" means that in a given volume of air, there are fewer molecules present, and this is what causes the feeling that there is not enough oxygen to breathe.

In our everyday lives, we worry about air pollution and express concern when we observe smog obscuring the blue sky, even though we know that the frequency of such phenomena depends on geography and the proximity of industrial activities to the places where we live. During smoggy days, many of us have the impression that dangerous smog saturated with pollution particles is far away from us because the exact distance from the polluted air is impossible

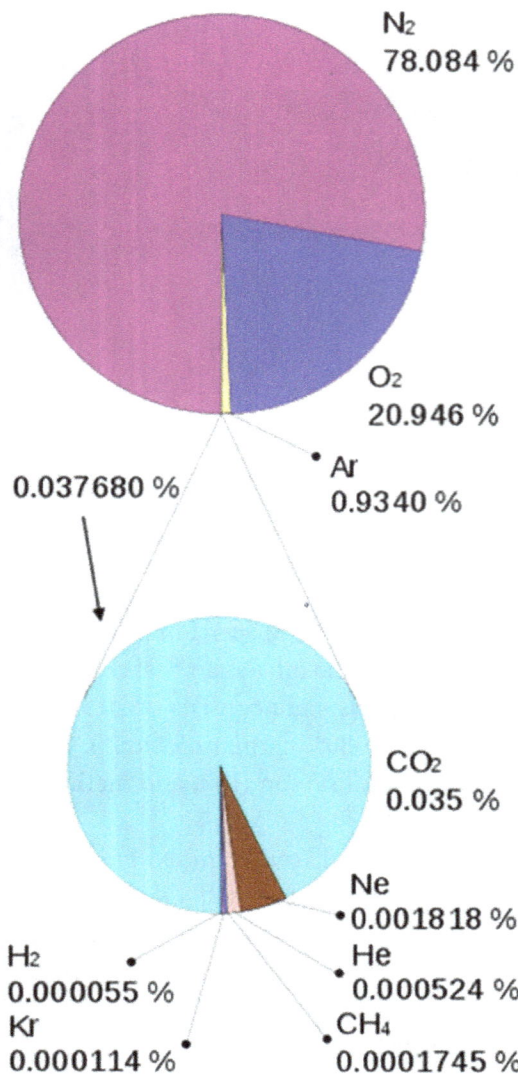

FIGURE 40 Composition of the Earth's atmosphere

to determine by simple observations. Small and invisible pollution particles float everywhere, including our classrooms, homes, highways, streets, at ground level, and spread further into higher regions of the atmosphere. How should one define whether the air is polluted? Air pollution is characterized by a high concentration of particulate matter (PM) that can cause harm to humans and the surrounding environment.

Particulate matter (sometimes also called particle pollution or PP) is a complex mixture of extremely small particles of different states, compositions, and origins. Particulate matter is more often classified in three categories based on its physical state properties (Fig. 41):

i.   Solid particles that consist of rocks, minerals, metals, and organic matter, including bacteria;
ii. Gases;
iii. Aerosols that are presented by liquid droplets of different chemistries.

The U.S. EPA recognizes two categories based partly on the size of the particles and partly on their origin:

i.   Inhalable coarse particles with diameters in the range 2.5–10 μm ($PM_{2.5}$–PM_{10}), such as the particles that compose the dust occurring in the vicinity of roadways, construction areas and dusty industries.
ii. Fine particles with diameters of 2.5 μm ($PM_{2.5}$) or smaller ($PM_{<2.5}$) that can be found in smoke and haze originating from forest fires, power plants and other industries, and automobile engines that emit gases into the air.

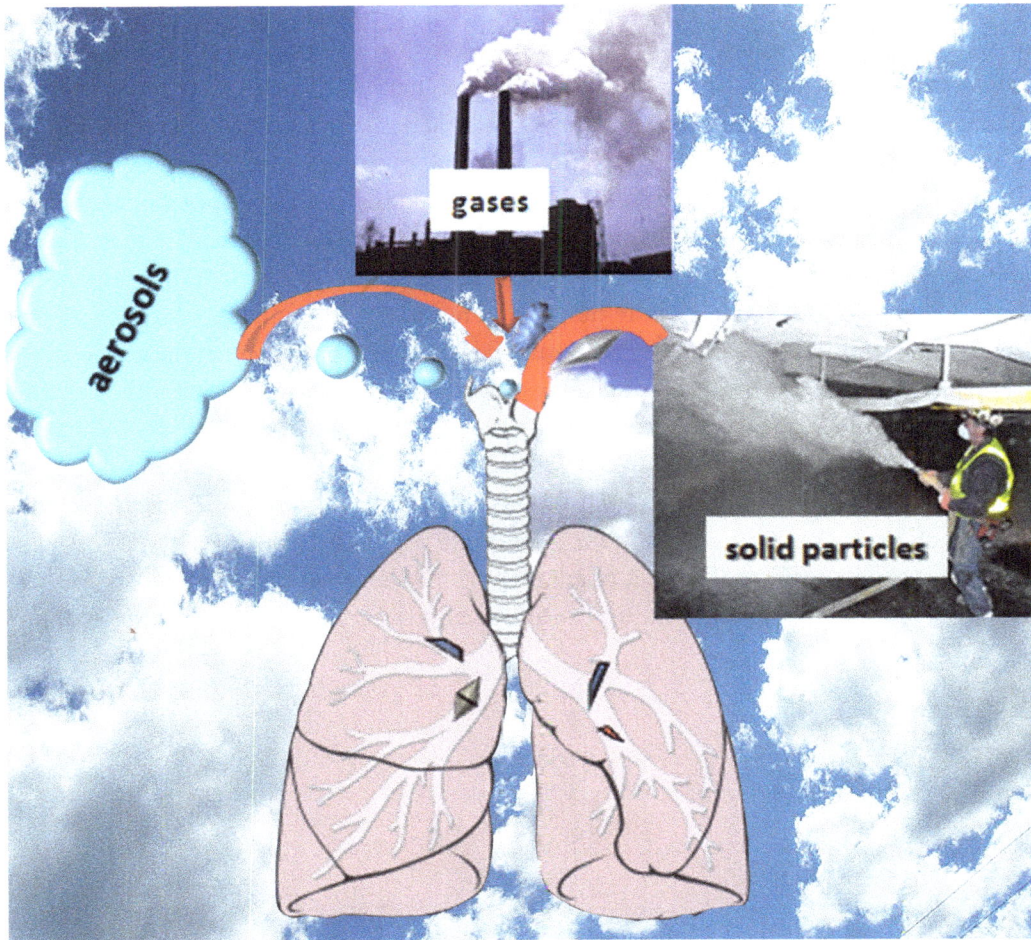

FIGURE 41 Three components of air pollution: solid particles, gases, and aerosols

The harmful pathway of air pollution is easy to understand. As we inhale air, we introduce pollution micro-components into our respiratory system, and some of the solid particles may travel deeper into the lungs and stack up among the lung tissues. If they reside there for a long time, the solid particles of minerals and organic matter can cause serious lung diseases such as asthma and cancer, as well as many other occupational illnesses such as farmer's lung, malt worker's lung, mushroom worker's lung, silicosis, coal worker's pneumoconiosis, asbestosis, and many other inflammatory lung diseases (e.g., Peters et al., 2012; Malo et al., 2013). Observations show that after two to three hours of exposure to dust of organic origin containing some fungus or mold, many agricultural workers develop a dry cough and breathlessness. Repeated, uncontrolled exposure may lead to severe conditions followed by fever and breathless rest (e.g., Rom and Markowitz, 2007).

In many cases, air pollution is caused by both geologically hazardous processes and industrial processes related to human activities or by one of them at a time. These processes include mechanical disintegration, mixtures, and chemical reactions that occur between solid, liquid, and gaseous particles in different combinations. During only mechanical disintegration, larger solid particles are broken down into smaller fragments without any compositional changes in the materials. For example, during intensive weathering processes (see Chapter 1) in mountain areas, solid rocks and minerals disintegrate into fragments and particles of different sizes that result in dust wind or dust storms. Activities around industrial and construction areas, open pits and quarries in mining territories, agricultural fields, coal and oil combustion, and forest fires, among others, produce many of the smaller-to-larger or coarse particles. Gas emission and the burning of fossil fuels are considered to be contributors to the chemical process because their small particles can further react with water droplets or combine with other compounds available in the atmosphere to produce complex pollutant particulate matter. A simplified scheme of these interactions is shown in Fig. 42. It includes:

(i) The greenhouse effect. This is a phenomenon that is formed in the atmosphere due to elevated concentrations of gases that can absorb and emit infrared radiation. The most abundant

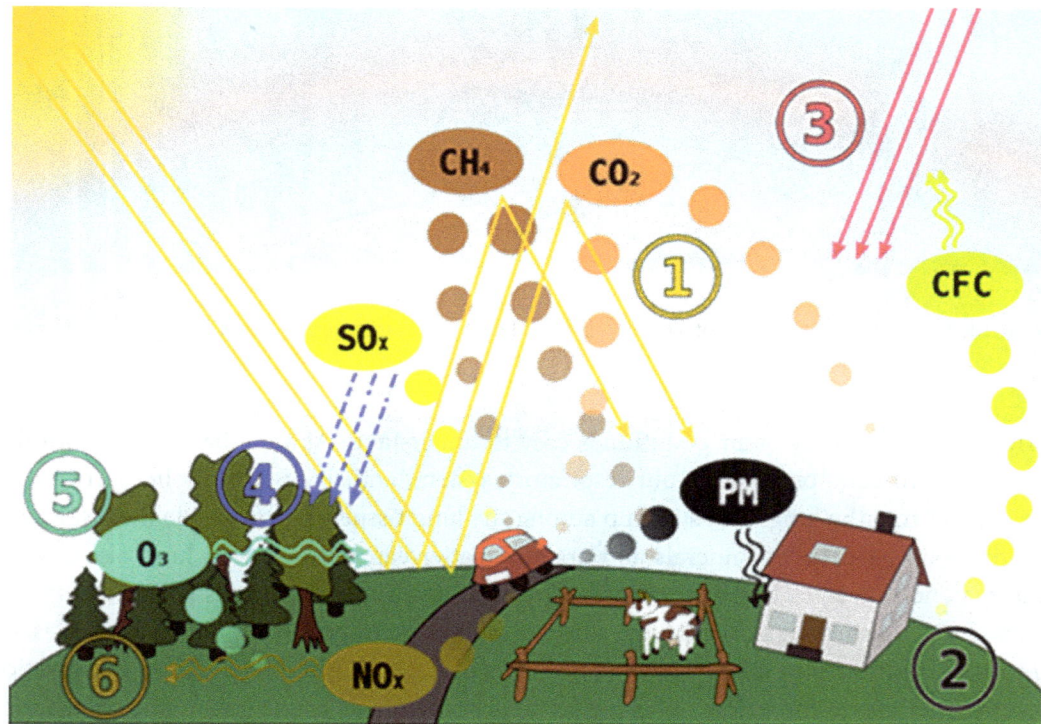

FIGURE 42 The diversity of the matter and phenomena that contribute to air pollution: 1, greenhouse effect; 2, particulate matter from land; 3, ultraviolet radiation (UV); 4, acid rain; 5, ground level of ozone concentration; 6, ground level of nitrogen oxides

gases that cause the greenhouse effect are water vapor (H_2O), carbon dioxide (CO_2), methane (CH_4), nitrous oxide (N_2O), ozone (O_3), and chlorofluorocarbons (CFCs).

(ii) Particulate matter from agricultural lands, and geological and atmospheric processes (volcanoes, earthquakes, tornados, wind storms, etc.). These consist of a mixture of solid particles, gases, and aerosols.

(iii) Ultraviolet (UV) radiation from the sun. The EPA classifies UV radiation from the sun and from tanning beds as a human carcinogen (http://www2.epa.gov/sunwise/health-effects-uv-radiation).

(iv) Acid rain. This is a broad term for rain that contains higher than a standard concentration of nitric and sulfuric acids. Acid rain is formed by both natural sources (e.g., volcanic eruptions and the decay of vegetation) and industrial activities (e.g., fossil-fuel combustion and byproducts from chemical/metallurgic plants). All of these natural and industrial sources are accompanied by the emission of sulfur dioxide (SO_2) and nitrogen oxides (NO_x) into the atmosphere, which react with droplets of water and precipitate down to the Earth as acid rain. The acidity of rain can be monitored by measurements of pH—i.e., the hydrogen ion concentration that characterizes the acidity or alkalinity of a solution. For comparison, recall (Chapter 3) that normal, pure water has a pH of 7.0, whereas a normal rainfall is usually slightly acidic (pH = 5.6) because some CO_2 that is available even in clean air is dissolved into H_2O and forms weak carbonic acid (H_2CO_3). Usually the pH of normal rain has been given a value that varies from a high of 5.6 to a low of 4.5 with the median value of 5.0. The most acidic rain (pH = 4.3 – 5) was recorded in the Eastern United States in 2000.

The man's contribution to the increasing/decreasing acidity of rainfalls will never be estimated, because the "natural background" of the rain acidity cannot be known. For example, in 1960 at the headlands of the Amazon River, in areas remote from urban activities, the monthly 100 rain events used to have an acidity ranging from pH = 4.3 – 5.2. On the island of Hawaii in the remote areas from industrial pollutions the median value of rainfalls pH was calculated as 5.3 over 4 years period of observations with a minimal value of pH = 3.8.

Acid rain precipitations may destroy forests, land, and other external environments, as well as buildings, historical monuments, and statues, especially those made from carbonate rocks such as marble and limestone (Fig. 43), and it can cause corrosion of metals, etc. It can also affect human health by causing lung problems such as asthma and bronchitis (Jie et al., 2013).

(v) Ozone (O_3). Though ozone is usually concentrated in the troposphere (e.g., 12–20 km above the surface of the Earth), it may be collected partly by the ground layers of soils/land from products of industrial combustions. Even very small concentrations of ozone may have a harmful effect on human health, and it is particularly known to worsen asthma conditions among children (Nawahda, 2013).

(vi) Nitrogen oxides (NO_x). About 1% of nitrogen oxides are formed naturally in the atmosphere by lightning, and they are also produced by plants and soils. The much larger volume (~80%) of the nitrogen oxide gases emitted into the atmosphere come from the burning of

FIGURE 43   (a) Forest and (b) sculpture damaged by acid rain

fossil fuels (coal, oil, and gas), such as exhaust emissions from vehicles. Other sources are metal refining and manufacturing industries and food processing. Breathing air with a high level of nitrogen oxides—where ozone may also be present—can cause serious respiratory problems (Mauzerall et. al., 2005).

Since the 1970s, the U.S. has continued to make substantial efforts to improve air quality. Such initiatives include using alternative sources of energy that do not produce and emit carbon dioxide into the atmosphere, as do coal, oil, and gas (Nemet et al., 2010). Natural gas is one of the most carbon-reduced fuels because it has less CO_2 emission than oil and coal. Alternative sources include solar power, wind, geothermal energy, low-head hydropower, hydrokinetics (e.g., wave and tidal power), and nuclear power—all of which are carbon-free. As to whether to consider biofuels as alternative, carbon-free or carbon-reduced sources of energy depends on the composition of the primary biomass source and the technology that is used to generate energy. Finally, another way to reduce air pollution is legislations and regulations by the EPA and other organizations governing the air quality control.

Dust   4.2.

Dust consists of solid particles of different sizes and shapes, and it can originates from hundreds of different minerals, rocks, volcanic ashes, meteoritic fragments, and any solid industrial or domestic materials and chemicals. Dust can be divided into many different categories based on the different physical and chemical properties of the constituent particles, the sources of origin, and the dynamics of the wind flow.

Primary and secondary particles. Particles of dust are divided into primary and secondary based on their physical and chemical characteristics. The primary particles are emitted into the air directly from their sources, such as construction sites, unpaved roads, freeways, agricultural fields, outcrops of "soft" geological formations, volcanic eruptions, earthquakes, tornados, smokestacks, and fires. The secondary particles form in the atmosphere from chemical compounds/elements (e.g., C, H, S, N, O, SO_x, NO_x, CO_x, and others), and their composition varies widely depending upon the location and time of year, emissions from industrial processes, smoke from the burning of wood and coal, and motor vehicle and other engine exhausts.

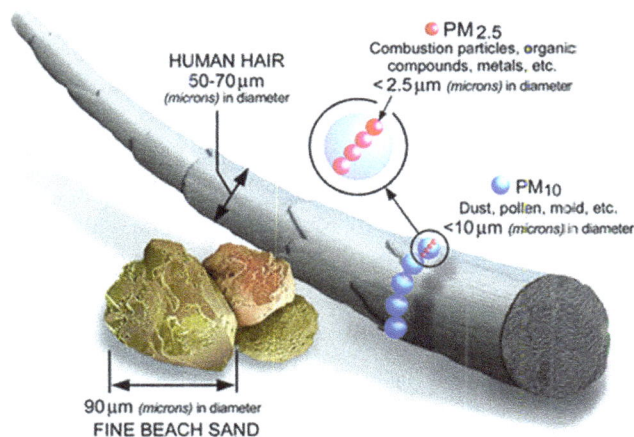

PM2.5
Combustion particles, organic compounds, metals, etc.
<2.5 μm *(microns)* in diameter

HUMAN HAIR
50-70 μm
(microns) in diameter

PM10
Dust, pollen, mold, etc.
<10 μm *(microns)* in diameter

90 μm *(microns)* in diameter
FINE BEACH SAND

FIGURE 44 Sizes of particulate matter PM2.5 and PM10 in comparison with human hair and fine beach sand

TABLE 6:   Lifetimes of Dust Particles

Particle Diameter (µm)	Time to Fall (Altitude - 1000 m)
0.02	228 years
0.1	36 years
1.0	328 days
10.0	3.6 days
100	1.1 hours
1000	4 min
5000	1.8 min

Size of dust particles. The size of dust particles (Fig. 44) is a very important parameter for understanding the dynamics of their transportation and lifetime when they participate in wind erosion and dust storms. Fig. 44 shows the size of a dust particle in comparison with a human hair and a fine-grain sand particle. Smaller particles of <10 µm in diameter are too cohesive (sticky) to be blown by the wind, and larger particles of >500–1000 µm are too heavy for the wind to lift, which explains why both the smaller and larger particles suspend in rivers. Particles of intermediate size that fall within the range 100–500 µm can easily "fly" for short to moderate distances. However, if there is a strong wind, the situation will be different for the smaller particles, which, despite their cohesiveness, will be transported over long distances. Once the small particles become airborne, they can go from one continent to another, forming intercontinental dust clouds. The atmospheric lifetime of dust particles depends on their size (Table 6).

Large particles with diameters from 1000 to 5000 µm quickly fall down from the 1000-m altitude due to gravitational settling; their lives in the atmosphere are about 1.8 and 4 minutes, respectively. The atmospheric lifetimes of sub-µm sized (i.e., <1 µm) particles can be up to 328 days, particles of 1.0-µm diameter can have lifetimes of up to 36 years, and extremely small particles of 0.02 µm in diameter can have lifetimes of 228 years or more (Table 6).

4.3.   Domestic (house) dust

All of us are familiar with domestic dust and how it looks when we clean our desks, surfaces of TV sets, furniture, and floors in our houses. The house dust is well known as a trigger of different allergy conditions and asthma diseases that affect millions of people around the world. Although allergy to house dust does not cause any direct threat to human life, asthma is a serious respiratory dysfunction that affects the airways that carry air to and from the lungs. Many scientists now believe that the presence of specific organisms (e.g., mites—*Dermatophagoides pteronyssinus*) in domestic dust is the greatest cause of allergic reactions and asthma, although immunologic mechanisms are complex, and understanding them continues to be a focus of much researches (e.g., Gregory and Lloyd, 2011). House dust mites are very small (0.1–0.6 µm) micro-organisms, but they are not insects, and therefore, they cannot be killed by any insect-killing chemicals. Their period of life lasts two to five months and strongly depends on the environments in which they are found.

What kind of environments do they prefer? Mites feed on skin scales shed by humans and their pets, which accumulate microfungi, yeasts, and other micro-organisms, and they also consume organic detritus from food available in houses (Platt-Mills and de Weck, 1989). Statistical data show that during the past few decades, the prevalence of house-mite-dust allergies and asthma has increased worldwide. The exposure to house dust allergens may be increased due to improved insulation and heating of houses, which now are much better than a decade ago. Furthermore, ventilation of buildings is internal and is not supported by the opening of windows and doors anymore, and there is more carpeting on the floors in homes and offices—both of which are factors that collect dust inside buildings, making ideal habitats for dust mites (e.g., Hart, 1995). The number of factors may grow even further as more studies are undertaken. For more information, see http://housedustmite.com.

Atmospheric mineral dust and intercontinental dust storms   4.4.

Atmospheric dust takes its origin from both geological and anthropogenic sources. Geological processes leading to weathering and anthropogenic processes such as agricultural activities destabilize the land, and these processes are often followed by movement of hard particles of different sizes over uneven surfaces in turbulent flows of air or water. Once the airborne dust is collected in the atmosphere, it may generate strong local forcing that, together with wind, may transport dust clouds for distances >5000 km from their source.

Where are the major sources of mineral dust on the Earth? Significant areas of the world's continents are covered by deserts, which are major dust-producing regions (Fig. 45). The map of the different climatic zones (Fig. 45) includes updated climate classification data showing hot deserts with arid climates, and high-elevation cold deserts (see also Fig. 46); (Peel et al., 2007). The largest desert regions are in Africa, where, for example, the Sahara Desert occupies ~8.6 million km^2. In total, the desert areas of the world cover about 16.5 million km^2 (without Antarctica), which corresponds to 11% of all continental land area—and about 20% if Antarctica is included.

Many calculations have shown that the global emission of mineral dust is 100–500 million tons per year. The Sahara alone, which is a hot desert with an extremely arid climate, produces 60–200 million tons of mineral dust per year. The transport of the dust across oceans, from one continent to another, or even around the globe has a great impact on both global and local climate change, and studies of mineral dust and aerosols are in a focus of many disciplines.

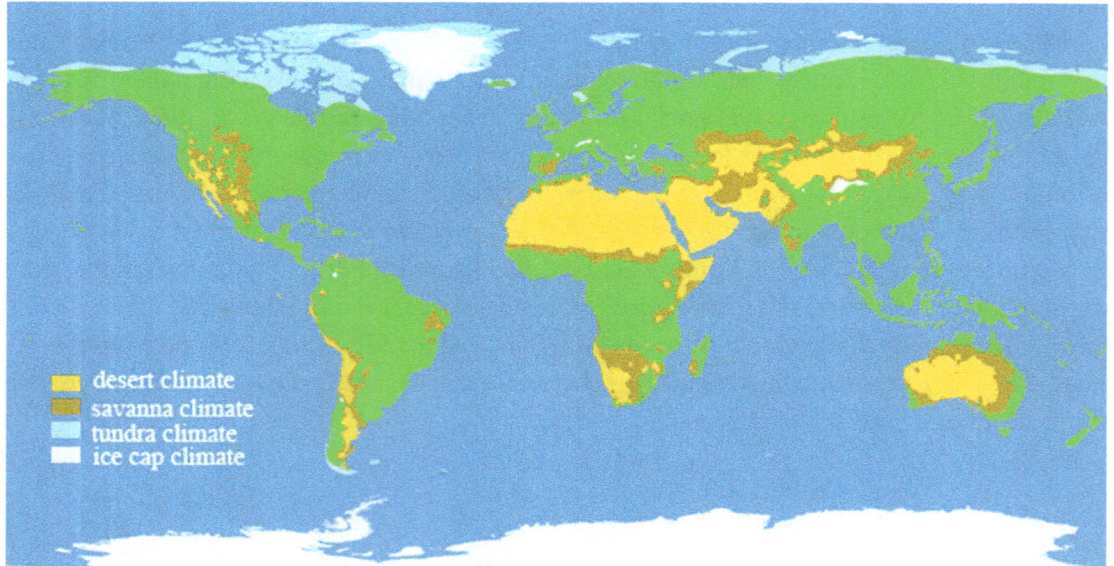

FIGURE 45   Schematic map of the world showing the climate zones (yellow: hot desert areas)

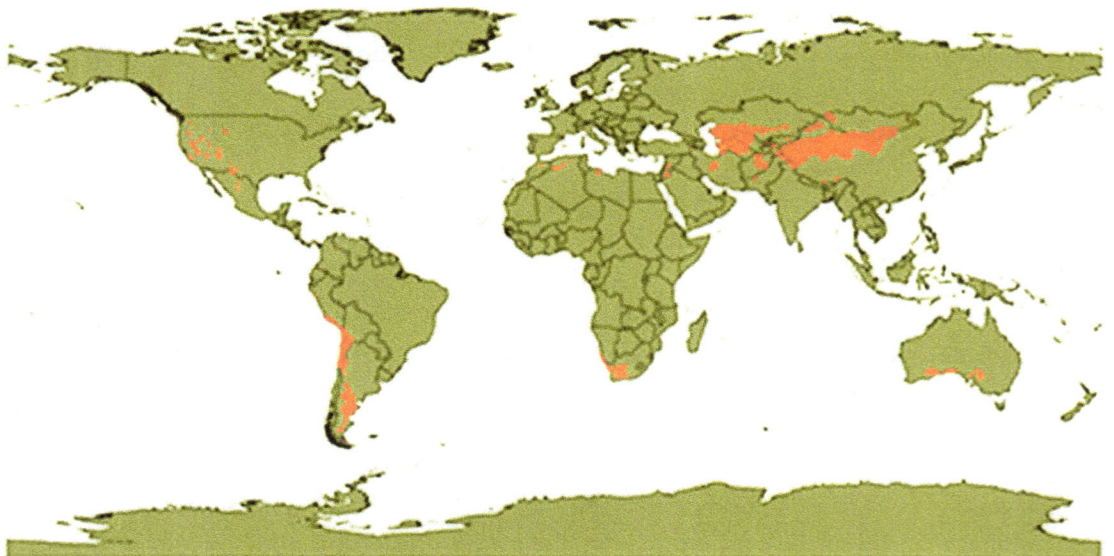

FIGURE 46   World map of the cold deserts

4.4.1 Intercontinental dust storms

As was already mentioned, the atmospheric lifetimes of mineral dust particulates correlate with the size of the particles (Table 6), though, of course, the length of time that particles fly also depends on wind power and direction, rain and clouds that cause dust precipitation, and many other factors. Systematic observations of different dust types and their classifications are based on chemical, physical, and mechanical properties of their mineral constituents. These observations allow developing of criteria that help scientists to distinguish both the geographic and geological sources of the dust precipitated in a particular region of interest. Direct observations of desert dust transportation from one continent to another and over the huge expanses of the Pacific and Atlantic oceans are well supported by remote-sensing satellite technologies. These data are delivered every day to monitoring stations and stored there for further study. There are many satellite observational systems, such as the Total Ozone Mapping Spectrometer (TOMS) and the Advanced Very High Resolution Radiometer (AVHRR), and coupled together with surface-monitoring networks, they provide high-resolution, global-scale monitoring of dust emissions, the directions of the transport of dust clouds, and their spatio-temporal features.

These data are used or the reconstruction of dust-cloud morphology, spatial configurations of dust clouds, dust-particle concentrations, and their evolution over time. The data are also used for recording spectral characteristics of mineral/aerosol particles to provide better understanding of their chemistry and physics. The combination of remote-sensing imaging and spectral analyses, together with the chemical-physical characteristics of the local dust collected from the land, has allowed scientists to determine, for example, that dust from the African Sahara Desert was once precipitated on the southeastern areas of the United States (e.g., Prospero, 1999). The chemical composition of the Sahara dust deposited during dust storms in 1901 (Fig. 47) showed that the mineral particulates of the dust clouds were rich in Si, Al, Ca, and Fe, and that aerosol particles also contained organic components (Barac, 1901). Moreover, traces of Sahara mineral dust have shown that it was not only transported to the U.S., but also to other regions across the Mediterranean, including Europe and the Middle East, across the Atlantic to the Caribbean, and to the western North Atlantic, among other places. A Sahara dust movement across the Atlantic on March 2, 2003, is shown in Fig. 48, which is a high-resolution satellite image showing that a storm released a massive dust plume moving northwestward over the Atlantic Ocean and extending more than

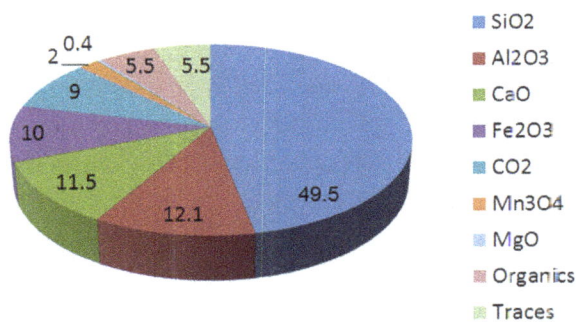

FIGURE 47 Composition (mass %) of the Sahara dust particulates deposited in March 1901

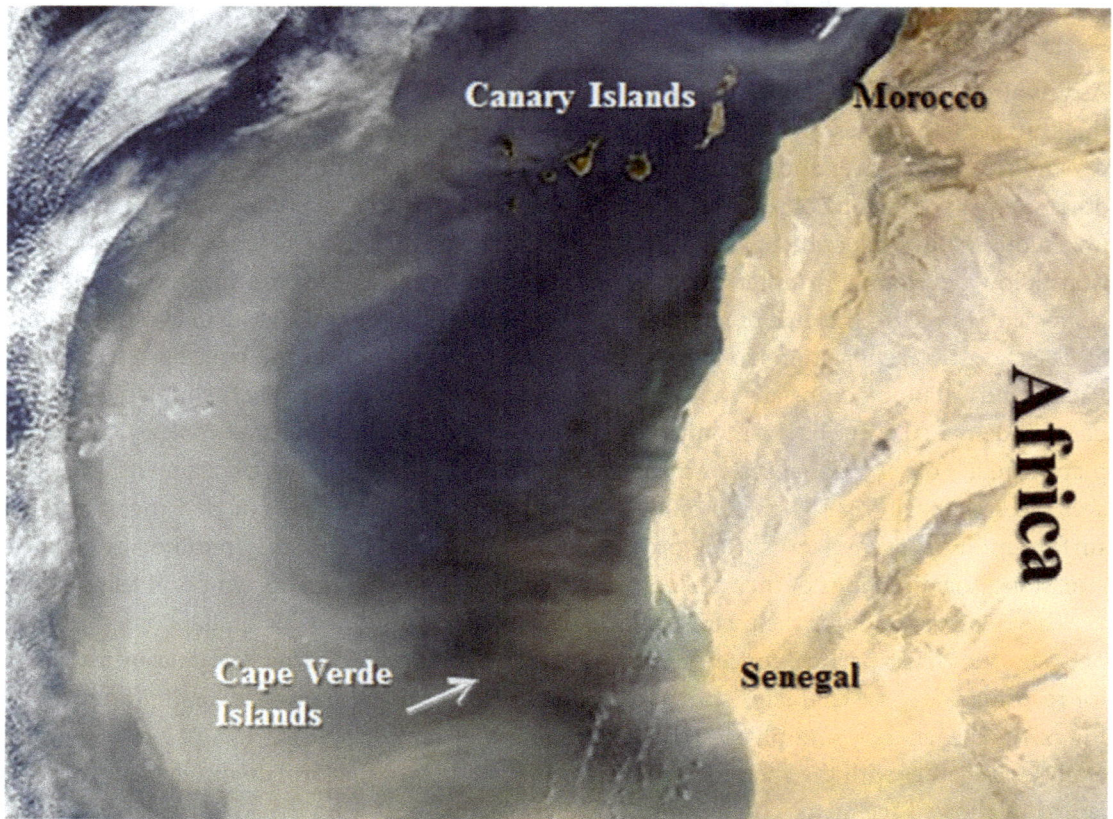

FIGURE 48 Sahara dust storm moving off Africa over Atlantic Ocean toward North America; the image was acquired by the Moderate Resolution Imaging Spectroradiometer (MODIS) aboard NASA's Terra satellite

1600 km. The image reflects a true-color scene in which the light brown contrast corresponds to the thick dust plume that is seen blowing westward from the coast of Senegal and then being redirected northward by strong winds blowing from the south. Other large sources of intercontinental dust storms are Asian deserts, which transport and spread mineral particulates across the Pacific Ocean. Calculations showed that $6–12 \times 10^6$ tons of Asian dusts are transported annually to the North Pacific (Uematsi et al., 2012). The high altitudes of the dust transportation is suggested by observations of the Saharan dust layer that have been recorded by satellite imagining at 1–6km above the marine layer of the Atlantic Ocean (e.g., Choobari et al., 2014).

One more interesting fact related to intercontinental dust storms is that together with mineral particles and aerosols, they transport fungi and bacteria. It has been observed since the 1970s that the coral reefs in the Caribbean have become progressively degenerated. There have been numerous hypotheses that overfishing, anthropogenic pollution, diseases, and bleaching caused by rising global temperatures have been responsible for coral's demise. Studies of Saharan

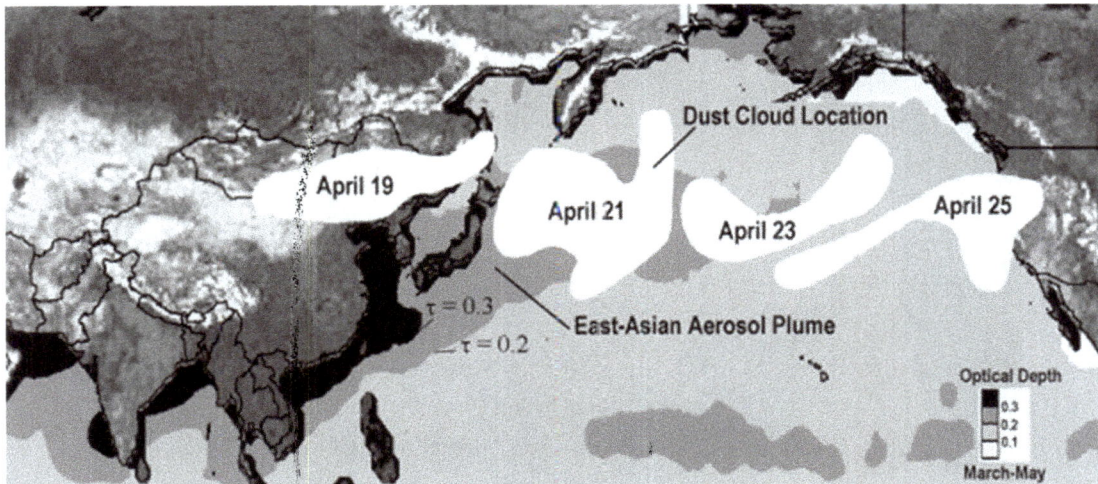

FIGURE 49   Intercontinental Asian dust storm, April 1998

mineral dusts precipitated in the Caribbean showed that besides solid mineral particulates, they contain fungi and bacteria that may be behind the coral reefs' degradation (Shinn et al., 2000).

Many studies show that intercontinental dust storms have seasonal and spatial activities and that their directions of movement and intensities are determined by predominating winds and many other factors, including geological formations from which dust particulates originate. For example, in the Southern Hemisphere, dust is transported from the Australian continent across the Indian Ocean in austral (i.e., southern) summer and spring. However, dust particles from East Asia are characterized by more active movements in spring, when they are transported across the Pacific Ocean to the West Coast of North America and precipitated there. The dust storms in North Africa are more intensive during summer, when they have a higher concentration of mineral particles and follow a more northward trajectory.

One of the most severe Asian dust storms was monitored in April 1998 (Fig. 49). It swept through Mongolia and the northern part of central China on April 15, and on April 19, it passed over the Gobi Desert and the Loess Plateau in the Gansu province of China. By April 21, the gigantic yellow dust cloud had stretched more than 1000 km over the Pacific, and by April 25, it reached the West Coast of North America and was observed inland as far as Salt Lake City, Utah. The sky was discolored in such a way that the usual blue color became milky-white due to redistribution of solar radiation interacting with flying dust particulates. The altitude of the dense aerosol/dust layer was estimated to be as high as 2 km. This Asian dust storm had an extremely hazardous impact on human health and welfare. According to CNN, on April 15, in the Xinjiang region of China, twelve people died because of the storm, and yellow, muddy rain precipitated over Korea and regions of eastern China. In the U.S., when the dust storm passed over the states of Washington, Idaho, and Oregon, the air pollution advisory agencies

FIGURE 50   Loess deposit exposed in a bluff (Vicksburg, Mississippi)

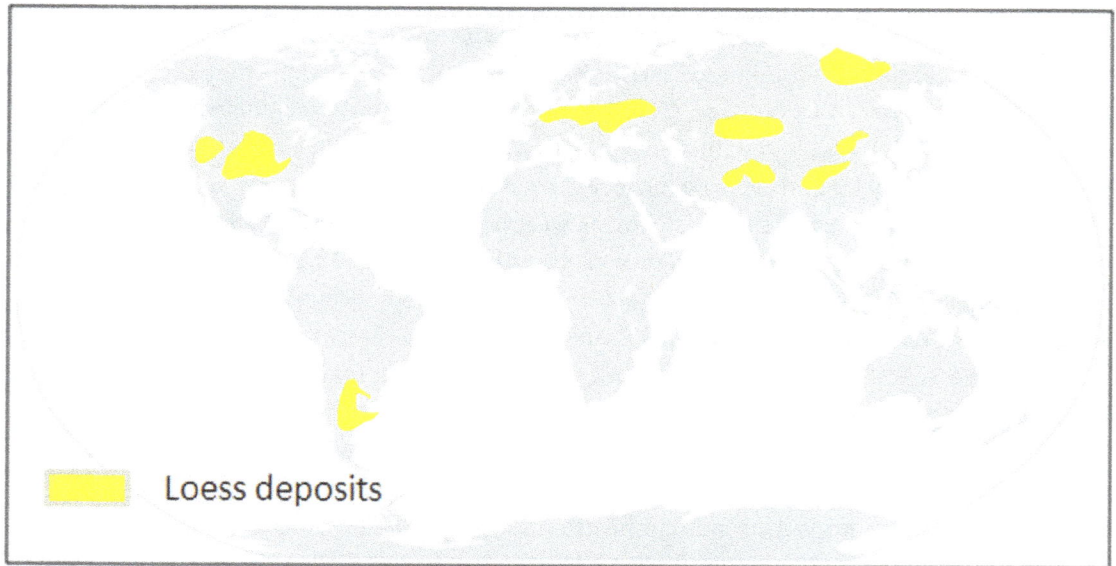

Loess deposits

FIGURE 51   World scheme of loess deposits

sent warnings to the general public and supplied constant information about the dust storm through a special website (Husar et al., 2001).

The monitoring of intercontinental dust storms and studies of dust products and their distribution over inland areas and oceans are important for understanding the complexity of dust-plume formation and transportation. Such observations provide valuable knowledge about how mineral dust and aerosols influence the physics of atmospheric clouds, how they change the radiance level of the sky, and how they may cause destruction of ecosystems and biological diversity and affect local climate systems.

4.4.2. Loess and dust storms

Loess is a rock which consists of fine-grained sediments (Fig. 50) that are mostly formed by the accumulation of wind-blown dust consisting of silt and sand of diverse compositions [quartz–SiO_2, feldspars–$KAlSi_3O_8$, $NaAlSi_3O_8$, $CaAl_2Si_2O_8$, micas–$K(Mg,Fe)_3AlSi_3O_{10}(F,OH)_2$, $KAl_2(AlSi_3O_{10})(F,OH)_2$], carbonates (dolomite–$Ca,MgCO_3$, siderite–$FeCO_3$, and calcite–$CaCO_3$), and clay minerals.

FIGURE 52   Loess landscape near Hunyuan, Shanxi Province, China

The rock-forming minerals of loess have good aeration because of their small sizes (silt particles are ~20–60 μm, and sand particles are ~80 μm) and water-holding ability, so they can be easily eroded by rivers and winds. Loess sediments were formed on all continents except Africa and Australia during the Quaternary period, which spans from 2.588 ± 0.005 million years ago to the present time (Fig. 51). Loess deposits are readily formed on continental shields beyond the limits of ice sheets, in high mountain ranges, and on the semi-arid margins of some lowland deserts (Fig. 52). These deposits may be situated as deep as 100 m or more, which is one of the main characteristics of loess formation in the arid climate zones of China and Central Asia. The remarkable thickness of loess deposits is due to a rapid uplift of the Tibetan Plateau, followed by a high rate of sediment production because of the existence of strong winds and effective dust traps downwind of the source regions. Such conditions lasted about two to three million years, during which thick loess sediment layers were accumulated (Crouvi et al., 2010).

Other gigantic areas of loess deposits are in Europe, North America (the Midwest of the U.S.), and some areas in the southwestern part of South America (Fig. 51). Intensive agricultural activity by humans is closely connected to loess regions because loess soils are extremely fertile and easily tilled, and they mostly are situated within the mid-latitude belt of the Earth, where weather is also favorable for agricultural development. However, under specific weather conditions, loess soils may be easily destroyed and blown away from agricultural fields by strong winds. For example, from 1931 through 1939, active farming in the Midwest and Southwest regions of the U.S. suffered under a long period of drought that led to a decrease in the moisture of the soil. The drought destroyed crops and loosened the soil, and strong winds raised enormous clouds of dust that covered everything, including houses. Dust suffocated many domestic animals, and many children became ill with pneumonia.

4.5. The hazardous effects of mineral dust

Dust storms of any scale have mostly adverse effects on human activities and health, though they also have some positive effects on the environment. The adverse effects include a haze that can reduce visibility to 300–500 m, which is a threat to safety of drivers, especially on freeways. The haze is characterized by a high concentration of fine particulate matter ($PM_{2.5}$, particle size = 2.5 μm) and an elevated amount of nitrogen oxides (NO_x) and sulfur dioxide (SO_2). These pollutants are a serious threat to the health of people who suffer from chronic respiratory diseases, and they may even cause premature death. In addition, dust storms may spread viruses and microbes that may be carried along with inorganic and organic particulates. Also, solid dust particles can be precipitated into buildings and businesses, where they may get inside computers and telecommunications equipment and ruin delicate technologies. The positive effect of dust storms is that they may provide iron and other micronutrients available in mineral particles rich in essential components to marine and terrestrial ecosystems.

4.5.1. Mineral dust and pneumoconiosis/silicosis in High Asian regions

Pneumoconiosis is one of the lung diseases caused by long-term exposure to mineral dust of different types through inhalation of tiny solid particles of minerals. Pneumoconiosis destroys the lung ventilation cycle, followed by a decrease in lung volume, inadequate consumption of oxygen, and, in severe cases, shortness of breath.

Silicosis is also a lung disease caused by exposure to and inhalation of crystalline silica particles. The inhalation of silica dust can cause local accumulation of fluid and scar tissue in the lungs that also, like pneumoconiosis, may seriously affect a person's respiratory system and his or her ability to breathe normally. Chronic silicosis usually takes place after ten or more years of exposure to silica at low levels of concentration in the air. Acute silicosis symptoms can be noticed after only weeks or months of exposure to very high levels of concentration of silica in the air. Acute silicosis develops very rapidly and can be fatal within months. Both pneumoconiosis and silicosis are primarily occupational diseases (i.e., they are caused by work in a dust-producing industry), but it will be shown below that there are many cases in which

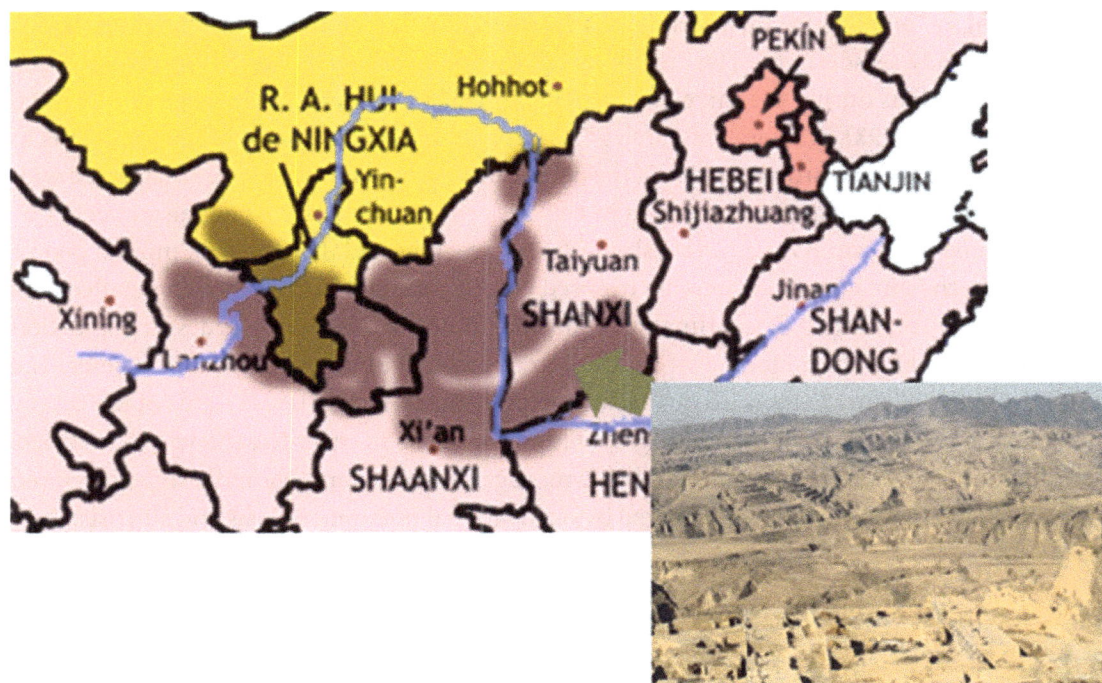

FIGURE 53   Loess Plateau (shaded area) in China

these diseases have nothing to do with industrial and/or occupational activities but instead are caused by geological processes.

High Asia, a non-industrial part of northern China, is situated within the Loess Plateau, where the loess formation occupies more than 600000 km² (Fig. 53). Beginning about 200 BCE and continuing through modern times, there are written records indicating that dust storms have often been active in this area. Some dust storms precipitated huge amounts of dust, which covered the landscape with up to 300–500-m-thick layers of tiny particles. Therefore, without any industrial impacts, populations living in these remote territories of High Asia (e.g., North China) were sometimes exposed to dust throughout their lives. The High Asia territories are characterized by a substantial tectonic uplift taking place during the Quaternary period up to the present. The uplift has created high mountain plateaus, and active plate tectonics continue to cause high-speed erosion of the loess formations. The 300–500-m thickness of the loess formations in High Asia is an undisputable fact of sustained and continuous deposition of airborne dust during at least the last 2.6 billion years of the Quaternary period of the Cenozoic era (see Table 1).

Loess-originated dust storms usually contain fine (<2-μm) particles of silts, clay minerals, and angular grains of quartz (SiO_2) derived from defragmentation (erosion) of previously formed loess deposits and loess soils. Observations of dust storms during the period 1982–1984 in the vicinity of Lanzhou (Xining Basin) in China showed that dust concentrations deposited indoors were ~20 mg/m³ and that the dust contained 61% free silica—e.g., mineral quartz. Observations during the windy season in the Hexi region of northwestern China in April 1991 revealed that dust concentrations during the three days of samples collection were in the range of 8.25–22.0 mg/m³ (Xu et al., 1993). The same research group also reported that among the 395 residents in the desert area of Gansu province, 21% of the people over forty years of age were diagnosed with silicosis. Radiographic studies of animals in this area showed that camels are also affected by silicosis. Some publications (e.g., Derbyshire, 2013) have indicated that about 24 million people in the northern China desert regions are exposed to loess-originated dust, though detailed data of exposure are still limited. The estimation of the magnitude of the High Asia desert population affected by inhalation of mineral dust should be continued with interdisciplinary studies. The research should be directed toward quantification of the respiratory disease status of the people living within the same geologic and geographic environments in order to achieve better understanding of the disease mechanisms, and risk-management programs should also be proposed for people who live in urban areas affected by occupational pneumoconiosis and silicosis.

4.5.2. Manhattan dust, New York, September 11, 2001

On September 11, 2001, after a terrorist attack that targeted the World Trade Center (WTC) in New York, the two tower-like buildings collapsed, releasing into the atmosphere a huge volume of airborne particulate materials and chemically toxic aerosols (Fig. 54). Several thousand

FIGURE 54 Picture of the dust cloud around the World Trade Center, New York, on September 11, 2001, shortly after the second tower collapsed

people lost their lives, and thousands of others were injured in the collapse of the buildings. The concentration of dust particulates produced by the tons of pulverized debris of masonry, glass, electronic and electrical appliances, and different kinds of building materials was >5000 µg/m³ during the first hour after the collapse of the second tower, but it decreased thereafter to <1600 µg/m³ (Landrigan et al., 2004). The dust was dispersed over lower Manhattan and Brooklyn and several kilometers beyond, and eventually, it precipitated on the streets and the roofs of nearby buildings and entered into residential areas, offices, schools, etc. In addition to the tremendous traumatic consequences of the terrorism attack for New Yorkers and for the entire nation, this was also the most dramatic environmental catastrophe known in the twenty-first century.

Extensive scientific studies of the particulate materials and their impact on human health were undertaken by a consortium of six research centers under the auspices of the National Institute of Environmental Health Sciences (NIEHS), and these studies were extended to include collaboration with the New York City Department of Health, the EPA, the USGS, and other organizations.

As part of these efforts, samples of dust were collected in the vicinity of the disaster during days 1 and 2 after the buildings collapsed, and this was followed by studies of dust settled down from day 3 through December 20, 2011. Researchers reported that almost all samples contained particles of <10–53 µm in diameter, and they were represented by fibrous-shaped particles of asbestos, glass fibers, wool, wood, paper, and cotton. Organic pollutants included

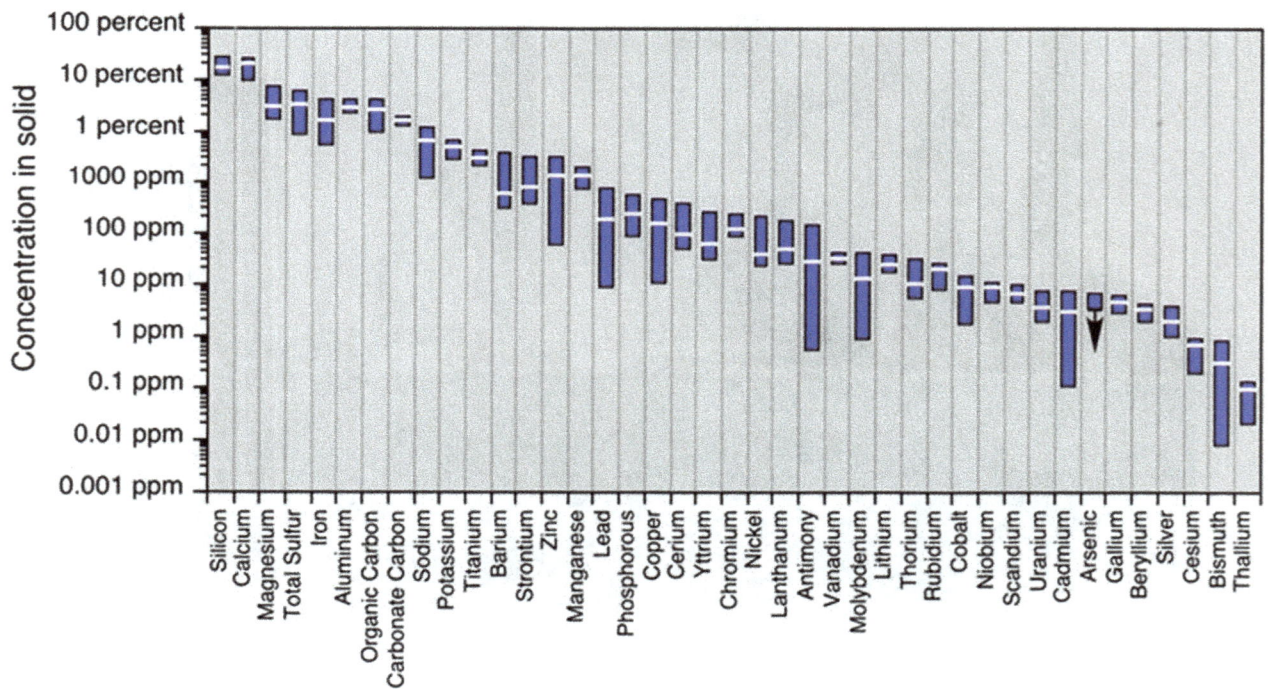

FIGURE 55   Composition of the WTC dust

toxic byproducts of burned plastic materials such as dioxin, polycyclic aromatic hydrocarbons, biphenyl, and many other chlorinated components with concentrations that were several orders of magnitude above their background levels (i.e., the levels that are usually present in urban areas). More than forty-one chemical elements were detected in the WTC dust (Fig. 55), of which fifteen elements (Pb, Cs, Cd, Cr, Be, Yt, Ta, Ce, As, Th, U, Sb, Sc, La, and Nb) are known to be extremely toxic. It was also found that the average concentrations of toxic heavy metals (such as Mo, Zn, Cu, Pb, Cr, Mg, Ni, and Ba) in the WTC dust were very high compared with their mean concentrations in natural soils from the eastern part of the United States.

When the short-term stage of the studies was completed, the conclusions were that environmental exposure after the 9/11 attack caused significant adverse effects on the health of New York's citizens. The NIEHS group determined that firefighters who were on duty on the day of the WTC collapse, cleanup workers, and other persons who helped as volunteers were at the highest risk of exposure to organic toxins, hazardous minerals, metals, and other solid particles. The 10116 firefighters who participated in the rescue efforts received a medical evaluation, and almost all of them showed exposure-related coughing (known now as "WTC cough"), bronchial hyperactivity, and wheezing, and community residents reported suffering from new onsets of coughing, wheezing, and shortness of breath.

Increased risk of developing asthma and even mesothelioma (a special form of lung cancer), especially among the firefighters and volunteers exposed to asbestos-rich dust, can be expected in years to come. Because many environmental hazards do not exhibit their adverse impacts on human health immediately or even in the short term (several months or years), it is important to continue as much follow-up as possible to determine the long-term consequences of the WTC disaster within the groups that were at higher risk of exposure.

Volcanic eruptions and human sickness from inhalation of volcanic gases and ashes 4.6.

According to the main concept of plate tectonics, volcanoes are associated with tectonic structures such as convergent and divergent plate boundaries and hot spots (see Chapter 1). Volcanic activities occur in places where large quantities of internal heat below the Earth's surface cause melting of the solid rocks, followed by eruption—e.g., the release of hot melt (magma) from a volcanic vent (Fig. 56). Some volcanoes erupt violently, releasing gigantic volumes of gases and ashes and ejecting pyroclastic materials called tephra (solid rock debris) up to tens of kilometers into the atmosphere, which, when they fall, can cause extensive destruction of property and loss of life. Volcanoes with explosive eruptions usually produce magma of andesitic-to-rhyolitic composition, which has high gas content and therefore is characterized by high viscosity.

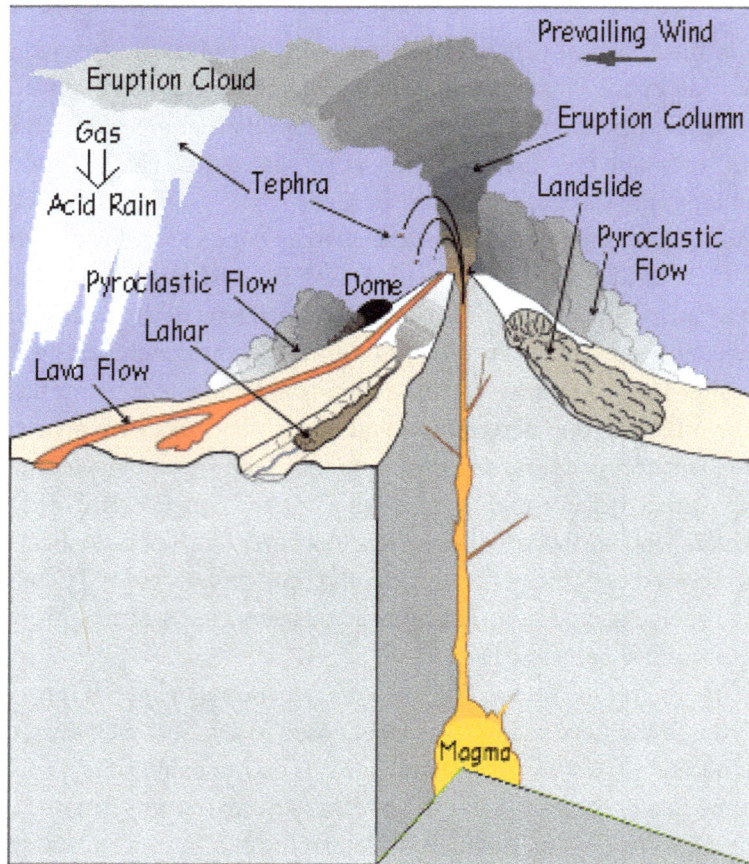

FIGURE 56   Schematic model of a volcanic eruption

Another type of volcanic eruption is characterized by non-violent (i.e., non-explosive) activity, and it only pours out streams of hot lava, which is typically magma of basaltic-to-andesitic composition, with a low gas content and low viscosity.

Volcanic hazards related to dangerous eruptions include the following (Fig. 56):

(i)   Direct blast;
(ii)   Tephra (ash) flows;
(iii)   Volcanic gases;
(iv)   Lava flows;
(v)   Lahars (mud and debris) flows;
(vi)   Pyroclastic (lapilli and tuff, ash) surges and flows;
(vii)   Landslides;
(viii)   Acid rain.

Although volcanic activity is a vivid witness to the fact that the Earth is a dynamic planet, the exact number of volcanoes in the world is not known. This is because some of them are dormant for long time, or they are not easily recognized without geological mapping because their specific "dome-like" configuration has been modified through geological time, or because they are situated at the bottoms of oceans, especially the Pacific. According to the Smithsonian Institution database "Volcanoes of the World," there are 1511 volcanoes on the Earth, and 177 of them are situated in the U.S. (http://www.volcano.si.edu). The Smithsonian and the USGS, through a cooperative project called the Volcano Hazards Program, provide weekly summaries of information on volcanic activity in the world, which can be found on the website "The Weekly Volcanic Activity Report" (http://volcano.si.edu/reports_weekly.cfm). Monthly reports on various volcanoes may be found in the *Bulletin of the Global Volcanism Network*.

In the past 500 years, over 200000 people have lost their lives during volcanic eruptions, and the loss of property and destruction of land and biodiversity are immeasurable. In addition to the visible hazards such as explosive blasts, lava flows, avalanches, tephras, lahars, and

FIGURE 57 Formation of volcanic aerosols, acid rains, and contamination of surface waters with leachates

pyroclastic material flows, the greatest threat to human health is the mixture of tiny, invisible particulates, aerosols, and gases ejected into the atmosphere. Because of their small size, these substances remain in the atmosphere for longer periods, and since they are spread by wind, they may act in ways similar to intercontinental dust storms, which can travel across oceans and continents, contaminating environments far from the primary volcanic vents.

A simplified model of volcanic aerosols and particulate formation (Fig. 57) includes, at least, three stages:

(i) Volatile materials that consist of water vapor (H_2O), carbon dioxide (CO_2), sulfur dioxide (SO_2), hydrochloric acid (HCl), hydrofluoric acid (HF), ash, and tiny particles of volcanic glass shards are injected into the atmosphere during eruptions. Volcanic ash consists of tiny fragments of the melt (magma) enriched in gas bubbles that instantly solidify during eruptions.

(ii) SO_2 and CO_2 gases slowly react with vapor, forming aerosols of sulfuric acid (H_2SO_4) and carbonic acid (H_2CO_3). Together with HCl and HF, they may form acid rain that can fall in the vicinity of the eruption and up to tens of kilometers away from the active volcano.

(iii) Part of acidic aerosol droplets can be adsorbed (i.e., gathered onto a surface) by ash and volcanic glass particles. The volcanic glass particles, being pitted with many cavities (known as vesicles), have a large surface energy, and therefore, each such particle may accommodate a substantial amount of acidic aerosols on its surface. These aerosols, which wrap large, solid particles, will reach the ground more rapidly due to the faster deposition velocities of the heavier particles composed with volcanic ash and glass. Furthermore, rainfall or surface waters that flow onto freshly deposited ash will leach the aerosols into the environment. In the other scenario, if a huge volume of ash particles wrapped by aerosols fall into water reservoirs, they will cause changes to local water chemistry and hence impact the water quality. The products of volcanic aerosol leaching (leachates) may also cause chemical changes in the underlying soil, although the results can be either adverse or beneficial. The outcome depends on the composition and alkalinity of the primary soil of the region before the eruption and ash falls. For example, if the underlying soil is too alkaline, the acidic components added from volcanic eruptions may balance the alkalinity to a neutral level.

4.6.1. Toxic effects of volcanic materials ejected into the atmosphere

4.6.1.1. Gas and acidic aerosols

Sulfur dioxide (SO_2). This is a gas with a specific and irritating smell, which may be identified when the concentration is 0.3–1.4 ppm, and it is clearly noticeable when the concentration reaches 3–5 ppm. Sulfuric acid (H_2SO_4), which is a product of sulfuric gas reaction with H_2O, is known as a severe irritant of the eyes and skin. In many volcanic plumes, the SO_2 concentration is <10 ppm at a distance of around 10 km from the vent, which is very high compared with the SO_2 content in the troposphere (0.00001–0.07 ppm). The health effect of SO_2 depends on

TABLE 7: Health Effects of Volcanic Gas SO_2 (the USGS Data)

Concentration Limit (ppm)	SO_2: Health Effects
1–5	This level is threshold for respiratory response in humans upon exercises or deep breathings.
3–5	These concentrations decrease lung function at rest and cause increasing of the airway resistance.
6	Causes immediate irritation of eyes, nose, and throat.
101–105	Represents threshold of toxicity for prolonged exposure.
20 and >20	Extended exposure may cause paralysis or death.
150	Extremely dangerous. Healthy individuals can tolerate this concentration for a few minutes.

the exposure limit. According to the USGS data (Table 7), the SO_2 gas is noticeable when its concentration reaches 3–5 ppm. At a concentration of 6 ppm it causes immediate irritation of the eyes, nose, and throat, and at concentrations >20 ppm, paralysis or death may occur after extended exposure. At a concentration >150 ppm, SO_2 can be withstood for only a few minutes by a healthy person.

Carbon dioxide (CO_2). This is a gas that has no color and no odor. Its concentration in volcanic plumes ejected by different volcanoes may vary from 1 ppm to hundreds of ppm. The CO_2 background in the troposphere is ~360 ppm. Carbon dioxide partly reacts with H_2O vapor to produce the acid H_2CO_3, but the larger part of erupted CO_2 spreads down-slope, where it may collect at low elevations by filling small topographic depressions on the land. This is because CO_2 gas is 1.5 times heavier than air. Because the CO_2 gas remains in the lower atmosphere for approximately four years after an eruption, it can be considered to be a long-term, active hazard. If a volcano eruption has occurred in a populated region, the emitted CO_2 gas becomes very dangerous for people because it can self-collect in natural topographic depressions, excavation quarries, open-pit excavations, buildings, and basements. The harmful effects of CO_2 gas on human health are summarized in Table 8.

TABLE 8: Summary of the CO_2 Adverse Effects on Human Health (the USGS Data)

CO_2 Concentration in Air (%)	CO_2: Health Effects
2–3	Shortness of breath may go unnoticed at rest, however during exertion may have a marked increase.
3–5	Causes acceleration of breathing rhythm, and if exposure is repeated it provokes headache.
5	Breathing becomes difficult followed by headache, sweating and bounding pulse.
7.5	Increasing heart rate, shortness of breath, headaches, sweating and sensation of ringing in the ears. Also causes dizziness, drowsiness, muscular weakness and loss of mental abilities.
8–15	In addition to headache, vertigo and vomiting, the individual may lose consciousness followed by possible death if oxygen is not given immediately.
10	After the loss of consciousness a respiratory distress develops rapidly within 10–15 minutes.
15	Exposure to levels above 15 cannot be tolerated by humans, and may be lethal for many individuals.
>25	Rapid loss of consciousness and convulsions after a few breaths followed by death if level of this concentration is maintained.

Hydrochloric (HCl) and hydrofluoric (HF) acids. Vapors and tiny droplets of HCl and HF are common within products of volcanic eruptions, and both have high solubility in water, which means that their absorption in the human respiratory system can happen very quickly. Both are known as severe irritants to the eyes and the upper respiratory system, causing inflammation of the nose, throat, and lungs that may result in chronic coughing if exposure was prolonged. Table 9 shows the concentration thresholds of HCl and HF that cause threats to human health.

TABLE 9: Hazardous Concentrations of HCl and HF in Volcanic Aerosols and Their Effects on Human Health (the USGS Data)

HCL (ppm)	HCL: Health Effects	HF (ppm)	HF: Health Effects
>5	Cause coughing.	<3	Some individuals feel irritation of nose and eyes.
35	After a short time of exposure people feel throat irritation.	>3	After 1 hour most individuals have sensations of burning eyes and throat, and cough.
>35	Cause severe difficulties of breathing, inflammation of skin and/or burns.	30	Worsen respiratory system symptoms which can be tolerated only for several minutes.

10–50	These concentrations represent maximum level that can be sustained for several hours.	50–250	Extremely dangerous even for very brief exposure.
>100	Cause lungs swelling and throat spasm.	120	Severe irritation of the respiratory system, smarting of the skin and conjunctivitis. Most individuals can tolerate this concentration only for 1 minute.
50–1000	Very harmful for humans, though some individuals may tolerate this concentration during 1 hour exposure.		
1000–2000	Extreme concentrations which are very dangerous even for a very short exposure.		

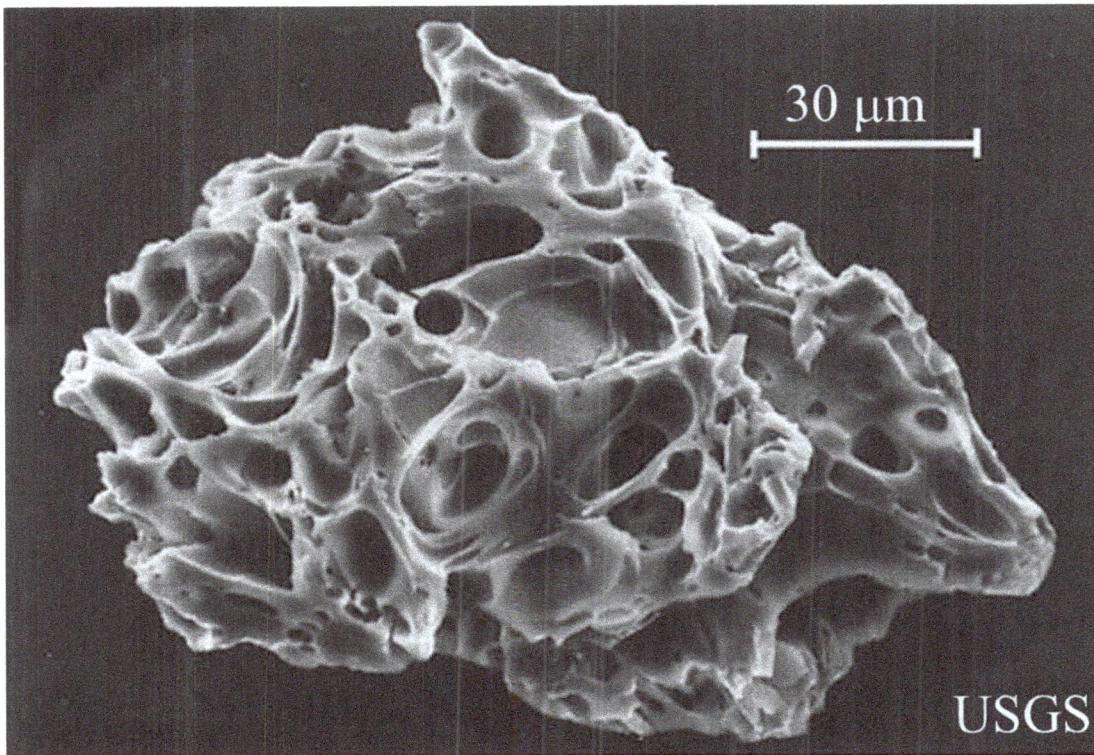

FIGURE 58 Scanning electron microscope images of a highly vesicular glass shard of ash ejected by Mount St. Helens volcano, USA: eruption, 18 May 1980

FIGURE 59 (a) Small explosion in 1997 produces light ash fall at Soufriere Hills volcano, a Caribbean island of Montserrat; (b) NASA satellite photo of an eruption on 11 October 2009, during which a plume of ash reached ~6 km into the sky

4.6.1.2. Volcanic ash

Volcanic ash consists of solid, sharp particles of minerals, rock fragments, and volcanic glass that are represented by rapidly solidified droplets of melt (Fig. 58). When a volcano erupts, a huge volume of volatile components, including gases, vaporized H_2O, and melt, are violently ejected into the atmosphere (Fig. 59). Due to the extremely rapid release of energy, the escaping gas shatters solid rocks from the walls of the vent, shreds cooling magma, and produces a mixture of solid particles of different shapes and compositions—what is called *ash* (Fig. 59a). Volcanic ash, therefore, is not the same ash that is familiar as a product of combustion. The size of volcanic ash particles can vary from 1 μm to 2 mm.

The ash particles themselves are not soluble in water unless they adsorb the aerosols, which create a thin film around each solid particle. Due to the sharp and curvilinear configuration of each particle, ash has strong abrasive properties that make it extremely dangerous for aviation because small particles can be collected inside a plane's engine and cause catastrophic failure. The smallest particles of ash can travel hundreds to thousands of kilometers downwind from a volcano vent, similar to what we learned in the cases of intercontinental dust. Their speed of

FIGURE 60 May 18, 1980 Mt. St. Helens volcanic eruption, Washington, USA; in the lower corner, map of the volcanic ash distribution over the United States

transportation and longevity in general will depend on the volume of ash erupted, the size of the ash particles, the height of the eruption column, and the wind speed. At the local scale, both wind and human activities may bring ash into areas that were previously clean, which can stir up ash for weeks to years following an eruption.

Volcanic ash inhalation, especially small particles that are <10 μm (PM_{10}), can cause lung and respiratory diseases similar to those caused by mineral dust and aerosols, which we have already discussed above in Sections 4.4 and 4.5. People who are in the ash-fall zone during a volcanic eruption should wear masks and protective eyeglasses, cover their bodies with clothing as much as possible, and avoid strong exposure to the ash. Volcanic ash also causes poor visibility on roads and freeways, creates slippery road surfaces, and damages vehicles.

The great leap in understanding the scale, reality, and danger of volcanic hazards in the U.S. followed the May 18, 1980, eruption of the Mount St. Helens volcano situated in the Cascade Mountain Range in Washington state. Prior to the eruption, there were a series of small earthquakes on March 16, followed by hundreds of additional earthquakes on March 17. These were followed by a magnitude 5.1 earthquake and, on May 18, the most massive eruption by this volcano in over 100 years (Fig. 60). The previous eruption, in 1856, was well-documented by mapping the traces of ash deposits in Montana and Alberta, Canada, as well as the fragments of ash deposits that made rich soils in eastern Washington.

During the Mount St. Helens eruption in 1980, a plume of ash, gases, and fragments of melt reached ~24 km into the sky. The ash was deposited in eleven U.S. states and some regions of Canada during the three days after the eruption, and smaller ash particulates continued to circle around the Earth for fifteen days. The eruption caused melting of the snow, ice, and several entire glaciers situated near the top of the volcano, and a series of volcanic mudslides streamed downward to the Columbia River, as far as 80 km from the point of the eruption.

The largest ash volume fell on the city of Yakima, Washington, ~110 kilometers east of Mount St. Helens, and created a 5–8-cm-thick layer of ash and debris in the first twenty-four hours following the explosion. City Manager Dick Zais provided the following overview: "We estimated that several million tons of ash was deposited on the entire region." He continued, saying:

> "By noon, the city was engulfed in darkness, and communications by home telephone were impossible. It was like an eclipse of the sun that lingered, a blinding blizzard, and an electrical storm all in one. Light-sensitive street lights came on automatically, traffic stopped, and a strange quiet fell on our community; and everywhere a talcum-like sandy gray powder kept accumulating. Cars, trucks, buses, and trains all stopped, and planes were re-routed away from the ash cloud... From noon on May 19 until 6:00 a.m. the following morning, the City was in total darkness. Three types of ash fell alternatively on the City: dark gray sand, medium gray sand, and a

light gray cement-like dust. All three grades were gritty and light, difficult to sweep or shovel, especially when dry. To make matters worse, shifting winds blew the ash everywhere, severely impairing visibility and driving in our area. It was exceedingly harmful and abrasive to mechanical and electrical equipment, especially motors of vehicles, aircraft, and electronic systems. Unlike snow, however, this precipitation was not going to melt!" (see more at http://volcanoes.usgs.gov/ash/dickzais.html).

The Mount St. Helens eruption killed fifty-seven people and destroyed 250 homes, 47 bridges, 24 km of railways, and 298 km of highways. Although now the Mount St. Helens volcano is silent, its "behavior" has been monitored continually since the 1980 eruption. About seventeen minor explosive events and/or lahars, accompanied by small earthquakes, were recorded between 1981 and 1986, and from 1996 through 1999, about thirty brief, but intense seismic tremors were accompanied by small explosions from the dome that caused minor pyroclastic and lahar flows. In the period 2004–2008, a strong seismic shake continued, although no eruptions were recorded that could be connected to the seismic signals (see reports by the Cascade Volcano Observatory at http://volcanoes.usgs.gov/observatories). Other sources of the seismic energy release were proposed, such as possible deformation of the rocks and/or gas geochemistry in the area of the volcanic conduit.

Earthquakes 4.7.

4.7.1. What is an earthquake?

An earthquake is a geological phenomenon that is characterized by shaking or trembling of the ground caused by a sudden release of energy from inside the Earth. Earthquakes mostly occur along tectonic plate boundaries or inside the plates along the deep faults in the Earth's crust called rift zones (Fig. 61). Earthquakes begin with slipping of the blocks of the crust along the fault, but the

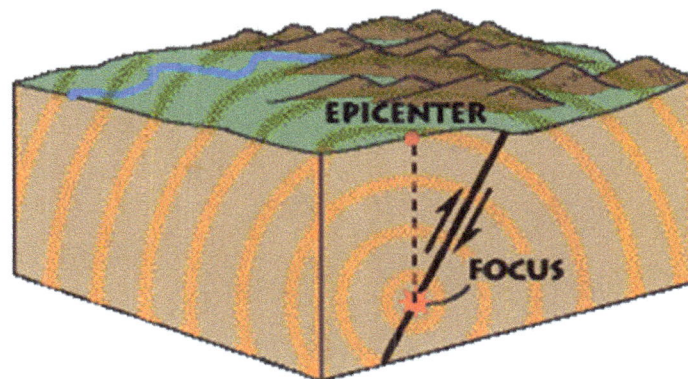

FIGURE 61   Schematic model of the epicenter of an earthquake

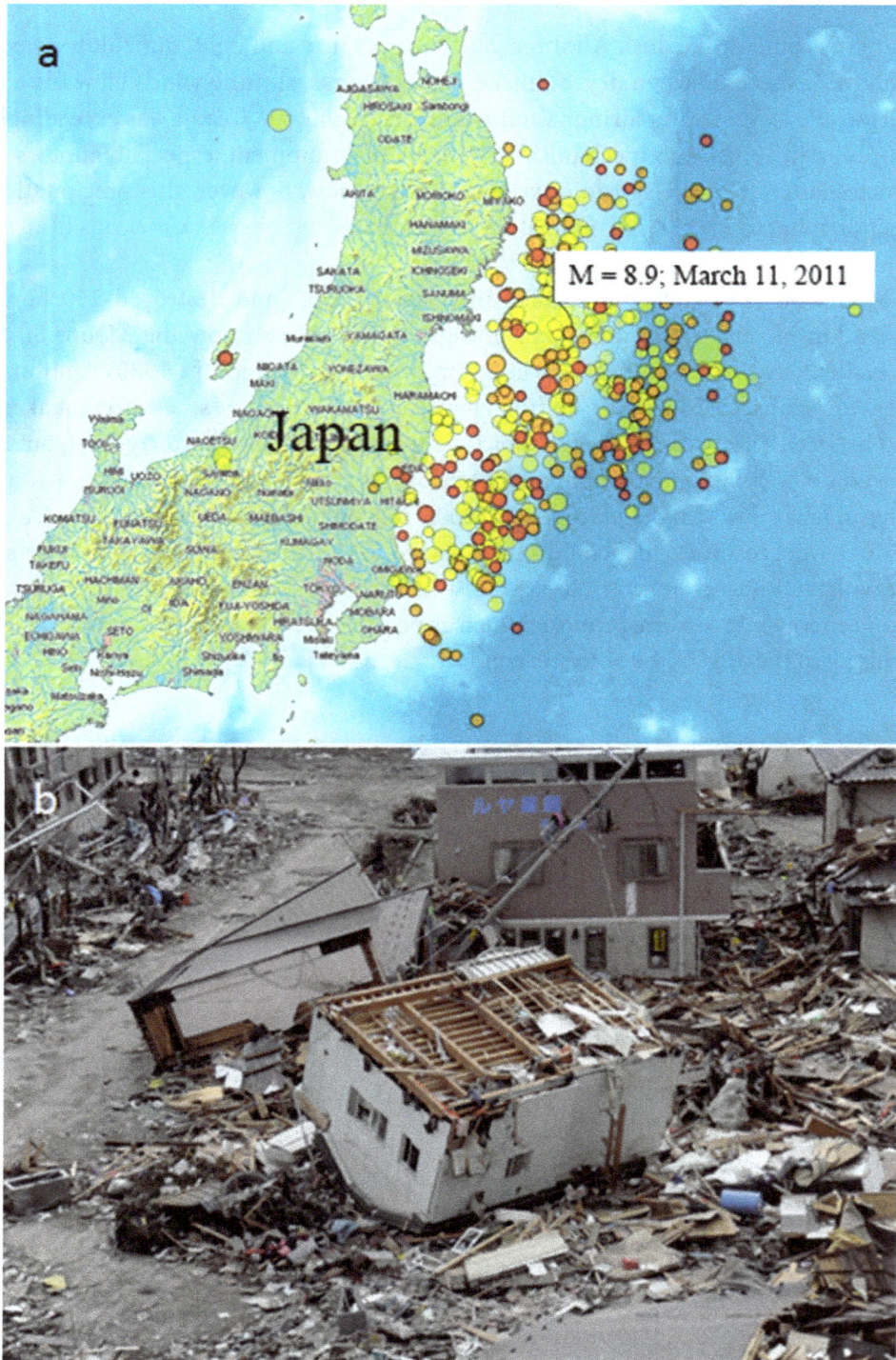

FIGURE 62 (a) Tohuku (Sendai) earthquake, Japan, March 11, 2011; (b) an upended house is among debris in Ofunato, after tsunami

released energy (seismic waves) radiates in all directions from the initial source, which is called a hypocenter or focus (originating from the word *foci*, which means "a point"); (Fig. 62). Slippage along the fault causes breaking of the rocks, and the continuing adjustment of the moving blocks' positions results in aftershocks (e.g., repeated minor tremors).

4.7.1.1. How do we measure the power of earthquakes?

Historically, the power of earthquakes was measured only by description of the severity of the shaking and the results of destruction. In 1857, after the great Italian earthquake, a first attempt was taken to evaluate its power in a scientific way. The results of damages were plotted on a topographic map, and lines were used to connect places of equal damage, which were "calibrated" as lines of equal intensity of the earthquake. Later, in 1902, G. Mercalli developed an intensity scale from I to XII, which is summarized below:

I. The earthquake is "not felt except by a very few under especially favorable circumstances";

II–VIII. This range of intensities is based on many details of human feelings, the appearance of cracks, details of properties damage, which progressively increases from II to VIII (the details of those long descriptions are intentionally omitted here);

IX. "Damage is considerable in specially designed structures. Buildings shifted off foundations. Ground cracked conspicuously."

X. "Some well-built wooden structures and most of the masonry and frame structures were destroyed. Ground badly cracked."

XI. "Few if any (masonry) structures remain standing. Bridges are destroyed. Broad fissures in the ground."

XII. "Damage total. Waves seen on ground surface. Objects thrown upward into air."

As you can see, the Mercalli earthquake-intensity measurements are based on the description of the damage severities to buildings and bridges, which does not provide any quantitative characteristics. Instead, it depends on the materials from which the buildings and bridges were made, the types of construction design, and the types of rocks (sedimentary or metamorphic and/or igneous) on which the urban regions were settled up. Therefore, according to Mercalli's evaluations the destruction caused by earthquakes may not reflect the real power of an earthquake and its actual magnitude.

In 1934, Charles F. Richter, a scientist from the California Institute of Technology, proposed a new earthquake magnitude (M) measurements scale using actual seismic records. The Richter magnitude is calculated from the amplitude of the largest seismic wave recorded on a seismogram for the earthquake. The Richter magnitudes are based on a logarithmic scale, meaning that for each whole number you follow up the scale, the amplitude of the ground motion recorded by a seismograph will go up ten times. The Richter scale of earthquake magnitudes is very convenient because the size (power) of an earthquake is described only as a single number calculated directly from a seismogram. It is based on a scientific approach that includes a measurement of physical phenomenon—e.g., seismic wave amplitudes. The Richter scale is a powerful tool that can be applied to any region of the world (Table 10).

TABLE 10   Earthquake Magnitudes on the Richter Scale (modified from the USGS Earthquake Hazard Program, http://earthquake.usgs.gov/learn/topics/mercalli.php).

Richter Magnitudes	Damage Effects Near Epicenters
<2.0	Very weak earthquake. Recorded only by seismographs. May not be noticeable to persons because shaking is not felt, except of rare individuals on upper floors of buildings.
2.0–2.9	Weak earthquake. No damage to buildings. Most people felt vibrations similar to the passing of a truck and more noticeably felt only by individuals on upper floors of buildings.
3.0–3.9	Light earthquake. People awakened at night, and in day time they have sensation similar to that if a heavy truck would strike a building. Shaking of indoor disturbs dishes, windows, doors; walls make cracking sound, however damage of buildings rarely caused.
4.0–4.9	Moderate earthquake. Felt by nearly everyone; many awakened. Causing noticeable shaking of indoor objects, some dishes and windows can be broken. Unstable objects overturned. Pendulum clocks may stop. Usually causes no damage or only minimal damage.
5.0–5.9	Strong earthquake. Usually it felt by everyone. Some heavy furniture moved indoors; a few instances of fallen plaster. Damage to poorly constructed buildings; at most, slight or no damage to slight damage in buildings of good design and construction.
6.0–6.9	Very strong earthquake. Damage in moderate to negligible in buildings of modern earthquake-resistant construction; considerable damage in poorly built structures; some chimneys broken. Small to moderate shaking felt by persons up to hundreds of kilometers from the epicenter, and strong shaking in close proximity to the epicenter area. Death toll from none to 25000.
7.0–7.9	Severe earthquake. Damage to most buildings with partial collapse. Indoors (factories, plants) heavy furniture overturned. Fall columns, monuments, walls, some bridges and freeways are severe damaged. Shaking felt across to 250 km from epicenter. Death toll is in range from 0 to 250000.
8.0–8.9	Violent earthquake. Major damage to partial or full collapse of regular buildings and structures. Many well-designed modern buildings shifted out of foundations. Shaking felt over extremely large regions. Death toll is known to be as large as from 1000 to 1 million.
9– >9	Extreme earthquake. Some well-built wooden structures and buildings can be totally destroyed; most masonry and frame structures collapsed and destroyed with foundations. In railways extreme earthquakes cause rails bent, and permanent changes can be caused in ground topography. Death toll is >50000.

It needs only the presence of the seismometers, or seismic stations from which data may be digitally transmitted to any laboratory or computing network. The effects of damage and destruction near the epicenters of earthquakes can be easily added to the Richter scale (Table 10) using coherent observational data or data derived from comparisons with earthquakes that occurred centuries before the development of seismic instruments. The Richter scale does not have an upper limit of earthquake magnitude, but the largest known earthquakes measured by seismographs using Richter scale are a 9.5 earthquake in Chile in 1960, and the Great Alaskan Earthquake of 1964, which measured 9.2 on the scale.

4.7.1.2. Economic and societal impacts of earthquakes

From a geological point of view, earthquakes are natural phenomena that reflect the Earth's dynamics; they, together with volcanism, deep-mantle convection, radioactive decay, and internal heat generation, are the most important components keeping the Earth as a planet with sustainable life. According to the USGS, of the earthquakes that are recorded annually in the world by seismic stations—mostly in remote, non-urban territories—more than 1.3 million micro-earthquakes (M = 2–2.9) occur, along with 130000 small earthquakes (M = 3–3.9), 13000 medium earthquakes (M = 4–4.9), 1319 significant earthquakes (M = 5–5.9), 134 large earthquakes (M = 6-6.9), 15 very large earthquakes (M = 7–7.9), and 1 great earthquake (M = 8 or >8). During the past decade, there were several great earthquakes with magnitudes of 8.5–9.1 that caused almost complete destruction of bridges, commercial and industrial buildings, and residential structures and caused thousands of deaths. Among these earthquakes were the following:

  (i) West coast of Northern Sumatra, Indonesia, M = 9.1, 2004;
 (ii) Northern Sumatra, Indonesia, M = 8.6, 2005;
(iii) Southern Sumatra, Indonesia, M = 8.5, 2007;
 (iv) Offshore Maule, Chile, M = 8.8, 2010;
  (v) Tohoku (known also as Sendai earthquake), Japan, M = 9.0, 2011.

These earthquakes, which had epicenters near coastlines, triggered tsunamis that devastated coastal towns and villages and damaged shipping ports; bodies of dead and severe injured people were found mostly under buildings and inside cars after the earthquakes struck. The recent great earthquake at Tohoku (Sendai), Japan, M = 9.0, on March 11, 2011, occurred near the eastern shore of the island of Honshu, close to the Fukushima Daiichi Nuclear Power Plant in Fukushima City. Although the reactor was automatically shut down when the cooling system failed, radiation inside the plant exceeded 1000 times normal levels, and outside the plant, it was up to 8 times normal levels. Radioactive cesium, iodine, and strontium were detected in offshore oceanic waters and in the soil in some places close to the reactor area. The earthquake triggered a huge tsunami that caused extensive damage in many villages and cities situated along the coast, and the transport network also suffered severe disruptions (Fig. 62). About 51000 people were evacuated from the area close to the nuclear reactor, and the Japanese National Police Agency confirmed that 15883 people were killed, 6150 were injured, and 2651 people were still missing on November 8, 2013. Estimates of the cost of the damage range well into the tens of billions of the U.S. dollars.

4.7.2. Earthquakes and human health

Public health and human life are directly impacted by earthquakes. If the epicenter of a large earthquake occurs in an urban area, many people can be killed instantly due to severe crushing injuries to the head or chest, external or internal hemorrhaging, or drowning from earthquake-induced tsunami waves. Dust inhalation, chest compressions, or environmental exposure (e.g., to cold temperatures, causing hypothermia) can also result in rapid death. Many delayed deaths occur within several days due to dehydration, hypothermia, wounds infection, and sepsis. The so-called "crush syndrome" occurs when people are trapped in collapsed buildings and their limbs are under prolonged pressure, which can cause disintegration of muscle tissues, kidney failure, or fatal cardiac arrhythmias.

Dust inhalation creates a significant threat to the health of rescue and clean-up workers in earthquake destruction areas because it causes eye and respiratory-tract irritation followed by possible delayed chronic lung disease, depending on the toxicity of the dust. Burns and smoke inhalation from fires pose additional major hazards to humans because collapsed buildings usually contain broken remains of flammable wood treated with chemicals, electronics, and synthetic materials, some of which are extremely toxic. Chemical and petroleum products and storage facilities for hazardous materials that are available in all industrial cities might explode or spill out, and damage to buildings and operation rooms at a nuclear power plant could lead to widespread contamination by radioactive materials, as happened during the 9.0-magnitude earthquake of 2011 in Japan. In a major earthquake, pipelines carrying natural gas, water, and sewage can be disrupted, and drinking water reservoirs can be under threat of severe toxic contaminations.

4.7.2.1. Valley Fever: An epidemic disease caused by the 1994 Northridge earthquake in California

Few people know that epidemic illnesses can be triggered by geological hazards such as earthquakes. These events do not produce any pathogen bacteria, but they can blow them away from their primary sources and distribute them as far as tens to hundreds of kilometers. An unexpected consequence of the magnitude 6.7 earthquake (Fig. 63) that struck Northridge, California (in greater Los Angeles), on January 17, 1994, was an outbreak of Valley Fever (named after the San Fernando Valley, where Northridge is located), a respiratory disease contracted by inhaling airborne fungal spores. The medicinal name for Valley Fever is coccidioidomycosis,

FIGURE 63 Northridge earthquake (M = 6.7), California, 2004: (a) geographic position of the epicenter; (b) damage of a column in a freeway system; (c) landslide caused by earthquake; (d) dust storm triggered by landslide and wind after the earthquake

and it is caused by inhalation of dust containing a soil-inhabiting fungus, *Coccidioides immitis* (*C.i.*), which is unique to North, Central, and South America. Given proper conditions, infectious spores of *C.i.* are released into the atmosphere when the soil is disturbed by earthquakes, wind storms, or construction activities. People who have been exposed to earthquake-caused dust containing *C.i.* may develop, in a couple of weeks, the common symptoms of coccidioidomycosis, which start from reddish spots on the skin (Fig. 64) followed by chest pain, fever, headache, pain or swelling in the legs and ankles, cough with phlegm, loss of appetite, and stiffness in the joints. People with severe infection may also experience changes in mood, enlarged lymph nodes, severe lung infection, weight loss, and in some cases, it can be transformed into meningitis, a brain disease that can be fatal.

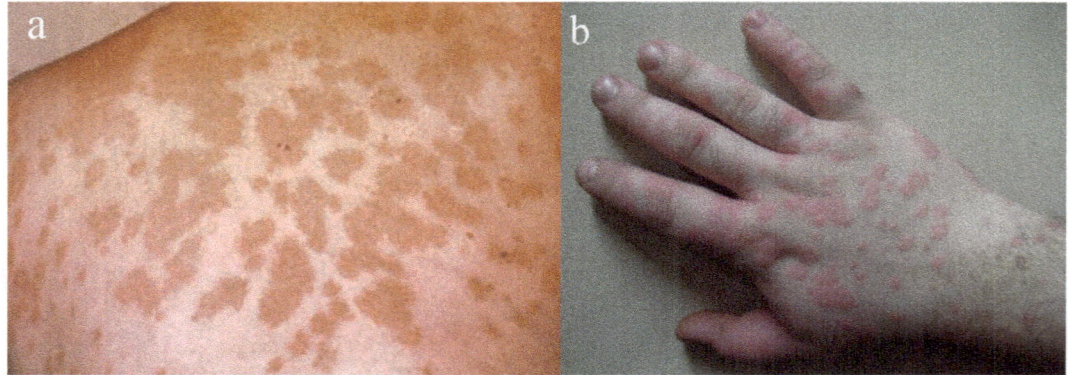

FIGURE 64 Valley Fever symptoms caused by coccidioidomycosis: the disease develops when *Coccidioides immitis* enters the human body from inhaling spores that have attached to dust particles

Within eight weeks after the Northridge earthquake, three people died and 203 were found to be ill due to contamination with *Coccidioides immitis*, although the onset of the disease for many of these cases took about two weeks after the earthquake. The three fatalities from the coccidioidomycosis epidemic accounted for 4% of the total earthquake-related fatalities. Almost all of the ill people were residents of Simi Valley, in Ventura County, north of Los Angeles County, where dust-generating landslides had been triggered by the earthquake and its strong

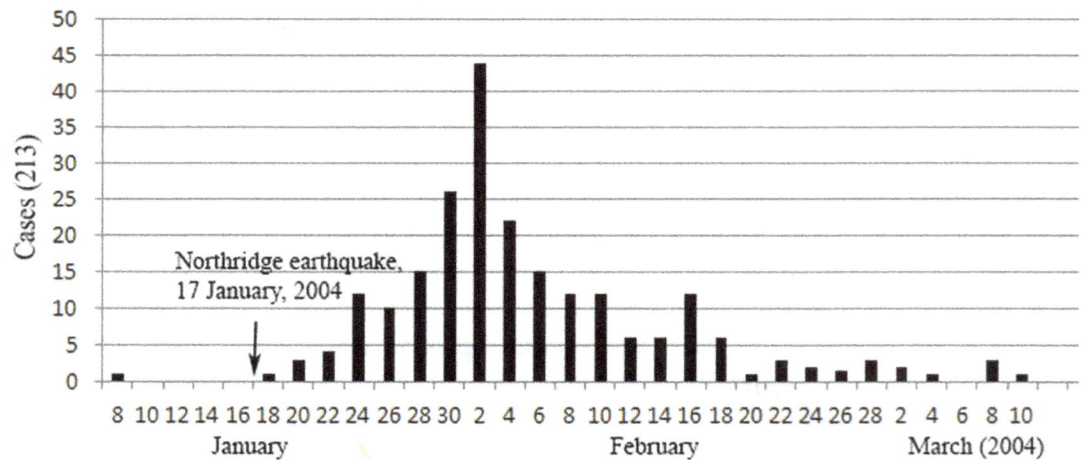

FIGURE 65 Valley Fever cases reached maximal range three weeks after the 6.7 magnitude earthquake in Northridge, California

aftershocks in the Santa Susana Mountains (see Fig. 63c,d). Prevailing winds after the earthquake and the aftershocks spread *Coccidioides immitis* in dust clouds generated by the earthquake to Simi Valley and other communities. Although the case of the epidemic of coccidioidomycosis caused by geological hazards was the first ever recorded, the statistical analyses of existing data clearly indicated that the maximal reported cases of the disease occurred in the third week after clouds of dust with the infectious fungus were scattered over the territory of landslides caused by the earthquakes (Fig. 65).

The first record of the coccidioidomycosis epidemic illness triggered by geological hazards is a great lesson for creating an awareness system that could prevent populations from similar geological environments where earthquakes may disturb soils containing *Coccidioides immitis*. The renewal of interest in coccidioidomycosis is important also because of the massive migrations of people in the U.S. to Phoenix and Tucson, Arizona; Bakersfield and Fresno, California; and El Paso, Texas, where formerly sparsely populated areas have now become large urban centers, thus putting the populations at risk of contracting the Valley Fever disease. Agricultural regions where soils contain *Coccidioides immitis* are also on the list of potential health threats for humans, so farmers should be made aware of the infectious nature of the fungus that silently lives in their fields.

References

Barac, M. 1901. The red dust of March 1901. *Monthly Weather Review* July 1901:316–317.

Choobari, O. A., P. Zawar-Reza, and A. Sturman. 2014. The global distribution of mineral dust and its impact on the climate system: A review. *Atmospheric Research* 38:152–165.

Crouvi, O., R. Amit, Y. Enzel, and A. R. Gillespie. 2010. Active sand seas and formation of desert loess. *Quaternary Science Reviews* 29:2087–2098.

Derbyshire, E. 2013. Natural aerosolic mineral dusts and human health. In: O. Selinus (Ed.), *Essentials of medical geology: Revised edition*. New York: Springer, pp. 455–475.

Gregory, L. G., and C. M. Lloyd. 2011. Orchestrating house dust mite-associated allergy in the lung. *Trends in Immunology* 32:402–411.

Hart, B. J. 1995. The biology of allergenic domestic mites, an update. *Clinical Reviews in Allergy and Immunology* 13:115–133.

Husar, R. B., D. M Tratt, B. A. Schichtel, S. R. Falke, F. Li, et al. 2001. Asian dust events of April 1998. *Journal of Geophysical Research* 106:18317–18330.

Jie, Y., Z. M. Isa, X. Jie, and N. H. Ismail. 2013. Asthma and asthma-related symptoms among adults of an acid rain-plagued city in Southwest China: Prevalence and risk factors. *Polish Journal of Environmental Studies* 22:717–726.

Landrigan, P. J., P. J. Lioy, G. Thurston, G. Berkowitz, L. C. Chen, et al. 2004. Health and environmental consequences of the World Trade Center disaster. *Environmental Health Perspectives* 112:731–739.

Malo, J-L., M. Chan-Yeung, and D. I. Bernstein (Eds.). 2013. *Asthma in the work place.* 4th Ed. Florence, KY: CRC Press, Taylor and Francis Group, 454 pp.

Mauzerall, D. L., B. Sultan, N. Kim, and D. F. Bradford. 2005. NO_x emissions from large point sources: Variability in ozone production, resulting health damages and economic costs. *Atmospheric Environment* 39:2851–2866.

Nemet, G. F., T. Holloway, and P. Meier. 2010. Implications of incorporating air-quality co-benefits into climate change policymaking. *Environmental Research Letters* 5:1–9.

Nawahda, A. 2013. The association of $PM_{2.5}$ and surface ozone with asthma prevalence among school children in Japan: 2006–2009. *Health* 5:1–7.

Peel, M. C., B. L., Finlayson, and T. A. McMahon. 2007. Updated world map of the Köppen-Geiger climate classification. *Hydrology and Earth System Science* 11:1633–1644.

Peters, S., H. Kromhout, A. C. Olsson, H.-E. Wichmann, and I. Brüske, et al. 2012. Occupational exposure to organic dust increases lung cancer risk in the general population. *Thorax* 67:111–116.

Platt-Mills, T. A. E., and A. L. de Weck. 1989. Dust mite allergens and asthma—A worldwide problem. *Journal of Allergy and Clinical Immunology* 83:416–427.

Prospero, J. M. 1999. Long-range transport of mineral dust in the global atmosphere: Impact of African dust on the environment of the southeastern United States. *Proceeding of the National Academies of Science of the USA* 96:3396–3403.

Rom, W. N., and S. B. Markowitz (Eds.). 2007. *Environmental and occupational medicine.* 4th Ed. Baltimore: Lippincott Williams &Wilkins, 1884 pp.

Shinn, E., G. W. Smith, J. M. Prospero, P. Betzer, M. L. Hayes, V. Garrison, and R. T. Barber. 2000. African dust and the demise of Caribbean coral reefs. *Geophysical Research Letters* 27:3029–3032.

Uematsu, M., R., A. Duce, J. M. Prospero, L. Chen, J. T. Merrill, and R. L. McDonald. 2012. Transport of mineral aerosol from Asia over the North Pacific Ocean. *Journal of Geophysical Research: Oceans* 88:5343–5352.

Xu, X., Z., X. G. Gai, and X. S. Men. 1993. A study of siliceous pneumoconiosis in a desert area of Sunan County, Gansu Province, China. *Biomedical and Environmental Sciences* 6:217–222.

Web resources

http://www.epa.gov/visibility/action.htlm
http://www2.epa.gov/sunwise/health-effects-uv-radiation
http://housedustmite.com
http://www.volcano.si.edu

http://volcano.si.edu/reports_weekly.cfm
http://volcanoes.usgs.gov/ash/dickzais.html
http://earthquake.usgs.gov/learn/topics/mercalli.php
http://volcanoes.usgs.gov/observatories

Image Credits

Figure 40: Mysid, "Atmosphere gas proportion," http://commons.wikimedia.org/wiki/File:Atmosphere_gas_proportions.svg. Copyright in the Public Domain.

Figure 41a: Copyright © Patrick J. Lynch (CC by 2.5) at http://commons.wikimedia.org/wiki/File:Lungs_diagram_simple.svg.

Figure 41b: National Institute for Occupational Safety and Health, "Coal miner spraying rock dust," http://commons.wikimedia.org/wiki/File:Coal_miner_spraying_rock_dust.jpg. Copyright in the Public Domain.

Figure 41c: Alfred Palmer, "AlfedPalmersmokestacks," http://en.wikipedia.org/wiki/File:AlfedPalmersmokestacks.jpg. Copyright in the Public Domain.

Figure 41d: Copyright © Xuanxu (CC BY-SA 2.0) at http://commons.wikimedia.org/wiki/File:Cielo_(4690375817).jpg.

Figure 42: Copyright © Chris (CC by 3.0) at http://en.wikipedia.org/wiki/File:Air_Pollution-Causes%26Effects.svg.

Figure 43a: Nipik, "Acid rain woods1," http://commons.wikimedia.org/wiki/File:Acid_rain_woods1.JPG. Copyright in the Public Domain.

Figure 43b: Copyright © Nino Barbieri (CC by 3.0) at http://commons.wikimedia.org/wiki/File:Pollution_-_Damaged_by_acid_rain.jpg.

Figure 44: Environmental Protection Agency, "Sizes of particulate matter PM2.5 and PM10," http://www.epa.gov/pm/graphics/pm2_5_graphic_lg.jpg. Copyright in the Public Domain.

Figure 45: Copyright © LordToran (CC BY-SA 3.0) at http://commons.wikimedia.org/wiki/File:Trockenklimate.png.

Figure 46: Copyright © Peel, M. C., Finlayson, B. L., and McMahon, T. A. (CC BY-SA 3.0) at http://en.wikipedia.org/wiki/File:Koppen_World_Map_BWk.png.

Figure 48: Jaques Descloitres, MODIS Rapid Response Team, NASA GDFC, "Saharan Dust off West Africa," http://en.wikipedia.org/wiki/File:Saharan_Dust_off_West_Africa.jpg. Copyright in the Public Domain.

Figure 49: "Intercontinental Asian dust storm, April 1998," Husar, Tratt, Schichtel et al., 2001. The Asian Dust Events of April 1998. Journal of Geophysical Research. Atmosphere. 106 (D16): 18317-18330. Copyright in the Public Domain.

Figure 50: Wilson44691, "LoessVicksburg," http://en.wikipedia.org/wiki/File:LoessVicksburg.jpg. Copyright in the Public Domain.

Figure 52: Copyright © Till Niermann (CC BY-SA 3.0) at http://en.wikipedia.org/wiki/File:Loess_landscape_china.jpg.

Figure 53a: Edescas2, "Meseta de Loes," http://en.wikipedia.org/wiki/File:Meseta_de_Loes.png. Copyright in the Public Domain.

Figure 53b: Copyright © Till Niermann (CC BY-SA 3.0) at http://en.wikipedia.org/wiki/File:Loess_landscape_china.jpg.

Figure 54: Copyright © Wally Gobetz (CC by 2.0) at http://commons.wikimedia.org/wiki/File:September_11_2001_just_collapsed.jpg.

Figure 55: United States Geological Survey, "Chemistry Figure 1," http://pubs.usgs.gov/of/2001/ofr-01-0429/chem1/wtcchemfig1new11-7.gif. Copyright in the Public Domain.

Figure 56: United States Geological Survey, "Types of volcano hazards usgs," http://commons.wikimedia.org/wiki/File:Types_of_volcano_hazards_usgs.gif. Copyright in the Public Domain.

Figure 57a: Copyright © Bidgee (CC by 3.0) at http://commons.wikimedia.org/wiki/File:Rapid_Creek_flooding_1.jpg.

Figure 57b: "MSH80 eruption mount st helens 05-18-80-dramatic-edit," http://commons.wikimedia.org/wiki/File:MSH80_eruption_mount_st_helens_05-18-80-dramatic-edit.jpg. Copyright in the Public Domain.

Figure 57c: warszawianka, "Tango Weather Showers Scattered," https://openclipart.org/detail/30151/tango-weather-showers-scattered-by-warszawiank. Copyright in the Public Domain.

Figure 57d: "Field with Water," http://nc.water.usgs.gov/projects/tile_drains/icons/ditch1.jpg. Copyright in the Public Domain.

Figure 58: United States Geological Survey, "Ash particle, 1980 Mount St. Helens eruption, magnified 200 times," http://volcanoes.usgs.gov/ash/properties.html. Copyright in the Public Domain.

Figure 59a: United States Geological Survey, "Small explosion in 1997," http://volcanoes.usgs.gov/ash/images/32424296-057_large.jpg. Copyright in the Public Domain.

Figure 59b: NASA, "Soufrière 2009 eruption," http://en.wikipedia.org/wiki/File:Soufri%C3%A8re_2009_eruption.jpg. Copyright in the Public Domain.

Figure 60a: United States Geological Survey, "MSH80 eruption mount st helens 05-18-80-dramatic-edit," http://en.wikipedia.org/wiki/File:MSH80_eruption_mount_st_helens_05-18-80-dramatic-edit.jpg. Copyright in the Public Domain.

Figure 60b: Murraybuckley, "1980 Mount st helens ash distribution," http://en.wikipedia.org/wiki/File:1980_Mount_st_helens_ash_distribution.svg. Copyright in the Public Domain.

Figure 61: United States Geological Survey, "Schematic model of epicenter of an earthquake," http://geomaps.wr.usgs.gov/parks/sunset/seismic.html. Copyright in the Public Domain.

Figure 62a: Heinz-Josef Lucking, "Map of Sendai Earthquake 2011," http://commons.wikimedia.org/wiki/File:Map_of_Sendai_Earthquake_2011.jpg. Copyright in the Public Domain.

Figure 62b: Matthew M. Bradley, "Upended House," http://commons.wikimedia.org/wiki/ File:US_Navy_110315-N-2653B-107_An_upended_house_is_among_debris_in_Ofunato,_Japan,_ following_a_9.0_magnitude_earthquake_and_subsequent_tsunami.jpg. Copyright in the Public Domain.

Figure 63a: United States Geological Survey, "Northridge, CA Earthquake Damage," http://gallery.usgs. gov/photos/01_15_2014_eja4DPo00W_01_15_2014_12#.UySwaGfn9D8. Copyright in the Public Domain.

Figure 63b: NASA, "Landslide caused by earthquake," http://earthobservatory.nasa.gov/Features/ Landslide/Images/golden.jpg. Copyright in the Public Domain.

Figure 63c: United States Geological Survey, "Dust Storm Triggered by Landslide and Wind after the Earthquake," http://pubs.usgs.gov/of/1995/ofr-95-0213/fig20vga.jpg. Copyright in the Public Domain.

Figure 64a: Copyright © James Heilman, MD (CC BY-SA 3.0) at http://commons.wikimedia.org/wiki/ File:Erythema_multiforme_minor_of_the_hand.jpg.

Figure 64b: Center for Disease Control, "Valley Fever symptom," http://www.cdc.gov/fungal/images/ phil-482_lores.jpg. Copyright in the Public Domain.

Review Questions

1. What is the composition of the atmosphere?

2. Is the oxygen amount in high mountain areas smaller than in the air at sea level?

3. Why do people say that the air in the high mountains is "thinner" compared with the air at sea level? What kind of scientific explanation is behind this?

4. What is air pollution? How should we define whether the air is polluted or not?

5. What is particulate matter, and what categories is it classified in if you take into account its physical state properties?

6. What is the "greenhouse effect"?

7. How do acid rains form?

8. Air pollution has adverse effects on human health. What is the harmful pathway of air pollution?

9. Are ozone and nitrogen oxides air pollutants?

10. How do we define atmospheric dust? What are the primary and secondary dust particles? What are the particle sizes, and why is it important to know their parameters? What are the sources of dust?

11. Do smaller dust particles of <10 μm diameter fly faster than larger particles of >500–1000 μm diameter?

12. What are the lifetimes of the dust particles: 5000 μm, 1000 μm, 100 μm, 10 μm, and 0.1 μm? How fast do 1-μm particles fall? How fast do 0.02-μm particles fall?

13. What is the origin of domestic dust? People often say that they "have an allergy to dust"; what is really behind such an allergy?

14. How many tons per year of mineral dust are emitted globally into the atmosphere?

15. What is intercontinental dust? Where does it usually form? How is it transported? On what timescale does it move?

16. What are the benefits and hazards represented by intercontinental dust?

17. What is loess? What does it consist of? Why is it potentially dangerous? In what parts of the world is loess abundant?

18. Did the dust of September 11, 2001—the Ground Zero site of the World Trade Center (WTC) in New York—contain toxins? What kinds of toxins? Where did they originate?

19. What was the most typical adverse effect of the WTC dust on the health of firefighters and volunteers who helped during the first hours after the terrorist attack?

20. What are pneumoconiosis and silicosis diseases, and what causes them? Who are at risk?

21. What are the various hazards presented by volcanoes, and how do they inflict damage?

22. Provide characteristics of the structure of a volcano, and define pyroclastic and lahar flows, ash falls, volcanic gases, and lava flows. How far from a volcanic vent can these substances be ejected and/or distributed?

23. When did the Mount St. Helens volcano erupt? Why are the Cascade Mountains the "right" place for volcanic activity?

24. What kind of magma do violent, explosive volcanoes produce? What kind of viscosity does this magma have?

25. What kind of magma do non-violent (e.g. non-explosive) volcanoes produce? What kind of viscosity does this magma have?

26. Taking in account the viscosities of violent and non-violent volcanoes, try to determine which of them will produce magma that will flow for a longer distance from the volcanic vent.

27. In your own words, provide a simplified model that includes the three major stages of volcanic aerosol and volcanic particulate formation.

28. What kind of toxic volcanic gases/aerosols are ejected into the atmosphere?

29. What concentrations of volcanic SO_2, CO_2, HCl, and HF ejected into the air can cause death after extended exposure?

30. What is volcanic ash? Is it soluble in water? Can volcanic ash particles transport aerosols? Can volcanic ash cause acid rain?

31. Using as an example the story of the city of Yakima, Washington, which is ~110 kilometers east of Mount St. Helens, describe what societal and economic damages can be produced by volcanic activity.

32. What is an earthquake? What is the epicenter of an earthquake?

33. In what geological situations do earthquakes occur?

34. Can earthquakes and volcanic activities be connected to each other?

35. What methods are used to evaluate the power of an earthquake?

36. How are Mercalli's and Richter's earthquake intensity measurements different?

37. List the magnitudes of the Richter scale, and provide some possible correlations with Mercalli's scale.

38. List the geographic names and magnitudes on the Richter scale of five great earthquakes that occurred from 1990 to the present.

39. How are earthquakes and tsunamis related to each other?

40. Why are earthquakes dangerous?

41. What is Valley Fever disease, and what causes it? Where is it a problem?

Quizzes (see answers on page 307)

1. Although very large dust particles settle out of the air very quickly, the finest dust particles (~0.02 µm) can remain suspended in the atmosphere for as long as...
   a. Days.
   b. Many months.
   c. Hundreds of years.
   d. Millions of years.
   e. Forever.

2. Which of the following is/are true about loess?
   a. It is made mostly from clay minerals.
   b. It is a fine-grained sediment formed by wind-blown silt combined with some sand and clay.
   c. It is easily eroded if not covered by vegetation, and it can cause dust storms.
   d. All of the above.
   e. b and c only.

3. Valley Fever is a fungal disease common in the Southwest. Arizona has some of the worst problems in the U.S. with this disease, but do we in California have to worry about this disease? Why or why not?
   a. Yes, even more of the fungus that causes it grows in California because of the wetter coastal climate.
   b. Yes, there is plenty of dust that carries the fungus responsible for Valley Fever. In addition, there is the added danger of earthquakes loosening sediments and creating more dust.
   c. Yes, the fungus contaminates ground water within several aquifers that are shared by Arizona and California.
   d. No. Although many conditions are similar, the fungus causing the disease does not grow in California.
   e. No. Although the fungus causing the disease can be found in California, the type of dust that can transmit the fungus to humans is not present.

4. Loess sediments, which were formed during the Quaternary period, which spans from 2.588 ± 0.005 million years ago to the present time, are found...
   a. Only on the Loess Plateau in China.
   b. In China and the U.S. Midwest.
   c. In China and the Amazon rainforest.
   d. In the U.S. Midwest and the Sahara Desert.
   e. On almost all continents except Africa and Australia.
   f. In Antarctica, the U.S. Midwest, and the Loess Plateau in China.

5. The oxygen concentration in the atmosphere is ~20.9%. When you are in the high mountains, you may feel a shortage of oxygen and become sick because...
   a. The concentration of oxygen varies with elevation.
   b. The concentration of oxygen in the air remains the same anyplace in the world, but the ratio of nitrogen and methane changes with the elevation.
   c. The air at high altitudes is less compressed and is therefore "thinner," which means that in a given volume of air, there are fewer oxygen molecules present.
   d. The concentration of oxygen remains constant (~20.9%) anyplace on the Earth's surface.
   e. Only c and d are correct answers.

Geological hazards: From Minerals to Chemical Disasters

Chapter 5

Essential and non-essential mineral elements for human health

5.1.1. Geochemical classification of elements: Major, minor, and trace

Ninety-two natural elements of the 118 total listed on the periodic table are the major building blocks of minerals (see Fig. 24). While the periodic table classifies elements based on their atomic weights, geochemistry and geology classify elements according to their concentrations. In rocks and minerals, major elements are those that have concentration levels >1% by mass (1% = 1 gr/100 gr), minor elements have concentrations in the range 0.1–1.0% by mass, and trace elements are those that have concentrations <0.1% by mass. For example, the chemical composition of the mineral albite ($NaAlSi_3O_8$) is Na_2O = 11.05%, K_2O = 0.03%, CaO = 0.95%, Al_2O_3 = 20.44%, SiO_2 = 67.52%, PbO = 0.003%, P_2O_5 = 0.007%, total = 100%. From this chemical analysis, one can easily determine that the major elements are Na, Al, Si, and O; the minor elements are Ca and K; and that the trace elements are Pb and P. The major, minor, and trace element categories are different for different materials. For example, in geological and biological materials, the same types of elements may belong to different categories (see examples in Table 11).

TABLE 11: Major, Minor, and Trace Element Abundances in Two Different Categories of Material: Rocks of the Earth's Crust and the Human Body.

Elements	Earth's Crust (Concentration, %)	Earth's Crust (Major, Minor, or Trace Elements)	Human Body (Concentration, %)	Human Body (Major, Minor, or Trace Elements)
Oxygen (O)	46.6	Major	65.4	Major
Silicon (Si)	27.7	Major	0.006	Trace
Iron (Fe)	5.0	Major	0.026	Trace
Calcium (Ca)	3.6	Major	1.4	Major
Sodium (Na)	2.8	Major	0.14	Minor
Potassium (K)	2.6	Major	0.34	Minor
Magnesium (Mg)	2.1	Major	0.5	Minor

5.1.2. Essential and non-essential mineral elements

The trace elements that are accumulated in human bodies are often classified as essential and non-essential. However, the understanding of which elements are essential and which are not is a philosophical question, especially when different disciplines try to consider the significance of the elements necessary for life. For example, the nutritional and toxicological literature considers a specific group of elements as toxic, though this term can be applied to any elements because they can all become toxic if they are present in significantly higher quantities than are required for normal biological functions. Geomedical and environmental publications use the definitions that are described below.

Essential elements are any chemical element required in large amounts (macronutrient) or in very small amounts (trace element) by an organism for healthy growth and sustainable life. The main role of essential elements includes the formation of skeletal structure and the maintenance of metabolic and biochemical processes that support the functioning of cells. Six elements (oxygen, hydrogen, carbon, nitrogen, calcium, and phosphorous) compose about 99% of the human body, and at least twenty-three other elements of the periodic table are known to be physiologically active, with eleven of them being trace elements.

Essential major and minor elements that are required by the human body in high quantities are considered to be macronutrients, the most important of which are sodium (Na), calcium (Ca), magnesium (Mg), potassium (K), and chlorine (Cl), among others. Essential trace elements are considered to be micronutrients, and the human body needs them in very small quantities (generally, less than 100 mg/day). They are represented by transition metals such as vanadium (V), chromium (Cr), manganese (Mn), iron (Fe), cobalt (Co), copper (Cu), zinc (Zn), and molybdenum (Mo), and by non-metals such as selenium (Se), fluorine (F), and iodine (I). Most of these trace elements are characterized by an optimal relationship between the sizes of their nuclei and electron availability in the outer orbits, which makes it easy for them to interact with the organic molecules available in biological systems.

The concentrations of essential trace elements in a healthy human body are constant during certain periods of human life, but they can be changed by the aging processes. Decreases of some essential elements in the human body may lead to abnormalities, such as some physiological symptoms, and uncontrolled increases in their concentrations due to environmental or occupational exposures may cause serious diseases. Trace elements that are essential to biological structures can be toxic if their concentrations are beyond the levels that are necessary for accomplishing normal biological functions. Therefore, in this context, *non-essential elements* are defined as those that are introduced into the human body in excess quantities from food and environments (e.g., Fraga, 2005). The toxicity of minerals (elements), however, is not as simple as just the excess of their concentrations—it also depends on the bioavailability of elements (e.g., how much of particular elements may be absorbed by the body to influence human health), their physical and chemical properties, and the places of their accumulation in the human body.

5.1.3. Human geophagy—eating "earth"

5.1.3.1 Geophagy and pica: Historical background

Human geophagy (*ji-o-fagi*), or geophagia (*ji-o-fagia*), is a practice of the deliberate eating of "earth," which is used here as a generic term for soil, earth dirt, chalk, and clay. There are many different explanations for why geophagy takes place, but there is no understanding of what first caused the deliberate consumption of the earth by humans. It is believed that geophagy is a variety of *pica* (ˈpaɪkə/py-kə), an abnormal desire to eat non-food substances (Waywodt and Kiss, 2002). Pica is a Latin word that means *magpie*, a bird noticed for its indomitable appetite for soil, dirt, and other non-nutrient substances and for carrying away all kinds of extraneous objects.

Geophagy was recognized in humans as early as 400 BC by Hippocrates. In one of his medicinal textbooks, *Ouvres Complètes d'Hippocrate*, which was translated into English by Potter (1995), he links geophagy (pica) to pregnancy conditions: "If a pregnant woman feels the desire to eat earth or charcoal and then eats them, the child will show signs of these things."

Many studies indicate that since that time, practically every culture has had an "earth-eating" stage in its development, and there is an agreement among many scientists that geophagy is a common habit of infants and young children, pregnant women, and, more rarely, of some adults. Geophagy has been traced through ancient Greece to European cultures, and it is well known in Africa and South America and has been observed in the American Indians and in rural areas of Mississippi in the U.S., as well as in Bangladesh and India (e.g., Young et al., 2011). It is believed that geophagy has some connection to cultural habits in the family, and it is deeply incorporated in some traditional cultures and religions. In many developing countries, or during wars and other catastrophic disasters, eating earth materials is often related to satisfaction of hunger pangs. Geophagy is widely disseminated in countries with low socioeconomic status, where poverty and malnutrition are common. Two photos on Fig. 66 show the tablets of white chalky earth from the Milk Grotto in the holy city of Bethlehem exhibited in the Science Museum, London, UK. It was said that these tablets originated from the site where Christians believe the Virgin Mary stopped to breastfeed Jesus as they fled to Egypt, and these circumstances would have given it a special significance (see for details: http://wellcomeimages.org). Cookies prepared from mixture of clay, mud, salt and vegetables are used as food in present-day Haiti, the poorest country in the Western Hemisphere, where 57 % of population lives in poverty. By eating such cookies, people make their stomachs feel "full," and they also satisfy part of their nutrient deficiency by receiving some microelements (mostly Fe) available in the mud. The only question that remains is whether they know how many toxic metals, non-essential chemical elements, and pathogen bacteria are hidden in the cookies and the kinds of adverse effects on their health that might be expected.

FIGURE 66   White chalky earth from Bethlehem, Palestine, 1920–1930; these tablets were intended to be eaten

Some studies show that in the tropical rainforest regions of South America, human geophagy was developed to reduce the toxicity of various fruits and plants grown in the local environments. In South Asian and African countries, geophagy is considered to be a response to nutritional deficiencies because the diets of poverty-stricken populations included mostly starch-rich, fibrous food. Starch is a white, tasteless, solid carbohydrate $(C_6H_{10}O_5)_n$ that is in high concentrations in sweet potatoes, cassavas, and other similar fiber-rich foods, but it is poor in essential elements such as magnesium (Mg), calcium (Ca), and zinc (Zn). Because these essential elements are important nutrients during motherhood, early childhood, and a transitional stage of human development from teenagers to adults, geophagy may be considered as a kind of self-regulation process. Indeed, geophagy delivers these elements to individuals whose diets are deficient in them.

5.1.3.2. Adverse effects of geophagy

Geophagy practice may have serious adverse effects on human health because it can result in exposure to toxic metals and pesticides as well as pathogen viruses that are widely present in soils, rocks, and minerals. Toxic elements in soils and clays usually originate from weathered minerals and rocks, and they may be concentrated due to industrial contamination of environments by man. Geophagy increases exposure to geohelminths (biological pathogens living in soils) and potentially to toxic elements that are abundant in mining and ore-processing areas and in polluted urban regions. Ingestion of soil, clay, or any other geophagy-related earth materials in excessive quantities may also cause intestinal blockage, excessive tooth wear, and hyperkalemia—an excess of potassium if the ingested soil is rich in potassium.

Although geophagy is not common today in Western countries, there are certain clays, powdered stones, and soils (mud) available as "over-the-counter" traditional medicine. Although the geographic sources of such products are now known, full chemical analyses of trace elements and their concentrations may not have been performed, and therefore, the harmful risk potential from ingestion of such "earth" is high. In Bangladesh, for example, "baked" natural clay rock (one local name is "sikor") is consumed in quantities of ~50 g per day by geophagic women in order to ensure healthy pregnancy and easy childbirth (Middleton, 1989). Geochemical studies of commercially traded sikor have revealed that it contains toxic elements such as arsenic (As) and lead (Pb), among others, in concentrations that are dangerous to human health. Further studies have shown that the consumption of 50 g of sikor is equivalent to ingesting 370 µg of arsenic, which exceeds its recommended maximum tolerable daily level by almost twofold (Al-Rmalli et al., 2010). The consumption of such clay during pregnancy may have adverse effects on the fetus because it is a well-known epidemiological fact that Bangladeshi women are concurrently exposed to As-contaminated drinking waters (Khan et al., 2003).

A literature review has also revealed many cases of dangerous geophagy practices existing among African women and schoolchildren (Njiru et al., 2011). In Kenya, for example, a pregnant woman consumes ~20 g of soil a day, even though according to the EPA, a non-harmful exposure to soil through hand-to-mouth contact is only 50 mg per day. This means that these Kenyan women consume an amount of soil almost 400 times the typical quantity ingested unintentionally through hand-to-mouth contact. Kenyan children five to eighteen years old ingest between 28 g and 108 g of soil per day, which is a dramatic "overdosing" if compared with the EPA's non-harmful dose.

Studies carried out in Iran in the 1960s showed that villagers with long-standing geophagy had profound anemia caused by iron deficiency, endocrine abnormalities, and dwarfism, and similar syndromes have been observed over the years among earth-eating people in Egypt, Turkey, and Portugal (Halsted, 1968).

5.1.3.3. Why do people eat earth?

There have been various hypotheses to explain geophagy, although as of yet, there is no consensus as to whether this is an atavistic trait (i.e., a hold-over) from human evolution, a practice designed to access valuable nutrients from the earth to support normal biological functions, a psychiatric disorder, or a mysterious cultural habit (Young et al., 2011). Indeed, human geophagy is a multi-component phenomenon that is associated with physiological, cultural, behavioral, and socioeconomic factors as well as an unexplained craving for soil and dirt among children, pregnant women, and even individuals with well-balanced nutrients in their diets. However, the existing concepts may be placed in three general categories: functional, behavioral, and cultural.

Functional concepts

(i)   **Detoxification.** This is a response designed to minimize noxious components in a human's diet and to target gastrointestinal distresses caused by ingestion of toxic substances. Statistics have confirmed that historically, geophagy has occurred in tropical regions, where soils, plants, vegetables, and fruits are more toxic and contain more pathogens than in other climatic zones.

(ii)   **Nutritional value.** The earth provides humans with essential elements such as iron, zinc, calcium, magnesium, phosphorous, and potassium, among other nutrients, and the fact, that children and pregnant women are the geophagy addicts, supports this hypothesis. Interestingly, many researchers have observed a positive correlation between mineral deficiency in a diet and geophagy, but at the same time, there are well-documented cases showing that such a correlation is negative.

(iii)   **A response to suppress hunger.** Geophagy has always developed in times of famine, when people have insufficient food, or during other periods of food insecurity caused by local wars, droughts, and other natural and geological hazards.

Behavioral

(i)   Eating disorders. These can occur as a result of pathological psychiatric conditions (i.e., mental illness), and they are often considered in the medical literature (e.g. Baheritibeb et al., 2008).

(ii)   Atavistic habits from human evolution. The history of anthropology shows that early humans living in extreme conditions often had scarcities of food, and so clay, rocks, minerals, and soils were used as the most easily accessible "food" and medicinal materials. One of the types of clay—kaolinite—was definitely used for fighting diarrhea in prehistoric times, and even in the present day kaolinite is a main ingredient in Kaopectate, the commercial, over-the-counter pharmaceutical product recommended for diarrhea and gastrointestinal distress.

Cultural

Geophagy is often associated with traditions, rituals, and religious practices. For example, in Central America (Honduras, Guatemala, Costa Rica, El Salvador, and Nicaragua), pregnant women eat clay tablets that have been blessed by the Roman Catholic Church (Hanter and de Kleine, 1984). The clay originates from a holy site associated with a cult of the "Black Christ," and it is believed that it has a healing power for easy pregnancy and healthy childbirth. In Africa, many young women from urban areas believe that eating clay or soil will make their skin lighter and thus make them look more attractive. In rural communities, geophagy is usually observed among people of lower education, children, and old people from families with poorer nutritional diets.

5.1.3.4. Negative vs. positive

Although the negative effects of geophagy are scientifically proven, historical and empirical facts suggest that the practice of eating soils is probably a necessity to satisfy a nutritional deficiency. Indeed, soil is a source of elements and minerals that can supply micronutrients for geophagic individuals. Some soils may be a good source for Fe in a quantity proportional to the daily intake recommended by nutritionists (Abrahams, 1997), and therefore, one may expect that soil-eating humans would never suffer from anemia, which historically was usually caused by an iron deficiency in diets. Surprisingly enough, this statement is in contradiction with observations published by physicians practicing in poor areas of Turkey (Okcuoglu et al., 1966), who concluded that many cases of geophagy in that area did indeed lead to anemia, the most common disorder of the blood, which causes feelings of weakness, fatigue, general malaise, and sometimes poor concentration.

The question that arises is how to draw a boundary between negative and positive effects of geophagy. Geophagy is still widely practiced in the world, and ingestion of soils as well as consumption of processed Earth materials, such as "holy clay tablets," definitely had some positive medicinal effects.

5.2. Arsenic—a toxic metalloid

5.2.1. Background on arsenic

On the periodic table, the element arsenic (As; number 33) is situated within the metalloids group, which also includes boron (B), silicon (Si), germanium (Ge), antimony (Sb), tellurium (Te), and polonium (Po), all of which have properties of both metals and non-metals. Arsenic can be found in soil, coal, volcanic emissions, hot springs, and "black smokers" (undersea hydrothermal vents). In nature, almost all arsenic-bearing minerals are associated with polymetallic ore deposits and hydrothermal veins and the products of their alteration hosted by different geological formations of the continental crust. The average concentration of arsenic in the Earth's crust is ~2 mg/kg. Table 12 shows that concentration of arsenic varies widely: it is minimal (0.001 mg/kg) in ultramafic rocks; it may be extremely high in sedimentary rocks such as shale/mudstones (up to 490 mg/kg), iron-rich formations (up to 2900 mg/kg), and bituminous shale (up to 900 mg/kg); and it reaches its maximum value in coal (up to 35000 mg/kg).

TABLE 12: Concentration of Arsenic in Different Rock Types of the Continental Crust (Data adopted from Smedley and Kinninburg, 2002)

Classification	Rock Type	Arsenic (mg/kg)
	Ultramafic rock (peridotite)	0.03–0.16
Igneous rocks	Basic (mafic) rock (gabbro)	1.5–110
	Acidic (felsic) rock (granite)	0.2–1.15
	Phyllite/slate	0.5–140
Metamorphic rocks	Schist/gneiss	<0.1–19
	Amphibolite	0.4–45
	Quartzite	2.2–7.6
	Shale/mudstone	3–490
	Sandstone	0.6–120
Sedimentary rocks	Limestone	0.1–20
	Iron-bearing formation	1–2900
	Bituminous shale	100–900
	Coal	0.3–35000

Arsenic in the form of arsenic trioxide (Ar_2O_3) dust/aerosol can be found in the terrestrial atmosphere as a byproduct of industrial smelting and coal burning. Arsenic has been detected in the atmospheres of Saturn and Jupiter, and its traces have been found in meteorites and in Moon-rock samples. Historically, arsenic has been known as a toxin and carcinogen if consumed in high concentrations, and poisoning from a single large dose can cause acute syndromes followed by death. In addition, small doses consumed repeatedly over a long period of time can result in chronic disorders that lead to skin changes (darkening or discoloration, redness, swelling) and hyperkeratosis (skin bumps that resemble corns or warts), as well as liver and kidney function problems. Arsenic is also classified as a carcinogen by the EPA. Because humans are routinely exposed to arsenic through food chains, water, air, and soil, knowledge about arsenic speciation in arsenic-bearing minerals, its solubility in water, its chemical reactions, and its migration from the rocks and minerals to soils and water reservoirs is crucial for environmental studies and for community education.

5.2.2. Minerals of the arsenic group

In geological environments, there are over 320 arsenic-bearing minerals that are presented by primary and secondary minerals; the latter are products of the alteration of the primary minerals. In addition, although it is not a major component, arsenic, as an element, occurs in many sulfide minerals and rocks in varying—usually very small—concentrations. The primary arsenic minerals are relatively rare, and their occurrences are limited mostly to polymetallic ore deposits. Below are examples of primary and secondary minerals more commonly found in the Earth's continental crust.

5.2.2.1. Primary arsenic minerals

1. *Native element* (As);
2. *Arsenic sulfides*: arsenopyrite (FeAsS), realgar (AsS or As_2S_2), orpiment (As_2S_3), lorandite ($TlAsS_2$), and many others;
3. *Arsenides*: algodonite (Cu_6As), niccolite (NiAs), oregonite (Ni_2FeAs_2), safflorite ($CoAs_2$), stibarsen or allemontite (SbAs), and others that contain varying amounts of Cu, Ni, Fe, and As.

Native arsenic is a very rare mineral (Fig. 67a) that is associated with quartz and calcite formed within hydrothermal veins. Arsenic sulfides are abundant within polymetallic ore deposits, and within those deposits, arsenopyrite (FeAsS), realgar (AsS or As_2S_2), and orpiment (As_2S_3) are the most important sources of arsenic for mining (Fig. 67b–d). Among others, sulfides rich in arsenic include mineral pyrite, which may contain more than 10% As. An arsenic-rich pyrite is called an arsenian pyrite—$Fe(S,As)_2$. In addition to the polymetallic ores, arsenian pyrite forms in low-temperature sedimentary environments in reduced conditions (e.g., where oxygen activity is limited), but it also may be concentrated in large amounts within river and lake sediments and in oceans. Neither arsenian pyrite nor arsenopyrite is stable in aerobic (high-oxygen activity) systems, and the reaction of oxidation leads to the release of free arsenic, sulfates, acids, and various trace elements into shallow rocks, soils, and aquifers and into the atmosphere. The presence of pyrite or arsenopyrite as a minor constituent in coal deposits is responsible for the formation of acidic rains and acid drainage in the vicinity of coal mines, and it also creates arsenic toxicity in areas of intensive coal burning. Other common sulfide minerals may contain 1% or more arsenic as an impurity. Realgar most commonly occurs as a low-temperature hydrothermal vein mineral associated with other arsenic and antimony minerals (Fig. 67c), but it is also found in hot-spring deposits in association with orpiment, arsenolite (As_2O_3, a secondary arsenic mineral), calcite ($CaCO_3$), and barite ($BaSO_4$). After long exposure to light, realgar decomposes into a reddish-yellow powder, and therefore, if one has this mineral in his or her collection, it should be protected from light exposure. Orpiment is an orange to lemon-yellow mineral (Fig. 67d) that is almost always associated with realgar formed in hot-spring deposits and as sublimated gases emitted from volcanoes.

FIGURE 67 Primary arsenic-bearing minerals: (a) native arsenic (As) with quartz and calcite (white crystals at the periphery of the black arsenic), St. Marie-aux-mines, Alsace, France; Romania; (b) arsenopyrite (FeAsS) from Panasqueira Mine, Portugal; (c) realgar (red crystals); (d) orpiment (yellow-orange crystals) with barite (white crystals)—(c) and (d) are from El'brusskiy arsenic mine, Elbrus, Northern Caucasus Region, Russia

5.2.2.2. Secondary arsenic minerals

4. *Arsenates:* Arsenates usually refer to the naturally occurring minerals possessing the $(AsO_4)^{3-}$ anion group and, more rarely, other arsenates with anions like $AsO_3(OH)^{2-}$ or $AsO_2(OH)_2^{1-}$. Examples of more typical arsenate-containing minerals include arsenolite or claudetite (As_2O_3), adamite (Zn_2AsO_4OH), alarsite $(AlAsO_4)$, erythrite $(Co_3(AsO_4)_2 \cdot 8H_2O)$, annabergite $(Ni_3(AsO_4)_{2}8H_2O)$; (see all in Fig. 68), and many others containing Cu, P, Mn, Pb, and Bi.

5. *Arsenites:* Arsenites are very rare, oxygen-bearing arsenic minerals containing the $(AsO_3)^{3-}$ anion. Examples of arsenite minerals are reinerite $(Zn_3(AsO_3)_2)$, finnemanite $(Pb_5Cl(AsO_3)_3)$, magnussonite $(Mn_5(OH)(AsO_3)_3)$, and others containing Pb, Ca, Fe, and Cu.

Arsenate and arsenite minerals containing the $(AsO_4)^{3-}$ and $(AsO_3)^{3-}$ anion groups, respectively, occur in oxidized environments among a variety of arsenic-rich soils and weathering products developed in the upper part of arsenic-bearing ore deposits. The latter often contain

FIGURE 68   Secondary arsenic minerals: (a) arsenolite (As2O3) from the White Caps Mine, Nye County, Nevada; (b) adamite Zn2AsO4OH from Ojuela Mine, Durango, Mexico; (c) erythrite (or "red cobalt") from Morocco; (d) annabergite Ni3(AsO4)2·8H2O (green crystals) from Greece

surface layers of oxidized and hydrated arsenate minerals that are a product of the weathering of primary sulfide minerals. Classical world localities in which such minerals occurred include the complex skarn-manganese deposit at Långban, Sweden, and the polymetallic deposit in Tsumeb, Namibia. Other notable deposits are located in Ontario, Canada; Germany; the Czech Republic; Cornwall, U.K.; Morocco; Sonora, Mexico; and Australia. In general, though secondary arsenic minerals occur in almost all rocks, sediments, and soils, the highest arsenic concentrations can be expected in rocks that have been enriched by sulfide and oxide minerals. A summary of the typical ranges of arsenic concentration in sedimentary rocks (Table 12) suggests that shale and mudstone, bituminous shale, iron formation, and coal have the highest concentrations of arsenic known among natural Earth materials (490–35000 mg/kg).

In the U.S., sandstones, shales, and coal located in Colorado, Utah, South Dakota, and Wyoming have high arsenic concentrations, and the phosphorite rocks in Arkansas, Idaho, Montana, South Carolina, and Wyoming are also rich in arsenic. Groundwater concentrations of naturally occurring arsenic are relatively high in Arizona, Idaho, Nevada, New Mexico, Oregon, Utah, Indiana, Michigan, Wisconsin, and Washington.

5.2.3. Main chemical reactions of arsenic minerals and solubility of arsenic in water

In natural conditions, arsenopyrite and other arsenic sulfides are slightly soluble in water, and only chemical reactions accompanied by weathering processes may transform them into highly soluble compounds (arsenic oxides).

The weathering processes turn primary arsenic minerals into arsenic oxide and arsenic acid compounds in which arsenic exists in the form of As^{3+} or As^{5+}. Such processes are influenced by the presence of moisture and/or aqueous fluid of a certain acidity/alkalinity (pH) circulating through the pores of the rocks/minerals, day-night and seasonal variations of temperature, the concentrations of CO_2 in the lower layers of the atmosphere and in the pores' water, and the oxygen activity of the surrounding environments. For example, the main weathering reaction of arsenopyrite occurs in the presence of oxygen and water, and it ends with the breakdown of FeAsS and the release of AsO_4. The breakdown of FeAsS is a complex reaction, which usually progresses through several stages, and a final compound of arsenic can also be presented by H_2AsO_3 if the reaction is controlled by redox conditions, in which the As^{5+} is not stable and usually transforms into As^{3+}. The summarized reaction of arsenopyrite breakdown may be expressed by the following equation:

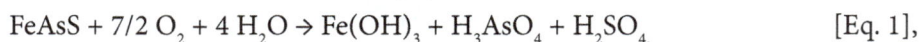

$$FeAsS + 7/2\ O_2 + 4\ H_2O \rightarrow Fe(OH)_3 + H_3AsO_4 + H_2SO_4, \qquad \text{[Eq. 1]},$$

and the next reaction [Eq. 2] expresses a transformation of As^{3+} and As^{5+}:

$$As^{3+} + 2Fe(OH)_3 + 6H^+ \rightarrow 2Fe^{2+} + 6H_2O + As^{5+}, \qquad \text{[Eq. 2]}.$$

The weathering reaction of arsenian pyrite breakdown is similar to that of Eq. 1; it transforms FeS_2 into $Fe(OH)_3$ and sulfuric acid, resulting in an overall reaction that is commonly written as follows:

$$Fe (As,S)_2 + 15/4\ O_2 + 7/2\ H_2O \rightarrow Fe(OH)_3 + 2\ H_2SO_4 + 2As, \qquad [Eq.\ 3].$$

Because the average arsenic concentration in mineral pyrite is ~0.02% to ~0.5%, and in some cases as high as ~6.5%, this reaction [Eq. 3] can also release free arsenic into the aqueous phase. In addition, the ability of arsenic to undergo conversion between As^{3+} and As^{5+}, depending on the variation of redox/oxidizing conditions, makes it more abundantly available in the environment.

Native arsenic is not soluble in water under ambient conditions, and it does not react with dry air, but it is easily oxidized and forms arsenous oxides through the reactions of Eqs. 4 and 5:

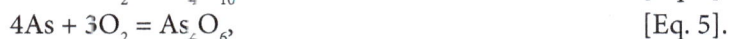

$$4As + 5O_2 = As_4O_{10}, \qquad [Eq.\ 4],$$
$$4As + 3O_2 = As_2O_6, \qquad [Eq.\ 5].$$

Mineral orpiment (As_2S_3) easily reacts with water, producing hydrogen sulfide and arsenous acid through the following reaction:

$$As_2S_3 + 6H_2O \rightarrow 3H_2S + 2H_3AsO_3, \qquad [Eq.\ 6].$$

Arsenic oxides, arsenic trisulfides, and arsenous acid compounds are usually found in natural groundwater, lakes, ponds, and other water reservoirs if they are situated close to arsenopyrite-bearing polymetallic ore deposits, products of ore weathering, or mining and ore-smelting operations.

Chemical reactions [Eqs. 2, 3] suggest that arsenic may be precipitated into water reservoirs under both oxidizing [Eq. 2] and reducing [Eq. 3] conditions, followed by precipitation of free arsenic and products of its oxidation by coagulation and adsorption. Adsorption of free arsenic on the surfaces of fine particles of sediments causes arsenic precipitation among the sediments, such as clay, sands, etc., which then become sedimentary and metamorphic rocks. Therefore, after some time, arsenic may dissolve once again from those rocks due to reducing or oxidizing chemical processes that accompany rock-weathering and rock-cycling processes.

The As^{3+} type of arsenic cations released from inorganic materials (e.g., minerals and rocks) is of more concern to environmental toxicologists than its more oxidized counterpart, As^{5+}. This is because As^{3+} is more difficult to remove from aqueous media than As^{5+} by using conventional physical and chemical As-removal techniques. The latter include oxidizing reactions with chlorine (Cl), ozone (O_3), and potassium permanganate ($KMnO_4$), all of which govern processes that convert As^{3+} to As^{5+}. There are also processes of biological removals of the As from a solution by micro-organisms (e.g., Katsoyiannis and Zouboulis, 2004).

5.2.4. Pathways of arsenic to humans

The pathways of arsenic from its natural geological occurrences to humans (Fig. 69) consist of two major routes that represent a combination of geological and atmospheric processes combined with industrial activities.

Round 1. This ncludes the release of inorganic As^{3+} and As^{5+} from sulfides, arsenides, arsenites, and arsenates and their other oxidation products into the atmosphere, the soil, and water reservoirs followed by their spreading out through biological ecosystems, including forests, plants, animals, fish, and other natural products of the food chain. Plants accumulate arsenic by root uptake from the soil, groundwater, rain, and irrigation waters, and/or by absorption of airborne arsenic particulates deposited on the leaves. In the atmosphere, As^{3+} and As^{5+} are mainly adsorbed on the surfaces of solid particles, which are dispersed as wind-blown dust and deposited on land and water reservoirs. The problems of As-contaminated drinking water are now solved in economically developed countries, and they are unfortunately increasing in many developing

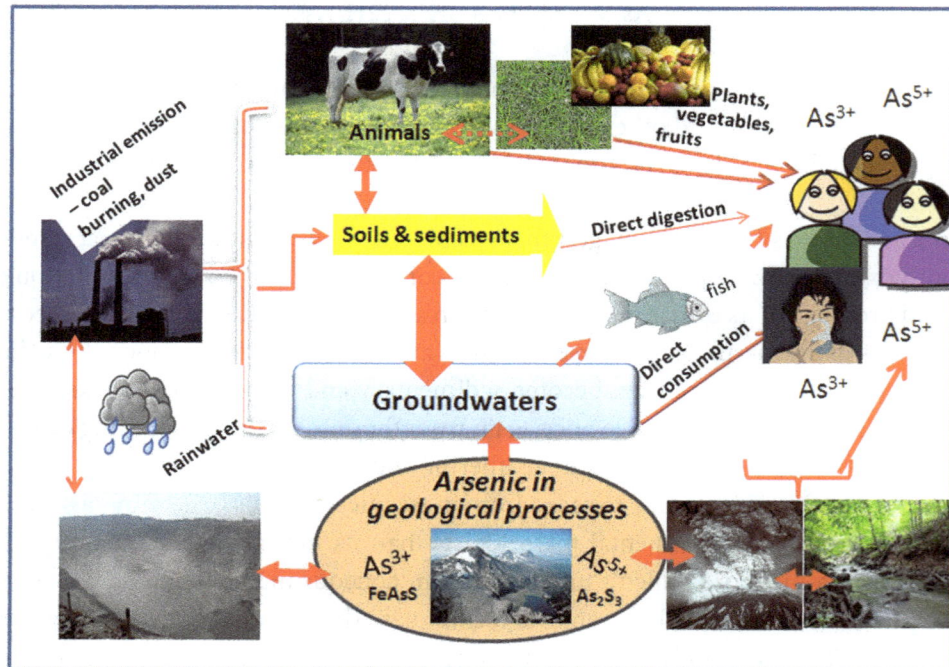

FIGURE 69 Pathway of natural arsenic to humans

countries, where As excess is a result of its remobilization through geological processes. If such naturally As-contaminated groundwater from non-drinking reservoirs (ponds, lakes, rivers) is used for agricultural irrigation, it creates a risk of additional accumulation of toxic As in soils and eventual human exposure through plants' uptake and meat, poultry, dairy, and grain product consumption. In other words, in many cases, non-occupational human exposure to arsenic is mostly through the ingestion of food and water.

Round 2. This refers to anthropogenic contamination through industrial activities. Historically, arsenic minerals with the general formulae $MAsS$—arsenic-bearing sulfides—and MAs_2—arsenides (where M represents metals such as Fe, Cu, Zn, Co, Pb, and others)—and native As ores have been used as the main commercial sources of arsenic. Arsenic has been used in agriculture as animal feed to promote growth (organic arsenic) and as an essential component of herbicides and pesticides (inorganic arsenic). In technology and engineering, it is used for lead batteries, metal alloys, semiconductors, wood and leather preservatives, sheep dips, paint pigments, antifouling paint, and as an additive component for glass manufacturing (Loebenstein, 1994). With the development of high technologies, the demand for arsenic-metal has increased for use in the electronics industry (e.g., germanium-arsenide-selenide alloys), for optical materials, for short-wave infrared technology (indium-gallium-arsenide), and for solar cells (gallium-arsenic semiconductors). Arsenic is also widely used for production of chromated copper arsenide (CCA) preservatives for antifungal wood and furniture treatments. However, with regard to the latter use, it has become clear that the CCA has the potential to leach out of the wood over time, causing contamination of the surrounding environment and creating a potential threat to human health. As a result, the wood-treating industry in many countries has eliminated all arsenical wood preservatives from residential use and now applies this treatment only for woods used for non-residential construction. Mining of polymetallic ores, metal smelting plants, the combustion of fossil fuel (coal and lignite), and the remobilization of historic sources such as mine drainage waters cause additional release of Ar^{3+} and As^{5+} into the atmosphere and onto water reservoir and land surfaces.

5.2.4.1. World production of arsenic

The total world production of arsenic in 2012 was ~44000 tons, and world reserves are predicted to be ~880000 tons. According to USGS data, in 2012, the world's leading producer of arsenic was China (26000 tons per year), followed by Chile (10000 tons) and Morocco (8000 tons). Since 1985, the U.S. has produced neither arsenic metal nor arsenic trioxide due to arsenic's adverse effects on human health and the environment. Arsenic is also no longer produced commercially in Canada, but it is still a significant byproduct of its gold and copper mining industries.

5.2.5. The toxicity of arsenic and its adverse effects on human health

Inorganic arsenic is extremely toxic, but human tolerance to arsenic varies greatly based on the form of the arsenic (inorganic or organic; As^{3+} or As^{5+}) and its compounds, the type of exposure (e.g., inhalation or ingestion), and the amounts and durations of exposure. Among the inorganic forms of arsenic, As^{3+} is highly soluble in water and acutely toxic to humans, and As^{5+} possesses only slightly less toxicity. Exposure to or intake of inorganic arsenic-bearing components over long periods of time can lead to chronic poisoning, called arsenicosis. Organic Ar^{3+} and As^{5+} (arsenobetaine), which are abundant in seafood and marine environments, are less harmful to health because the human body rapidly eliminates organic arsenic.

The problem with arsenic exposure is that many of the symptoms may result from other causes, so an individual who has been exposed often may not suspect arsenic. Furthermore, most inorganic and organic arsenic compounds have no smell (except of arsenopyrite, which has a specific garlic smell), and most have no special taste. Thus, one usually cannot recognize whether arsenic is present in one's food, water, or air. Exposure of humans to As leads to accumulations of As in skin tissues, hair, nails, and, for shorter periods, in the blood, all of which are used as biomarkers to determine the level of As concentration. However, blood analysis for an arsenic contamination is useful only in the case of acute arsenic poisoning because As is rapidly cleared from the blood. Measurements of arsenic in hair and nails provide much better results for evaluation of past arsenic exposure and can even be used to determine the length of time since the last acute exposure. If there are no special skin-damaged symptoms that can be easily attributed to arsenic poisoning, arsenic tests of hair and nails may be difficult to interpret because (1) health experts are not sure what "normal" levels of arsenic in the human body are, and (2) different people may react to arsenic in different ways.

Short-term exposure to large doses of inorganic arsenic through ingestion leads to acute gastrointestinal symptoms, disturbances of the cardiovascular and nervous systems, and eventually death. The immediate clinical symptoms are vomiting, abdominal pain, and diarrhea, followed by numbness and tingling of the extremities and cramping of the muscles.

Long-term exposure to arsenic through drinking water increases the risk of cancer in the skin, internal organs, and lungs, as well as other serious illnesses resulting in hyperpigmentation and keratosis (a characteristic lesion on the palms and soles) of the skin (Chen and Chiou, 2011). The first changes during long-term exposure appear as skin pigmentation, then skin lesions and hard patches on the palms of the hands (Fig 70) and the soles of the feet (Fig. 71). Other effects may appear as conjunctivitis, gastrointestinal illnesses, diabetes, renal-system symptoms, enlarged liver, bone-marrow depression, high blood pressure, and "black foot" disease/gangrene. The latter was specifically recorded in Taiwan, where drinking water was severely contaminated with inorganic arsenic. Interestingly, only a few cases of the "black foot" disease were identified in southwestern Taiwan in the early twentieth century, and at that time, nobody tried to connect this disease with arsenic pollution. After an intensive survey in 1990, it became clear that a total 2252 cases of the disease occurred in two

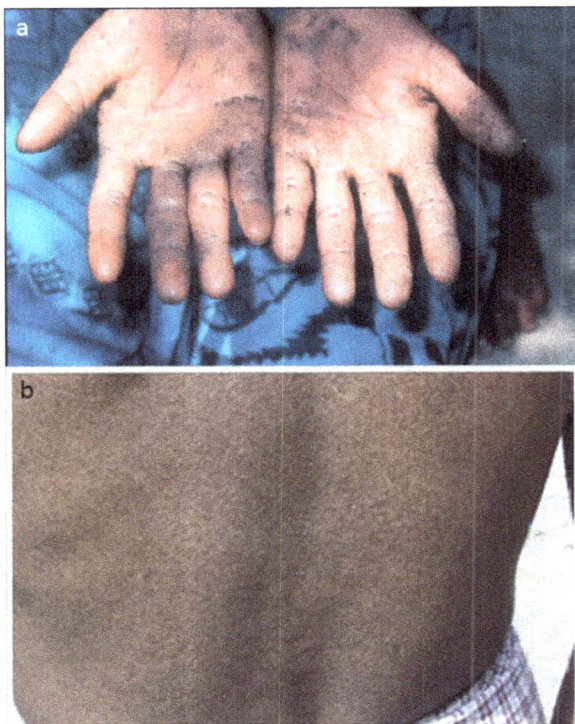

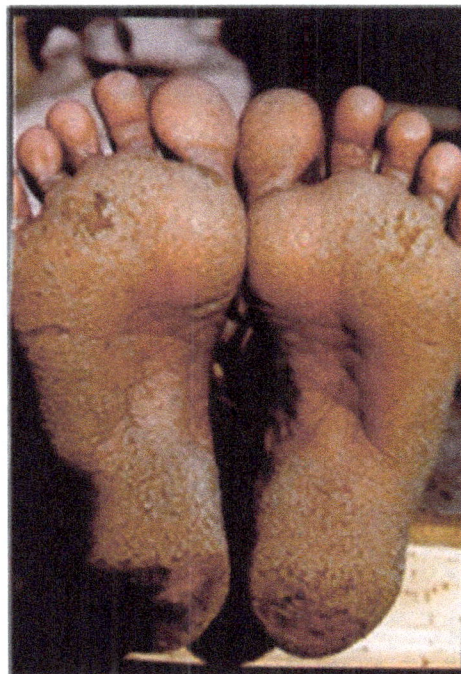

FIGURE 71   Soles of the feet of a patient with an arsenosis illness

FIGURE 70   Arsenic-caused diseases: (a) arsenic keratosis on the palms of a patient; (b) patchy skin hyperpigmentation—both conditions developed after the prolonged ingestion of arsenic-contaminated well water

arsenic-exposed areas in southwestern and northeastern Taiwan (Chen and Chiou, 2011). The arsenic concentration in drinking water of artesian wells that was consumed by citizens of those regions was as high as >350–300 μg/L, which is about 30–35 times more than the maximum concentration level of arsenic recommended by the WHO.

In 1993, the WHO established 0.01 mg/L (10 μg/L) as a provisional guideline for arsenic in drinking water. The EPA suggests the same arsenic standard (10 μg/L) for drinking water to protect consumers from the effects of long-term, chronic exposure to arsenic. Although this new permissible level of As in drinking water was established in 1993, there are many countries that still follow the older the WHO-EPA standard level of arsenic—50 μg/L. Now, it has become known that many cases of keratosis reported in China, Bangladesh, India, Argentina, Chile, Mexico, and Japan have a connection to high arsenic concentrations in drinking water.

The International Agency for Research on Cancer (IARC) has classified inorganic arsenic as a Group 1 carcinogenic substance due to the high carcinogenicity of inorganic arsenic to humans, followed by a statement that arsenic in drinking water increases the risk of cancer (http://www. inchem.org/documents/iarc/vol84/84-01-arsenic.html).

5.2.6. World arsenic-contaminated groundwater and environments

Arsenic is one of ten chemicals included in the WHO's list of the most toxic elements. The harmful effects of arsenic on human health and environments have been increasingly recognized in the past decade. The world map (Fig. 72) shows the most noteworthy areas of arsenic-contaminated groundwater, where arsenic concentration significantly exceeds 50 µg/L. Among them are Bangladesh and the neighboring territory in India (West Bengal), the western part of the U.S., Mexico, Chile, Argentina, Hungary, Romania, China, Nepal, Taiwan, Vietnam, and Thailand. The map also indicates areas where water and the environment are contaminated with arsenic originating from mining (Alaska, Canada, Mexico, U.K., Greece, Ghana, Zimbabwe, South Africa, and Thailand) and geothermal sources (the Aleutian Islands, the western part of the U.S. and Yellowstone National Park in Wyoming, El Salvador, Japan, France, Russia and the Far East, and New Zealand).

FIGURE 72 World map showing in yellow regions groundwater with arsenic contamination; red stars, arsenic contaminations related to mining operations; green stars, arsenic related to geothermal sources

5.2.6.1. Arsenic in drinking water

Bangladesh—extreme exposure. Historically, it was known that surface water reservoirs in

Bangladesh were severely contaminated with micro-organisms that caused a significant burden of gastrointestinal disease and mortality, especially among infants. In addition, the absence of appropriate knowledge of individual hygiene and the lack of sewage-system controls, aggravated by frequent flooding and seasonal monsoons, placed Bangladesh at the top of the world's highest infant mortality rates.

Then, in 1990, about 8 million new tube wells were installed in Bangladesh by an initiative of the World Bank and United Nations International Children's Emergency Fund (UNICEF) to provide "pure water" to the population in order to prevent morbidity and mortality from gastrointestinal diseases. This action brought down infant mortality and diarrheal illnesses by 50%. However, unexpectedly, it resulted in a new epidemic disease caused by elevated concentrations of arsenic in drinking water consumed from the new tube wells. Intensive studies showed that most citizens who consumed water from the new tube wells developed symptoms typical for arsenic poisoning: e.g., skin lesions, pigmentation changes in arms and legs, and keratosis of the palms and soles similar to those shown in Figs. 70–71. Later, research by many international organizations confirmed that arsenic contamination of Bangladesh's groundwater is caused by a naturally occurring high concentration of arsenic minerals in local rocks and sediments (Khan et al., 2003).

It appears that one out of five tube wells is still contaminated with arsenic, with concentration levels significantly exceeding Bangladesh's national standard of 50 µg/L. The water from the 8 million new tube wells was not tested for arsenic contamination, but only for the presence of pathogenic bacteria that caused gastrointestinal diseases. A British Geological Society study in 1998 found that many of Bangladesh's 2022 tube wells exceeded the standard limit by five times (e.g., 300 µg/L), and a few contained up to 1600 µg/L. Other studies showed that one out of ten people who drink water containing up to 500 µg/L for a long period may ultimately die from cancers of the lung, bladder, or skin. It has also been observed that skin cancer can take place within ten years after the first arsenic exposure, while lung and bladder cancer development may take about twice that long to be lethal (Khan et al., 2003).

In 2000, during the 4th International Congress on Arsenic in the Environment, the arsenic contamination of groundwater in Bangladesh was recognized as the world biggest arsenic catastrophe—a catastrophe greater than the accidents at Bhopal, India, in 1984 (a methyl

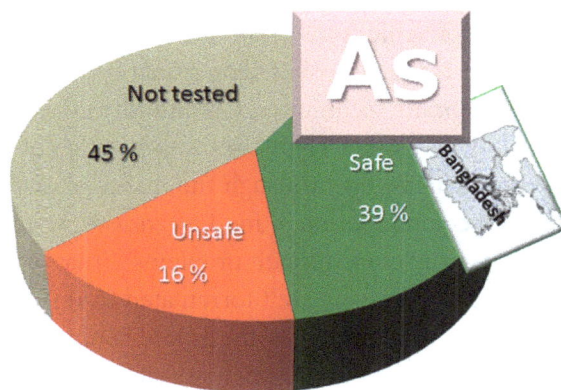

FIGURE 73 Diagram showing that among approximately 8 million total tube wells in Bangladesh, 45% were not tested for arsenic, 39% have safe levels of arsenic (e.g., < 50 µ/L), and 16% have elevated arsenic levels and their drinking water is unsafe

isocyanate gas leak at the Union Carbide India Ltd. pesticide plant) and Chernobyl, Ukraine in 1986 (a nuclear power plant explosion).

Despite efforts to decrease the arsenic exposure level, it is estimated that about 45 million people in Bangladesh are at risk of being exposed to high arsenic concentrations through drinking water. The statistical data revealed that 45% of tube wells existing in Bangladesh remained untested for arsenic levels, 39% are safe, and 16% are unsafe (Fig. 73). Those tube wells that have arsenic concentrations of >50 µg/L are painted red to warn people that the water there is not safe to drink, and the safe tube wells are painted green, but the unchecked wells remain unpainted. Still, this is an effective and low-cost means of discriminating between high-arsenic and low-arsenic sources that could help to rapidly reduce exposure to arsenic when accompanied by effective education of the citizens with regard to arsenic distributions in nature and its pathways to drinking water resources. Other prevention and control measures include installation of local filtering systems, but unfortunately, such systems are not always affordable by local communities.

Representatives of the UNICEF and other international organizations, with the support of local administrations, have emphasized awareness about arsenic poisoning in Bangladesh. They have educated people on how to blend low-arsenic water with higher-arsenic water to achieve an acceptable arsenic concentration level if water in the tube well was contaminated. It should also be noted that many arsenic-affected people in Bangladesh, in addition to their health problems, suffer from social stigma because arsenic-unaffected people often consider keratosis and other dermal disorders to be contagious diseases or a curse. Overall, however, the education and awareness about arsenic poisoning have now largely reached their target, and arsenosis patients report that this situation has improved significantly.

Improvement may also result from the increasing number of low-cost technologies for removing arsenic from small or household supplies that are being progressively established in Bangladesh, which are based on oxidation, coagulation–precipitation, absorption, ion exchange, and membrane techniques. Community education is continuing to help new generations understand the risks of high arsenic exposure not only through drinking water, but also through the food chain—e.g., through the intake of arsenic by crops of rice from irrigation water and through contamination of food by cooking water.

One of the most difficult problems is that while many approaches to solving the Bangladesh arsenic poisoning are well formulated in different documents and publications, no single solution is applicable across the whole country. This is because the distribution of the elevated As concentrations is extremely variable across the country and because in practice, every well (including untested) that is used now, and those to be built in the future, must be carefully tested. Thus, the large magnitude of As contamination of the national groundwater resources, coupled with the extreme poverty of the population, complicate the goals of mitigation.

Mexico. An origin of well-documented high As concentration (up to 624 µg/L) in the groundwater of the Lagunera region of north-central Mexico has been attributed to arid oxidizing environments in which arsenic exists in the form of As^{5+}. Around 400000 citizens of the Lagunera region are exposed to >50 µg/L As concentration in drinking water. In Santa Ana, one of the large cities in the area, drinking water contains 404 µg/L of As. In the Sonora state in northwestern Mexico, the highest arsenic concentration (~305 µg/L) was found in four towns—Hermosillo, Etchojoa, Magdalena, and Caborca—all of which are plagued by significant, chronic health problems related to arsenic exposure.

China. Elevated concentrations of As in drinking water were recognized as a health threat in the 1960s. Inner Mongolia and the Shanxi Province are considered epidemic areas in terms of arsenic poisoning, with the average rate of unsafe pump well water varying from 11% to ~52%. However, millions of groundwater wells remain to be tested in order to determine the magnitude of the As problem. In a survey conducted in 1991–1993, 4545 patients were recognized with symptoms of As-caused diseases, suggesting that chronic arsenicosis should be seriously taken into account as a newly emerging public health issue in China. By now, eight provinces and thirty-seven counties in China are included in epidemic areas of arsenicosis. According to a new statistical risk model (Rodriges-Lado et al., 2013), about 20 million people are at risk of being affected by the consumption of arsenic-contaminated groundwater.

United States. Arsenic in groundwater has been found in many places in the U.S., but no epidemic arsenic-related diseases are known because in most cases, arsenic occurs in small concentrations that are not dangerous to human health. There are, however, some exclusive cases in the southwestern U.S. where groundwater contains high levels of arsenic: the San Joaquin Valley in California (from 1 to 2600 µg/L) and in Fallon, Nevada (100 µg/L). In Nevada, the groundwater in at least in 1000 private wells were discovered to contain >50 µg/L of As, and in Maine, Michigan, Minnesota, South Dakota, North Dakota, Oklahoma, Texas, and Wisconsin, concentrations of arsenic were found to be slightly higher than 10 µg/L, which is the standard norm established by the EPA. In some areas of these states, elevated levels of arsenic have been reported, but they regionally coincide with agricultural activities, where arsenic may be leaching from pesticides. In northern Texas, for example, the As concentration in groundwater varies from <50 to <2500000 µg/L, and this extremely high concentration has been attributed to disposal of cotton gin waste, which contains arsenic levels up to 240 mg/kg (Welsh et al., 2000). However, Welsh et al.'s studies in Minnesota, North Dakota, South Dakota, and Wisconsin showed that groundwater is largely unaffected by the use of arsenical pesticides, and therefore, inorganic arsenic contributions to the groundwater may originate from geological sources.

5.2.6.2. Arsenic contamination from mining activity

Thailand. The most severe case of this type of arsenic poisoning is in Thailand, where 1000 people have been diagnosed with As-related skin disease in the territory close to the Sn-ore and Au mining operations (Williams et al., 1996). The concentration of arsenic in the shallow groundwater in that region has been measured at up to 5000 $\mu g/L$.

Ghana. In Ghana, which produces about one-third of the world's gold, the mining operations are associated with sulfide ore deposits rich in arsenopyrite. In the mining area and related ore-treatment localities, arsenic byproducts are specifically mobilized into the soil, and their concentration in shallow groundwater has been measured at 64 $\mu g/L$, which is not as high as in similar mining regions in Thailand and the U.S. (Smedley and Kinninburg, 2002).

Mexico. One of the most severe cases of As contamination from mining operations was reported in the Zimapan Valley, Mexico. In 1993, the National Water Commission of Mexico, during routine testing of water for cholera, found ~300 $\mu g/L$ As concentration in a municipal water reservoir and ~1000 $\mu g/L$ in the deep well of the town of El Muhi. Further studies showed that residents of the Zimapan Valley, and in particular the community of El Muhi, have constantly consumed arsenic-contaminated water for more than ten years, resulting in arsenic-related diseases, including cancer. More than 127000 people in eleven counties of northern Mexico have had long-term exposure to arsenic through drinking water and food crops contaminated though irrigation.

The U.S. The well-documented cases of As contamination include the metal-sulfur ore mining and processing industry in the Coeur d'Alene Pb-Zn-Ag mining region of Idaho and the Leviathan sulfur mine in Alpine County in northern California.

The Coeur d'Alene mines, Idaho. The region of the Coeur d'Alene River and its tributaries was the nation's largest producer of silver, lead, zinc, and other metals from 1800 through the twentieth century, including more than 100 mines and ore-processing smelting factories. The mining- and processing-related wastes containing toxic metals such as cadmium, arsenic, lead, and zinc were directly discharged, and runoff to the surrounding areas. More than 75 million tons of heavy metal-bearing sediments, including toxic arsenic, have been deposited into the beds of the local lakes since the late nineteenth century. Maximum arsenic levels in surface water ranged from 4.3–600 $\mu g/L$ (Mok and Wai, 1990), but much higher concentrations of arsenic have been found in the soil at a parking lot situated very close to the major Coeur d'Alene mine operation (1060 mg/kg) and in sediments in the vicinities of Thompson Lake (1.5 mg/kg) and Rose Lake (375 mg/kg). Discharged arsenic and other toxic metals in the area pose significant threats to people and wildlife. Since the early 1980s, after controversial discussions and the U.S. Congressional hearings, the Coeur d'Alene Basin Commission, with the help of the EPA, started clean-up projects to reduce exposure to toxic metals and return rivers to their pre-mining conditions so that they will be able to support healthy aquatic life, plants, and habitants in the areas. The Superfund Cleanup Implementation Plan of the Coeur d'Alene Basin is still active, but the remediation of the damaged eco-biological system and environment is a very slow process. The EPA continues to update its plans and keeps people involved in and informed about the status of the Coeur d'Alene Basin cleanup (for more information, see http://www.epa.gov/region10/pdf/sites/bunker_hill/cda_basin/imp_update_12-5-13.pdf).

The Leviathan Mine, California. The Leviathan Mine is an abandoned sulfur mine that was operated by the Anaconda Copper Mining Company in the 1950s and 1960s. Over 20 million tons of crushed rocks containing remnants of sulfur ore and associated toxic metals were spread over the site, where they remain today. Sulfuric acid and toxic components such as arsenic, copper, nickel, zinc, chromium, aluminum, and iron leached from the waste rock piles, causing significant contamination along Leviathan, Aspen, and Bryant creeks and the River Ranch Irrigation Channel. The Washoe Tribe of Nevada and California expressed concern that contaminated waters from the Leviathan Mine may have affected their lands and could cause adverse ecological, cultural, and health effects. Arsenic concentration levels as high as 49.9 mg/L in some ponds raised the most concern. The possibility of high levels of other toxic elements associated with As should also be considered because people can be exposed via inhalation of dust near the mine tails and through their contacts with surface water and sediments to which the toxic elements can migrate. Other risks of exposure include eating fish, plants, and animals raised near the Leviathan Mine. In 2006, the site was used for pilot studies of water treatment from toxic elements with lime and other neutralizing chemical compounds. Further clean-up activities were published, and recommendations for local people were formulated to minimize potential risks from consuming or using fish and plants from the area. These recommendations included the following:

1. Avoiding eating fish caught in the Leviathan, Aspen, and Bryant creeks and the River Ranch Irrigation Channel until it is determined that fish in these areas do not contain metals that could present health risks;
2. Collecting plants for consumption or other purposes as far as possible from the Leviathan Mine;
3. Washing collected plants with non-contaminated water to remove dust and dirt.

5.2.6.3. Arsenic in geothermal waters

Geothermal waters plotted on the map (see Fig. 72) exhibit maximum concentrations of arsenic, which are several hundred times higher than the standard acceptable level established for drinking waters. For example, in California, according to data published by Welsh et al. (2000), the level of As in geothermal waters is as follows: Honey Lake Basin: up to 2600 μg/L; Coso Hot Springs: up to 7500 μg/L; Imperial Valley: up to 15000 μg/L; and Long Valley: up to 2500 μg/L. The geysers and hot springs of Yellowstone National Park in Wyoming contain from <1 to 7800 μg/L. There are a total of 1283 geysers that have erupted in Yellowstone, 465 of which remain active during an average year. It is interesting that much of the high arsenic concentration that was found in groundwaters within the Madison and upper Missouri River valleys was accumulated from the Yellowstone geothermal system.

5.2.7. Arsenic removal from drinking water

Natural groundwaters contain As^{3+} if their chemical environment has a shortage of oxygen (e.g., reduced media), and they contain As^{5+} in oxidized conditions. Modern technologies for removal of inorganic As^{3+} and As^{5+} cations from water include four different processes, and a choice of an appropriate application depends on the ratios of As^{3+} and As^{5+} and the bulk chemistry of the water. The technologies are underlined by fundamental chemical processes such as the following:

1. *Purification by precipitation.* This includes coagulation and filtration (direct filtration, coagulation-assisted microfiltration, and enhanced coagulation), and enhanced lime softening. During these treatment processes, the physical and chemical properties of dissolved matter are altered, and the resulting particles settle out of solution by gravity or are removed by filtration.

2. *The adsorption process* (specifically, with activated alumina). This treatment uses the activated alumina prepared by dehydration of $Al(OH)_3$ at high temperature, which adsorbs both As^{3+} and As^{5+}.

3. *The ion-exchange process* (specifically, anion exchange). This treatment is based on the process by which ions in the solid phase are exchanged for ions in the feed water.

4. *Membrane filtration* (reverse osmosis and electrodialysis). This process is based on using a "membrane" as a selective barrier allowing one constituent to pass while blocking the passage of others. The driving forces are controlled by pressure (osmosis) and electrical potential-driven filtration (electrodialysis).

In the U.S., organizations that supply drinking water are obligated to notify their customers if the concentration of As exceeds a standard requirement—e.g., 10 μg/L—as soon as practical, or at least within thirty days after a violation is detected. The delivery of alternative drinking water supplies may be required to prevent serious risks to public health (for details, see http://water.epa.gov/lawsregs/rulesregs/sdwa/publicnotification/index.cfm). Those who consume water from private wells are expected to regularly check with the local water system administration for information on arsenic contamination in the area around the well. In addition, countertop arsenic-removal water filters for individual usage are available on the market.

References

Abrahams, P. W. 1997. Geophagy (soil consumption) and iron supplementation in Uganda. *Tropical Medicine and International Health* 2:617–623.

Al-Rmalli, S. W., R. O. Jenkins, M. J. Watts, and P. I. Haris. 2010. Risk of human exposure to arsenic and other elements from geophagy: Trace element analysis of backed clay using inductively coupled plasma mass spectrometry. *Environmental Health* 9:79.

Baheritibeb, Y., S. Law, and C. Pain. 2008. The girl who ate her house—Pica as an obsessive-compulsive disorder. A case report. *Clinical Case Studies* 7:3–11.

Chen C.-J., and H.-Y. Chiou. 2011. *Health hazards of environmental arsenic poisoning. From epidemic to pandemic*. Singapore: World Scientific Publisher Co. Pte. Ltd.

Fraga, C. G. 2005. Relevance, essentiality and toxicity of trace elements in human health. *Molecular Aspects of Medicine* 26:235–244.

Halstead, J. A. 1968. Geophagia in man: Its nature and nutritional effect. *The American Journal of Clinical Nutrition* 21:1384–1393.

Hanter, J. M., and R. de Kleine. 1984. Geophagy in Central America. *Geographical Review* 74:157–169.

Katsoyiannis, I. A., and A. I. Zouboulis. 2004. Application of biological processes for the removal of arsenic from groundwaters. *Water Research* 38:17–26.

Khan, M. M., F. Sakauchi, T., Sonoda, M. Washio, and Mori, M. 2003. Magnitude of arsenic toxicity in tube-well drinking water in Bangladesh and its adverse effect on human health including cancer: Evidence from a review of the literature. *Asian Pacific Journal of Cancer Prevention* 4:7–14.

Loebenstein, R. J. 1994. The materials flow of arsenic in the United States. *U.S. Bureau of Mines Information Circular* 9382:1–12.

Middleton, J. D. 1989. Sikor—an unquantified hazard. *British Medical Journal* 298:407–408.

Mok, W. M., and C. M. Wai. 1990. Distribution and mobilization of arsenic and antimony species in the Coeur d'Alene River, Idaho. *Environmental Science and Technology* 24:102–108.

Njiru, H., U. Elchalal, and O. Paltiel. 2011. Geophagy during pregnancy in Africa: A literature review. *Obstetrical and Gynecological Survey* 66:452–459.

Okcuoglu, A. A., V. Arcasoy, Y. Minninch, S. Tarson, O. Cin, et al. 1966. Pica in Turkey. I. The incidence and association with anemia. *American Journal of Clinical Nutrition* 19:125–131.

Potter, P. (Ed.). 1995. *Hippocrates*. Volume 8. Cambridge, MA: Harvard University.

Rodríguez-Lado, L., G. Sun, M. Berg, Q. Zhang, H. Xue, Q. Zheng, and C. A. Johnson. 2013. Groundwater arsenic contamination throughout China. *Science* 341:866–868.

Smedley, P. L., and D. G. Kinninburg. 2002. A review of the source, behavior and distribution of arsenic in natural waters. *Applied Geochemistry* 17:517–568.

Waywodt, A., and M. Kiss. 2002. Geophagia: The history of earth-eating. *Journal of the Royal Society of Medicine* 95:143–146.

Welsh, A. H., D. V. Westjohn, D. R. Helsel, and R. B. Wanty. 2000. Arsenic in groundwater of the United States: Occurrence and geochemistry. *Ground Water* 38:589–604.

Williams, M., F. Fordyce, A. Paijitprapapon, and P. Charoenchaisri. 1996. Arsenic contamination in surface drainage and groundwater in part of the southern Asian tin belt, Nakhon Si Thammarat Province, southern Thailand. *Environmental Geology* 27:16–33.

Young, S. L., P. W. Sherman, J. Lucks, and G. Pelto. 2011. Why do people eat earth? A test of alternative hypotheses. *Quarterly Review of Biology* 86:97–120.

Web resources

http://www.inchem.org/documents/iarc/vol84/84-01-arsenic.html;

http://water.epa.gov/lawsregs/rulesregs/sdwa/publicnotification/index.cfm;

http://www.epa.gov/region10/pdf/sites/bunker_hill/cda_basin/imp_update_12-5-13.pdf.

Image Credits

Figure 66: Copyright © Wellcome Images (CC by 4.0) at http://commons.wikimedia.org/wiki/File:White_chalky_earth_from_Bethlehem,_Palestine,_1920-1930_Wellcome_L0059009.jpg.

Figure 67a: Aram Dulyan, "Native arsenic," http://en.wikipedia.org/wiki/File:Native_arsenic.jpg. Copyright in the Public Domain.

Figure 67b: Copyright © JJ Harrison (CC BY-SA 3.0) at http://en.wikipedia.org/wiki/File:Arsenopyrite,_Panasqueira_Mine,_Portugal.jpg.

Figure 67c: Copyright © Kluka (CC BY-SA 3.0) at http://commons.wikimedia.org/wiki/File:Realgar,_1Rumunia1,_Baia_Sprie.jpg.

Figure 67d: Copyright © Rob Lavinsky (CC BY-SA 3.0) at http://commons.wikimedia.org/wiki/File:Orpiment-Baryte-149290.jpg.

Figure 68a: Copyright © Rob Lavinsky (CC BY-SA 3.0) at http://commons.wikimedia.org/wiki/File:Arsenolite-333170.jpg.

Figure 68b: Copyright © Rob Lavinsky (CC BY-SA 3.0) at http://commons.wikimedia.org/wiki/File:Adamite-282289.jpg.

Figure 68c: Copyright © Didier Descouens (CC BY-SA 3.0) at http://commons.wikimedia.org/wiki/File:Erythritemaroc1.jpg.

Figure 68d: Copyright © Didier Descouens (CC by 2.0) at http://en.wikipedia.org/wiki/File:Annabergite-Siderite-_Grece-1.jpg.

Figure 69a: Copyright © bengt-re (CC BY-SA 3.0) at http://commons.wikimedia.org/wiki/File:Green_green_grass_of_home_(5418533268).jpg.

Figure 69b: Copyright © Mortis (CC by 3.0) at http://commons.wikimedia.org/wiki/File:Forest-Creek-Eagleville-PA-USA.jpg.

Figure 69c: Cretep, "Jwaneng Open Mine," http://commons.wikimedia.org/wiki/File:Jwaneng_Open_Mine.jpg. Copyright in the Public Domain.

Figure 69d: warszawianka, "tango weather showers scattered," https://openclipart.org/detail/30151/tango-weather-showers-scattered-by-warszawianka. Copyright in the Public Domain.

Figure 69e: Alfred Palmer, "AlfedPalmersmokestacks," http://commons.wikimedia.org/wiki/File:AlfedPalmersmokestacks.jpg. Copyright in the Public Domain.

Figure 69f: Keith Weller/USDA, "Cow female black white," http://commons.wikimedia.org/wiki/File:Cow_female_black_white.jpg. Copyright in the Public Domain.

Figure 69g: Copyright © Bill Ebbesen (CC by 3.0) at http://commons.wikimedia.org/wiki/File:Culinary_fruits_front_view.jpg.

Figure 69h: nicubunu, "People," https://openclipart.org/detail/15048/people. Copyright in the Public Domain.

Figure 69i: OIKu, "night thirst," https://openclipart.org/detail/171676/night-thirst. Copyright in the Public Domain.

Figure 69j: johnny_automatic, "generic fish," https://openclipart.org/detail/1997/generic-fish-by-johnny_automatic. Copyright in the Public Domain.

Figure 69k: United States Geological Survey, "Three Sisters Image," http://vulcan.wr.usgs.gov/Volcanoes/Sisters/Images/framework.html. Copyright in the Public Domain.

Figure 69l: United States Geological Survey, "MSH80 eruption mount st helens 05-18-80-dramatic-edit," http://commons.wikimedia.org/wiki/File:MSH80_eruption_mount_st_helens_05-18-80-dramatic-edit.jpg. Copyright in the Public Domain.

Figure 70a: "Dirty Hands," http://www.atsdr.cdc.gov/csem/arsenic/images/full_arsenic_pic7.jpg. Copyright in the Public Domain.

Figure 70b: "Back," http://www.atsdr.cdc.gov/csem/arsenic/images/full_arsenic_pic6.jpg. Copyright in the Public Domain.

Figure 71: Richard Wilson, "Arsenic poisoning on foot," http://users.physics.harvard.edu/~wilson/arsenic/8.%20Diff.%20Keratosis%20Sole.jpg. Copyright © by Richard Wilson. Reprinted with permission.

Figure 72: Copyright © Jayarathina (CC BY-SA 3.0) at Adapted from: http://commons.wikimedia.org/wiki/File:Arsenic_contamination_areas.jpg.

Figure 73: Adapted from: Rarelibra, "Bangladesh administrative divisions," http://commons.wikimedia.org/wiki/File:Bangladesh_administrative_divisions.png. Copyright in the Public Domain.

Review Questions

1. How are elements classified in the periodic table, and how are they classified by geologists?

2. Provide definitions of major, minor, and trace elements.

3. What are essential vs. non-essential elements?

4. What is geophagy? What is pica? How is it related to geophagy?

5. Where does geophagy most often occur today?

6. What are the potential benefits to those who practice geophagy?

7. Why is it sometimes associated with famine? Or local wars? Or hazardous events?

8. What are the negative effects/potential risks of geophagy?

9. Why did humans develop geophagy in the tropical rainforest regions of South America?

10. How large is the U.S. EPA-permissible dose of soil that may be consumed by humans through hand-to-mouth contact?

11. Are there any connections between long-standing geophagy and anemia (the most common disorder of the blood, which causes feelings of weakness, fatigue, and general malaise)?

12. Why do people eat earth?

13. Describe the functional concept of geophagy development.

14. What are the behavioral and cultural aspects of geophagy?

15. What properties of clay are useful for geophagists?

16. Would you prefer to consume natural "earth" instead of pharmaceutically prepared analogues? Please explain your answer.

17. In what geological conditions do arsenic minerals form?

18. Is there any arsenic in meteorites or in any other planet than Earth?

19. Where does arsenic occur: in the Earth's mantle, in the crust, or in the inner core?

20. How do primary arsenic minerals differ from secondary arsenic minerals?

21. Which of the two arsenics (As^{3+} or As^{5+}) is more toxic?

22. Is organic arsenic more toxic than inorganic? Where does organic arsenic mostly occur?

23. How many minerals of the arsenic group are known in terrestrial conditions?

24. What symptoms follow human poisoning from a single large dose of inorganic arsenic?

25. What symptoms follow human poisoning from long-term consumption of small doses of inorganic arsenic?

26. Does arsenic-contaminated water have a specific smell?

27. How is arsenic released into water and onto land?

28. What is the pathway of arsenic to the human body?

29. What is the acceptable, non-harmful concentration of arsenic in drinking water in the U.S.? In Bangladesh?

30. How large is the arsenic contamination in tube wells in Bangladesh?

31. What country is the leading producer of arsenic in the world? For what purposes is arsenic used in industry and technologies? In the agricultural sector? In the wood-production industry?

32. How much arsenic did the U.S. mine during the period 2001–03?

33. What is better to use for arsenic exposure examination: a blood test or measurements of As in hair and nails?

34. What are the first noticeable changes that appear in the human body during long-term exposure to arsenic through drinking water?

35. What societal problems do arsenosis patients have in Bangladesh?

36. Where do high concentrations of arsenic occur in the U.S.? What kinds of populations are at risk?

37. Why did the Coeur d'Alene mines of Idaho and the Leviathan Mine of California cause trouble to local communities?

38. If you travel through the U.S., would you swim or drink water from a hot spring that you have found in a remote area?

39. List four fundamental chemical processes used for arsenic removal from water.

Quizzes (see answers on page 307)

1. A trace element is…
   a. An element with a low atomic mass.
   b. An element that can be detected and used for tracking molecules.
   c. An element that is in very low concentrations (less than 0.1% by mass) in a rock or mineral.
   d. An element in a rock or mineral that is toxic to humans.
   e. An element that is necessary for good health in humans.

2. Arsenobetaine is…
   a. An arsenic-bearing organic compound found in mushrooms and certain marine foods.
   b. An arsenic-bearing mineral formed in low-temperature hydrothermal deposits.
   c. A cell protein that absorbs arsenic and removes it from the body.
   d. A naturally occurring mineral that absorbs arsenic.

3. Geophagy is…
   a. A severe rash caused by arsenic exposure.
   b. A disease caused by deficiency of selenium.
   c. The practice of eating earthy or soil-like substances.
   d. The process that wears down rock and produces sediment.
   e. The study of geology and human health.

4. The toxicity of arsenic is in part related to:
   a. Organic or inorganic form.
   b. Valence state.
   c. Solubility.
   d. Rate of absorption and elimination.
   e. All the above
   f. a and d.

5. How does drinking water in tube wells in Bangladesh come to be contaminated with arsenic?
   a. Much coal was burned that released free arsenic into the atmosphere followed by its precipitation into the wells.
   b. They used a lot of arsenic-bearing pesticides that leached into wells.
   c. The natural geological formations of Bangladesh are enriched in arsenic.
   d. None of the above.

6. What is the acceptable concentration of inorganic arsenic in drinking water suggested by the U.S. EPA?
   a. 10 mg/L in drinking water from a private well.
   b. 10 µg/L in any source of drinking water.
   c. 50 µg/L in tap water in a house.
   d. 20 µg/L in any source of drinking water.

Geological and Anthropogenic Hazards from Toxic Metals

Chapter 6

Toxic metals **6.1.**

6.1.1. Natural heavy minerals and metals—background

Native metals and many metal-bearing sulfides and metal-bearing oxides belong to the group of heavy minerals that have specific gravity higher than 2.9 g/cm^3. Some of them are known as extremely toxic. Heavy minerals are also called accessory minerals because they usually occur in all types of rocks in very small concentrations (<1%). Under certain geological conditions, igneous and metamorphic rocks may segregate heavy metals and heavy minerals in suitable economic concentrations. They also may be accumulated in sediments formed by aquatic processes and in granite-pegmatite veins. Large or even gigantic masses of heavy minerals that have accumulated in rocks and veins form ore deposits, and when they are concentrated in aquatic sediments, they form so-called placer deposits.

Because of today's global demand for metals, their exploration and mining are of great economic importance. Industrial progress depends not only on understanding how gigantic metal deposits can be exploited at reasonable economic cost, but also on the risk management of their adverse effects on the environment and public health. The mining and refining of metals produces large amounts of waste because metals represent only a small fraction of the total volume of the rocks. The most important ore resources are those that contain industrially valuable metals such as iron (Fe), copper (Cu), aluminum (Al), lead (Pb), zinc (Zn), silver (Ag), gold (Au), chromium (Cr), nickel (Ni), cobalt (Co), cadmium (Cd), manganese (Mn), molybdenum (Mo), tungsten (W), vanadium (V), tin (Sn), mercury (Hg), magnesium (Mg), platinum (Pt), titanium (Ti), and uranium (U). Production of Cu, Pb, Cd, Cr, and Zn causes an enormous degradation of the environment because their raw ores require smelting operations that release toxic byproducts in the forms of both fumes and solid particulate matter emitted into the atmosphere and precipitated in soils and water.

Almost all metal pollutants remain in the soil from hundreds to thousands of years, which makes the rehabilitation of abandoned mining sites and their surrounding areas expensive and time-consuming processes. Moreover, if toxic metals have already entered the food chain, they can remain there for a long time as a potential poison for people who consume such foods (e.g., Dudka and Adriano, 1997). In this respect, it is important to know that some toxic metals are often included in pesticides, and exposure to the agricultural products grown in pesticide-treated soils can cause chronic, long-term diseases and acute intoxication or poisoning. The severity of intoxication depends on the dose; an individual's genetic vulnerability, age, and general health conditions; the length of exposure; environmental factors; and parallel ingestions of other toxic chemicals from different sources.

Humans have been exposed to heavy metals since prehistoric times, and such exposure is continuing to grow in developing countries, whereas over the past 100 years, toxic metal emissions have declined in economically more developed parts of the world. Three of the most toxic metals—mercury, cadmium, and lead—are considered below as examples of unprecedented environmental pollutants that represent real threats to human health, sometimes causing irreversible failure of biological functions of the human body and even degradation of intellectual development.

6.2. Mercury

6.2. 1. Mercury minerals and their natural occurrences

Mercury (Hg) is an extremely toxic element that belongs to the metals group in the periodic table. In geological environments, mercury associates with the rocks that compose the continental crust, where it can be found in concentrations as small as 0.08 ppm, but no mercury in analytically detected concentrations are known in the Earth's mantle. Overall, there are more than twenty mercury-bearing minerals, and native mercury (Hg^0) is the only mineral that exists in a liquid form, although other minerals contain Hg in their crystalline structures. Minerals of the mercury sulfides group such as cinnabar (HgS—mercury sulfide), corderoite ($Hg_3S_2Cl_2$—mercury sulfide-chloride), and livingstonite ($HgSb_4S_8$—mercury antimony sulfide) are the major industrial sources of mercury (Fig. 74). They usually occur together with other metal sulfides within polymetallic ore deposits or in hydrothermal veins that are often found in sedimentary and metamorphic rocks, as well as in hot springs and other volcanic formations.

FIGURE 74 Mercury minerals: (a) native mercury from Sonoma County, California; (b) cinnabar from one of the closed mines in Spain; (c) crodeirite from Humboldt County, Nevada; and (d) livingstonite from Guerrero, Mexico

Two mercury minerals—cinnabar and native Hg—were found in the meteorite Tieschitz that fell in 1878 in the territory of the modern Czech Republic; the age of this meteorite is ~4550 Ma.

6.2.2. Mercury mining and production

In the U.S., mercury has been mined since 1800, mostly as the minerals cinnabar and native mercury associated with silica-carbonate and other sedimentary and volcanic rocks in the Coast Ranges of California, in Nevada, Arkansas, Texas, and southwestern Alaska, and in hot spring waters of the sulfur mines in California and Steamboat Springs, Colorado. The Coast Ranges contained more than fifty mines, and until the 1990s, California was a leading national producer of mercury. Cinnabar in economic concentrations is also found in many other localities of the

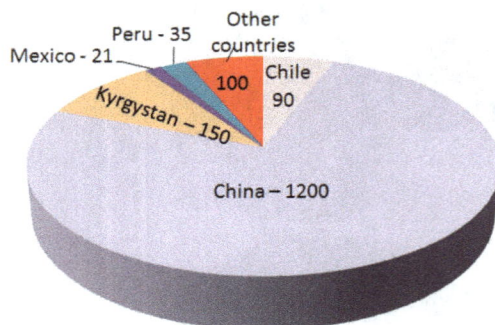

FIGURE 75 World production of mercury (in tons) in 2012 according to USGS data

world that yield mercury-sulfide ores, notably the Philippines, Spain, Slovenia, Egypt, Italy, Serbia, Peru, and China.

Other mercury-bearing minerals such as corderoite ($Hg_3S_2Cl_2$) and levingstonite ($HgSb_4S_8$) are widely distributed in the world, but in small quantities if compared with cinnabar and native mercury availability (Fig. 74c,d). Corderoite was first described in 1974 in the McDermitt Mercury Mine, Nevada. Levingsonite occurs in Mexico, Japan, Kyrgyzstan, and Spain, where it is found in low-temperature hydrothermal veins together with cinnabar, native sulfur, gypsum, and Fe-, Cu-, and Zn-sulfides.

The world's leading countries in mercury mining and production are China, Kyrgyzstan, and Chile, which in 2012 produced 1200, 150, and 90 tons, respectively, out of a total global production of 1600 (Fig. 75). Because mining and extraction of mercury are extremely hazardous processes for humans and the environment, large-scale mercury mining was first stopped in 1970 in Alaska, and later, in 1990, in California, followed by other places. This is an explanation of why the U.S. did not appear on the diagram of world mercury production (Fig. 75). Although the estimated world reserve of mercury is still very high (~600000 tons) and there is a large industrial demand for mercury, its production and export-import activities are decreasing every year because of its toxicity. In the U.S., mercury mining as a principle mineral commodity was discontinued in 1992, when the McDermitt mine, the largest mine in Nevada, was closed. Furthermore, the U.S. and the European Union have created new regulations and laws to control mercury pollution in their environments and to clean and remediate those areas that have already been degraded.

To better understand the mechanisms of mercury contamination, one should first be familiar with the industrial extraction of mercury and mercury chemistry. On an industrial scale, mercury extraction requires a mechanical grinding of rocks containing raw mercury ore followed by the roasting, condensation, and refining processes:

1.   *Ore crushing, grinding, and milling.* In the first stage, the ore of the mercury-bearing minerals is separated from the surrounding rocks by excavation followed by crushing and grinding by a series of mills, unless a fine-grained powder already exists.

2. *Roasting process.* This involves the heating of the powdered ore-concentrate in the presence of oxygen, during which free-vapor mercury and sulfur dioxide are released:

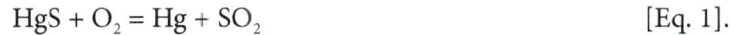

$$HgS + O_2 = Hg + SO_2 \qquad \text{[Eq. 1]}.$$

3. *Condensation process.* The goal of this process is to cool byproducts released by roasting—e.g., $Hg + SO_2$ (see Eq. 1). The cooling causes transformation of vapor mercury to liquid mercury—e.g., mercury metal. The mercury metal is then separated from SO_2 gas and other substances such as metylmercury (CH_3Hg), as well as any undesirable chemical impurities that accompany the condensation process.

4. *Refining process.* This is the last stage of the cleaning of the condensed mercury from impurities, and it may be repeated several times to reach higher purity. The refining includes mechanical filtering, oxidation, triple distillation, and possibly electrolysis (e.g., passing an electrical current through the mercury to remove the final impurities).

6.2.3. Where is mercury used?

Mercury has many applications, although due to its toxicity, some of them have been discontinued and others have undergone modification to replace the mercury with less toxic or non-toxic metals and composites.

Industrial applications. The primary use of mercury in the world was and continues to be as an important compound used by the artisanal and low-scale gold-mining industry. On a larger scale, mercury is used as a catalyst for vinyl chloride and chlorine-bearing synthetic products (chlorine-caustic soda industries) and for production of alkaline batteries, thermostats, medical monometers, and thermometers. Mercury vapor is used as an ingredient in traditional fluorescent lights and neon lights, though currently, more efficient lead lights are expelling them from the market. In addition, in the U.S., the chlorine-caustic soda industry is working now on replacing old Hg-cell technology with new, less toxic membrane-cell technology. Mercury is also no longer used for batteries and paints produced by the U.S. manufacturers, although it is still in production in some other counties.

Medical applications. Dental amalgams, fever thermometers, and older-generation blood-pressure manometers used to contain mercury. In 1991, the WHO estimated that 3% of world's mercury has being consumed for dental amalgams, which contained 50% metallic mercury, 35% silver, 9% tin, 6% copper, and traces of zinc. Many studies showed that such a high concentration of mercury in amalgams may be dangerous to human health. Since that time, the use of dental amalgams has declined and has been replaced by ceramic compounds, the quality

TABLE 13   Regulations Pertaining to the Distribution of Mercury-Containing, Non-Eye Cosmetic Creams and Soaps (A) and Eye-Area Cosmetic Products (B), N/A—Data Are Not Available.

Regulatory Source	Mercury Limit for Cosmetics – A	Mercury Limit for Cosmetics – B
European Union	Banned	≤ 0.0007% by weight
Many African Nations	Banned	N/A
U.S. Food & Drug Administration (FDA)	<1 ppm	≤65 ppm– Hg; ~100 ppm– phenylmercury acetate or nitrate
Health Canada	≤3 ppm	N/A
Philippine Food & Drug Administration	≤1 ppm	N/A

and safety of which are constantly improving due to ongoing studies of their chemical and mechanical properties. Dental ceramics are still sometimes under discussion as to whether they have better safety than the amalgams, but thermometers and monometers are now completely replaced by new, non-mercury, digitized devices. In thermometers, mercury has been replaced by a new product, "galinstan," which is a mixture of gallium, indium, and tin; this new product is not known to be as toxic as mercury. Thirty years ago, mercury was used in many western pharmaceuticals as an antiseptic and an antidiuretic, but production has been completely discontinued and replaced by alternative, less toxic substances.

Cosmetic applications. Mercury is often used as an ingredient for skin-lightening creams and soaps. If such products are applied to the skin, the mercury will restrain the formation of melanin, a complex polymer that is responsible for determining the color of the skin and hair. In some African and Asian countries and within dark-skinned populations in North America and Europe, such cosmetic products are used to achieve a lighter skin tone. Mercury has also been detected in some eye-cleansing cosmetics and in mascara that is used for making lashes longer and thicker. Although it is well known that mercury is a hazardous material, its concentrations in some cosmetics are 200000 times higher than the standard mercury limit. Many cases have been reported throughout the world with regard to mercury poisoning after using cosmetic beauty products. In response, the production and distribution of mercury skin-lightening cosmetics were banned in many countries, and the acceptable level of mercury was significantly lowered in some products to <1–3 ppm for non-eye cosmetics, and for eye cosmetics to 0.0007%—by weight, or to <65–100 ppm (Table 13).

The European Union forbids using both mercury and mercury components as ingredients for the production of soaps, creams, shampoos, lotions, and other skin-bleaching cosmetics.

Despite the ban on mercury in cosmetics and preventive measures to limit the production of the Hg-containing cosmetics, the unverified skin-lightening products (sometimes they even have no labeling) can be easily purchased on the Internet. One of the serious concerns has been the increasing cases of mercury poisoning in Minnesota, where women used skin-lightening products with high levels of mercury. Among the twenty-seven tested products, which included creams and soaps, 47% of them contained four to thirty times more mercury than the standard level, and some of the tested products contained a catastrophic level of mercury—up to 33000 ppm—according to the Minnesota Department of Health (Harrington, 2014). Such concentrations significantly exceeded the <1 ppm mercury limit established by the U.S. Food and Drug Administration (FDA); (Table 13). In 2010, in California, at least sixty people were reported to have been poisoned by the use of mercury skin-lightening creams manufactured in Mexico. The mercury poisoning symptoms were observed not only in those individuals who used the cream products, but also in six children and babies with no history of using the creams. Infants and small children are more vulnerable to mercury poisoning than adults, and in this case, they became ill from contacts through the skin with their mothers or other family members who used the products. After this case, the California Department of Public Health (CDPH) warned consumers to avoid using skin-lightening creams in unlabeled or hand-labeled containers originated from Mexico.

Traditional medicine. Mercury was historically used for traditional treatment in China, India, and other less developed countries. Cinnabar (HgS) was commonly used in China to treat illnesses from cardiovascular to nervous system disorders and insomnia. However, review publications on traditional medicine have shown that some patients have developed mercury poisoning symptoms such as inflammation of the intestinal system, kidney disease (nephrosis), muscular tremor, and bleeding of the gums (e.g. Zhou, 1986).

There have been many discussions of the Ayurvedic medicine of India, which is still practiced and has become very popular in Western countries. On the one hand, some publications say that traditional medicine is safe because the total concentration of mercury and other toxic metals (e.g., arsenic, lead, and cadmium) are insufficient to cause health problems, and the level of toxicity depends on the chemical forms of the mercury compounds. Ayurvedic medicine also believes that mercury sulfides are not as toxic as mercury vapor, mercury chloride, and methylmercury. Although many of the Indian Ayurvedic remedies have been used safely, there are not enough available statistical data to show the long-term effects. Some publications have shown that some people were poisoned by heavy metals after their use of traditional remedies, and more public awareness regarding the risks involved in the uncontrolled ingestion of traditional medications is needed to prevent health damage. For example, one-fifth of both the U.S.- and Indian-manufactured Ayurvedic medicinal products sold on the Internet were found to contain harmful concentrations of lead, mercury, and arsenic (Saper et al., 2008). The number of similar reports is growing as more scientific studies and examinations are applied to the traditional medicinal compounds.

Agricultural pesticides. Usage of mercury for pesticides is now banned in the U.S., and the EPA requires more than 100 different scientific tests to approve new pesticides for distribution and sale. The tests are supposed to confirm that the pesticide can be used with a reasonable certainty of no harm to human health and without posing unreasonable risks to the environment (http://www.epa.gov). However, in many other countries, some pesticides include natural rocks containing mercury minerals, so mercury poisoning through crop fertilization is not excluded.

6.2.4. Mercury toxicity and its adverse effects on human health

Mercury and its compounds are highly toxic to humans, and in particular, they are extremely harmful to developing fetuses and small children, with animals and ecosystems also being at high risk. The EPA considers mercury to be one of the top ten toxic chemicals that represent major threats to public health. High doses can be fatal to humans, but even relatively low doses can affect the nervous system and have been linked with harmful effects on the cardiovascular, immune, and reproductive systems.

In 2002, the EPA established that 0.1 µg/kg/day (micrograms per kilogram of body weight per day, which is equal to 0.1 ppm) is the maximum acceptable daily exposure to mercury to prevent harmful effects during a lifetime. The FDA has recommended a slightly higher level of mercury as a standard: 0.5 to 1 ppm, which is similar to the recommendations of the WHO.

When mercury is released into the air from industrial sulfide ore treatments or the burning of fossil fuels, it becomes highly mobile and soluble, and it can react with other elements cycling between the Earth's surface and the atmosphere. Mercury persists in the environment, where it combines with carbon and hydrogen to produce methylmercury:

$$Hg \rightarrow CH_3Hg, \hspace{4cm} [Eq. 2].$$

Methylmercury is called an "organic" mercury to distinguish it from the mercury incorporated in minerals. Methylmercury may also be formed in water and soil by bacteria acting on inorganic mercury compounds. Methylmercury can be easily bio-accumulated in fish, shellfish, and algae to levels that are many thousands of times greater than the levels in the water.

Although in some publications, one may find that methylmercury and native mercury are considered the most toxic among the mercury compounds, it is needless to say that exposure to high levels of any type of mercury, be it metallic, inorganic, or organic, can irreversibly damage the brain and kidneys, the cardiovascular, immune, and reproductive systems, and developing fetuses. Symptoms of mercury poisoning in adults include difficulty in concentrating,

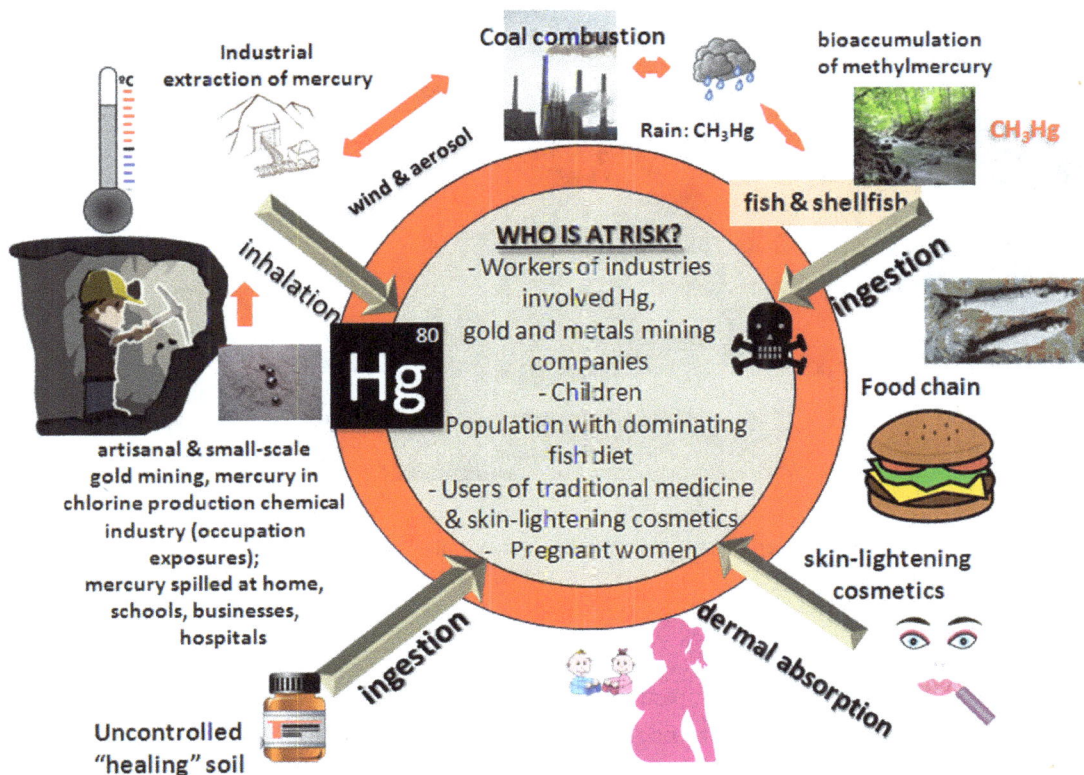

FIGURE 76 Anthropogenic sources and pathways of exposure to mercury

lack of coordination, impairment of speech and hearing, irritability, depression, insomnia, unexplained headaches, weight loss, feelings of exhaustion and fatigue, tremors, numbness or tingling in the hands or feet, and weakness in the extremities. Children poisoned by mercury usually lose their appetites and may have excessive thirst, irritability, poor muscle tone, leg cramps, or a rash.

6.2.5. Mercury pathways to the environment and humans

The pathways of mercury pollution to the environment and humans are complex (Fig. 76). The largest mercury contaminations take place through industrial activity, but mercury is also emitted into the air and the environment from volcanic eruptions, hot springs, rock weathering, and soil erosion.

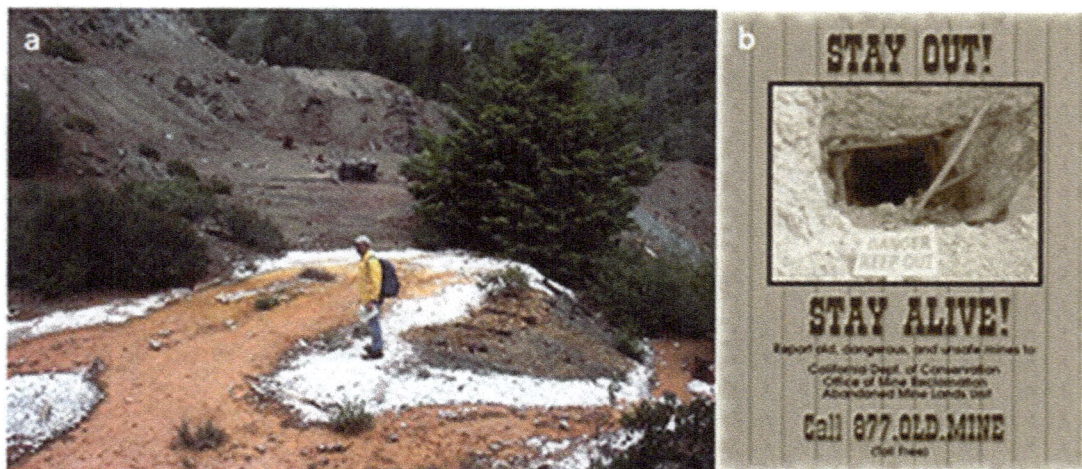

FIGURE 77   The abandoned Helen Mercury Mine, Lake County, California

6.2.5.1 Anthropogenic pollution

Anthropogenic pollution of the environment is produced by industrial mining and ore extraction, small-scale mining of gold, drainage of contaminated waters from abandoned mines and tailings, coal combustion and waste incineration, and domestic and other incidents of mercury spills in schools, hospitals, and businesses. The human exposure often occurs through ingestion of mercury-contaminated fish and other products in the food chain and through poorly controlled industrial and mining waste in workplaces.

(i) Industrial mining and ore processing. These processes create huge piles (tailings, also called mine dumps) of waste rocks that are left over after mercury separation from the non-economic portion of the ore. Although in 1992, mercury production was banned in the U.S., many abandoned mines continue to constitute hazards to human health. For example, at the site of the abandoned Helen Mercury Mine in Lake County, California (Fig. 77), the oxidized tailing material is "washed" away by wind every winter season into Dry Creek, a periodic stream tributary of the larger Upper Putah Creek. The drainage of the Helen mine tailings and the large volume of waste rocks that contain elevated levels of mercury (varying from 2.7 to 90 ppm) are potential sources of pollution for the ecosystem. The U.S. Bureau of Land Management (USBLM) works on removal of Hg-contaminated waste from the Helen mine and others as a means of reducing mercury transport to Dry Creek and protecting the ecosystem and the local biodiversity. Other precautions involve asking people who find abandoned mines or traces of mine-tailing drainage to stay away from them and report their locations to the California Department of Conservation and Mine Reclamation (CDCMR; Fig. 77b).

(ii) Gold mining. Mercury is widely used for gold recovery from alluvial sediments because mercury easily amalgamates with gold. Significant quantities of mercury were lost to the

environment and watersheds during these activities, followed by its entering the food chain and becoming a health hazard to humans.

(iii) Coal combustion. There are trace amounts of mercury in coal that are released into the atmosphere during coal combustion. Also, the coal industry and power plants that use coal to generate heat and electricity produce hundreds of millions of tons of waste products, including ash and sludge that contain mercury, uranium, thorium, arsenic, and other heavy metals. The emission factor of mercury from coal combustion by power plants is 0.1–0.3 g/t, and for residential and commercial boilers that use coal, it is 0.3 g/t.

(iv) Domestic spill of mercury. Spilled liquid mercury moves fast and evaporates at room temperature, producing an odorless, colorless vapor of extreme toxicity. Inhalation of mercury vapor as well as the swallowing of mercury or its byproducts is harmful to humans, especially children. Small-volume spills of mercury should be cleaned up and contained immediately in the following ways:

1. Drops of liquid mercury can be sucked up by any syringe or eye dropper, or they can be swept up into a dust pan and placed in any glass bottle or container, which should be immediately sealed to prevent evaporation.

2. Place sand around the spill of mercury to stop its movement and spreading out of the spill area. Then "amalgamate" mercury with silver or gold (if it happens at home, you can use a silver-bearing spoon) or with a special amalgamating powder that is available in mercury clean-up kits. In a few minutes, the mercury will be solidified, and then it can be easily picked up and placed in a safe container or plastic bag, which must be immediately sealed.

3. There are commercially produced "mercury indicating powders" that will change color over a twenty-four-hour period if mercury still remains in the contaminated area.

6.2.5.2. Human ingestion of mercury through the food chain

The principal route of human exposure is through consumption of fish and other sea products contaminated with mercury. The schematic pathway of both natural and anthropogenic mercury into aquatic systems, followed by the accumulation of methylmercury in the food chain, is presented in Fig. 78. Methylmercury is readily ingested by

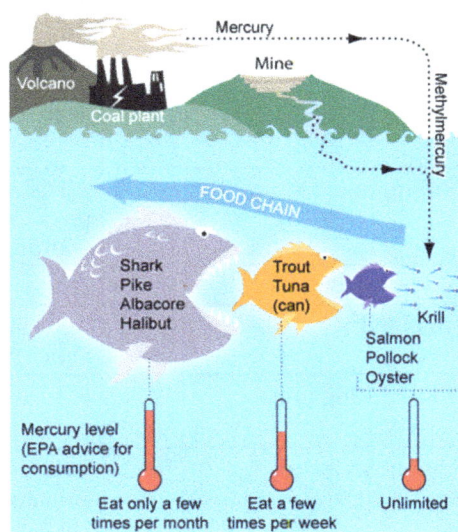

FIGURE 78 Schematic pathway of natural and anthropogenic mercury into aquatic systems and food chains

aquatic organisms and tends to accumulate itself in them in a higher concentration. Such a process, called biomagnification, can result in high mercury concentrations in predatory fish such as striped bass and sharks and in large, fish-eating birds and mammals.

TABLE 14   Maximum Allowed Levels of Mercury in Fish in Various Countries (UNEP Data; http://www.chem. unep.ch/mercury)

Country or Organization	Fish Type	Maximum Recommended Level of Mercury in Fish
Australia	Swordfish, tuna, shark, barramundi, ling, and rays. All other fish and mollusks.	1 ppm 0.5 ppm
Canada	All fish except shark, swordfish, and tuna. For Aboriginal people who consume a large amount of fish.	0.5 ppm 0.2 ppm
European Union	All fishery products with the exception of the following: Catfish, bass, ling, bonito, eel, halibut, rays, red, scabbard and sail fish, shark, mackerel, swordfish, sturgeon, marlin, and pike.	0.5 ppm 1 ppm
India	Fish	0.5 ppm
Japan	Fish	0.4 ppm 0.3 ppm (CH_3H_g)
Philippine	All fish except for the following predatory fish: shark, tuna, and swordfish.	0.5 ppm (CH_3H_g) 1 ppm (CH_3H_g)
S. Korea	Fish	0.5 ppm
Thailand	All seafood Other food	0.5 ppm 0.02 ppm
United Kingdom	Fish	0.3 ppm
United States	Fish, shellfish, and other aquatic animals. Locally caught fish (tribes, states).	1 ppm (CH_3H_g)—by FDA; 0.5 ppm (CH_3H_g)—by FDA; 0.1 ppm for all fish and sea products—by EPA
World Health Organization	All fish except of the following predatory fish: shark, swordfish, pike, tuna.	0.5 ppm (CH_3H_g) 1 ppm (CH_3H_g)

Fish and other seafood. Methylmercury accumulates in the tissues of fish, and people may be exposed to mercury poisoning by eating fish or shellfish as part of their regular diet. The consumption of mercury-contaminated fish is an especially serious threat to the health of pregnant and nursing women and young children. There are many national and several international regulations that urge people to avoid polluting the oceans, and there have been efforts to create a unified international standards system with regard to acceptable dosages of mercury in fish and other marine products. Such a system has not been developed yet, and as a result, the maximum allowed concentration of mercury in fish varies from country to country, from 0.1 to 1 ppm (for details, see Table 14).

Predatory fish such as shark, swordfish, tuna (Fig. 79), and other types of large-body fish have a tendency to accumulate the highest levels of mercury. Fish of smaller size (i.e., that can fit whole into a pan) and shellfish usually contain less mercury, and therefore, they can be safer for consumption. People who regularly

FIGURE 79 Predatory fish usually contain high levels of mercury: (a) Atlantic bluefin tuna; (b) striped marlin from Australia

fish for subsistence may not be aware that they fish in mercury-polluted aqua-systems, and the probability of such a situation is quite high. In the U.S., through a long history of gold and mercury mining, ~18 million acres of lakes, estuaries, and wetlands and 1.4 million river miles were contaminated with mercury. In 2008, the U.S. issued fish-consumption advisories and also warned citizens to limit consumption of certain types of fish caught in the local waters that have a history of mercury contamination. The U.S. Natural Resources Defense Council (NRDC) has classified fish based on their potential level of mercury concentration (Table 15) and has recommended that people not eat fish with the highest levels of mercury (>0.5 ppm), that they eat fish with high levels of mercury (0.3–0.49 ppm) not more than three times per month, and that they eat six servings or less per month of moderate-mercury-level fish (0.09–0.29 ppm). Fish containing less than 0.09 ppm of mercury are considered safe, and they are placed in a category of "least-mercury-level fish." See the types of fish by categories in Table 15.

FIGURE 80   Cadmium-bearing minerals: (a) greenockite (CdS); (b) otavite (CdCO3)—both are from the Tsumcorp Mine, Otjikoto Region, Namibia

TABLE 15   Classification of Fish and Shellfish Based on Their Potential Level of Mercury, and Recommendations for Consumption

*Highest-Mercury Fish: Hg = >0.5 ppm (Avoid Eating)	*High-Mercury Fish: Hg = 0.3–0.49 ppm (Eat Three Servings or Less per Month)	*Moderate-Mercury Fish: Hg = 0.09–0.29 ppm (Eat Six Servings or Less per Month)	*Least-Mercury Fish: Hg = <0.09 ppm (Enjoy This Fish!!!)
Mackerel (king), marlin, shark, swordfish, tilefish, tuna (bigeye, ahi).	Bluefish, grouper, mackerel (Spanish, Gulf), sea bass (Chilean), tuna (canned albacore), tuna (yellow-fin).	Bass (striped, black), carp, cod (Alaskan), croaker, halibut Atlantic and Pacific), silverside, lobster, mahi-mahi, monkfish, perch (freshwater), sablefish, skate, snapper, tuna (canned chunk light), tuna (skipjack), weakfish (sea trout).	Anchovies, butterfish, catfish, clam, crab, flounder, haddock (Atlantic), hake, herring, mackerel (North Atlantic, chub), mullet, oyster, perch (ocean), plaice, salmon (fresh), salmon (canned), sardine, scallop, shrimp, sole (Pacific), tilapia, trout (freshwater), whitefish.

(*All data are adapted from the U.S. NRDC: http://www.nrdc.org/health/effects/mercury/guide.asp)

6.2.6. The Minamata disease—an epidemic of methylmercury poisoning in Japan

An example of epidemic mercury exposure is the Minamata disease that was first recognized in 1956 within the population of the city of Minamata, Kumamoto Prefecture, Japan. The disease was a result of massive contamination of fish and other marine products by methylmercury (CH_3Hg) discharged, together with wastewaters, from a chemical factory into Minamata Bay. The chemical factory, which was built by the Chisso Corporation (Japan), manufactured plastic (acetaldehyde) materials from 1932 to 1968, and the acetaldehyde production progressivly grew, starting from 210 tons in 1932 and reaching 6000 tons per year by 1951. This increasing production spurred the local economy and created a quarter of all jobs in the region. A decade after the factory's production of acetaldehyde began, some Minamata residents began to feel unusual symptoms such as numbness and unsteadiness in their hands and legs, ringing in the ears, slurred speech, and awkward movements, but at the beginning, these symptoms were not officially recognized as an epidemic disease. In 1956, doctors from the local hospital examined a five-year-old girl who had severe difficulties in walking and speaking and who experienced convulsions. Soon, her sister and another girl from a neighboring house, as well as eight adults, were hospitalized with similar symptoms, which at that time were classified as "an unknown disease of the central nervous system." Three years later, the Japanese Ministry of Health and Welfare's Minamata Food Poisoning Subcommittee officially recognized the origin of the Minamata epidemic. They wrote: "The Minamata disease is a poisoning disease that affects mainly the central nervous system and is caused by the consumption of large quantities of fish and shellfish living in Minamata Bay and its surroundings, the major causative agent being some sort of organic mercury compound" (Lessons from Minamata Disease, 2013). The technological process of the Chisso Corporation's chemical factory was based on using mercury sulfate as a catalyst to govern the chemical reactions required for the acetaldehyde production. The highly toxic chemicals had been discharged into Minamata Bay over a period of almost thirty-six years, during which time they bioaccumulated in fish and various shellfish, which resulted in severe mercury poisoning of those for whom the local fish represented a major part of their diet.

Many published documents have shown that after the Chisso Corporation recognized that the Minamata disease was a result of mercury pollution from its factory, it did nothing to stop the operation or to change its technology. The marine products in the Minamata Bay region contained from ~5.61–35.7 ppm of Hg, which significantly exceeds the provisional level (0.1–1 ppm), and such a fact indicates the corporate negligence with regard to human lives. A second outbreak of the Minamata disease occurred in Niigata Prefecture in 1964, and it was also caused by toxic methylmercury wastewater that was being discharged into the Avano River by the Showa Denco Corporation's factory.

In 1968, the government of Japan officially determined that the Minamata disease was a result of the methylmercury pollution caused by the Chisso Corporation's operations. Also in 1968, the toxic mercury sulfate technology for production of plastic materials was officially discontinued. According to the Japanese statistics, 2955 people contracted the Minamata disease, and 1784 people have died from the disease. There were also many patients with chronic symptoms such as headaches, tiredness, weakness of memory, and loss of sense of smell and taste that are not considered severe illnesses but make human life difficult. Although the Minamata disease is not contagious and cannot be transmitted through air or handshaking, there are people who were born with deformities after being affected by mercury during their mothers' pregnancies because the mothers ate contaminated fish. Minamata disease patients have also suffered from ostracism by their neighbors, who have prejudicially thought that the illness might be contagious. Finally, the Minamata disease has no cure today, so patients may only receive treatment and physical rehabilitation therapy to lessen their symptoms.

6.2.7. Measures to prevent mercury contamination

The Minamata disease is one of the largest and most severe epidemic diseases caused by mercury pollution from industrial hazards, and it teaches us that mercury pollution can be a long-term chemical disaster if industrial operations neglect measures of environmental and human health safety. About 17000 people have applied for certification as Minamata disease victims, and the Chisso Corporation has had to pay $86 million in compensation to 10353 claimants. Chisso has also been ordered to clean up the contaminated environment and other lawsuits and claims for compensation of different scales continue to present. Although the toxic chemical discharge was stopped in 1968, it took approximately thirty years for the concentration of Hg and CH_3Hg in fish and marine products to decrease to the provisional level (0.4 ppm total Hg and 0.3 ppm CH_3Hg). In 1997, the government of the Kumamoto Prefecture officially announced that the Minamata Bay region was clean and safe.

Similar damage to humans and the environment from anthropogenic mercury pollution has been recorded in recent years in China, Canada, Tanzania, and the Amazon River region. In the U.S., Congress authorized the EPA and other government organizations to create and enforce regulations for protecting the air and water from mercury contamination, including mercury emissions from industrial coal combustion and local discharging from waste deposits and unsafely stored mercury products. In addition, the Clean Air Act, the Clean Water Act, and the Resource Conservation and Recovery Act were established in order to control anthropogenic sources of mercury contamination.

In 2011, the European Union banned mercury exports, and mercury use in the EU chlorine-alkali chemical industry was discontinued. The EU also suggested some measures for safely storing the secondary mercury products collected from the purification of natural gas and the processing of non-ferrous metals, including placing them in underground salt mines that would serve as toxic waste depositories. In 2005, a United Nations initiative established a Global

Mercury Partnership to achieve reductions of emissions and use of mercury worldwide. The U.S. plays a leading role in the activities of this partnership.

References

Dudka, S., and D.C. Adriano. 1997. Environmental impacts of metal ore mining and processing: a review. *Journal of Environmental Quality* 26: 590–602.

Harrington, R. 2014. "Minnesota's Mercury Concerns Bring New Test for Newborns." http://www.startribune.com/lifestyle/health/249993501.html

Lessons from Minamata Disease and Mercury Management in Japan. 2013. Environmental Health and Safety Division and Environmental Health Department, Japanese Ministry of the Environment.

Saper, R.B., R.S. Phillips, A. Sehgal, N. Khouri, R.B. Davis et al. 2008. Lead, mercury, and arsenic in US- and Indian-manufactured Ayurvedic medicines sold via the Internet. *Journal of American Medical Association* 300:915–923.

Zhou, T.Z. 1986. Analysis of the toxicity of Chinese mineral medicines. *Zhejian Journal of Traditional Chinese Medicine* 4:354–356.

Web resources

http://www.chem.unep.ch/mercury.
http://www.nrdc.org/health/effects/mercury/guide.asp.

Image Credits

Figure 74d: Copyright © Rob Lavinsky (CC BY-SA 3.0) at http://commons.wikimedia.org/wiki/File:Livingstonite-sea46a.jpg.

Figure 76a: Artmaker, "makeup," https://openclipart.org/detail/154303/makeup-by-artmaker. Copyright in the Public Domain.

Figure 76b: mlumen, "pregnancy silhouet," https://openclipart.org/detail/2017/pregnancy-silhouet-by-molumen-2017. Copyright in the Public Domain.

Figure 76c: Pippi2011, "Baby boy and girl," https://openclipart.org/detail/181638/baby-boy-and-girl-by-pippi2011-181638. Copyright in the Public Domain.

Figure 76d: kubitus, "thermometer," https://openclipart.org/detail/19459/thermometer-by-kubitus-19459. Copyright in the Public Domain.

Figure 76e: tzunghaor, "miner," https://openclipart.org/detail/165301/miner-by-tzunghaor. Copyright in the Public Domain.

Figure 76f: Giltesa, "Mercurio2," http://commons.wikimedia.org/wiki/File:Mercurio2.JPG. Copyright in the Public Domain.

Figure 76g: nicubunu, "RPG map symbols Mine 2," https://openclipart.org/detail/11481/-by--11481. Copyright in the Public Domain.

Figure 76h: Bogdangiusca, "Pollution de l'air," http://commons.wikimedia.org/wiki/File:Pollution_de_l%27air.jpg. Copyright in the Public Domain.

Figure 76i: warszawianka, "tango weather showers scattered," https://openclipart.org/detail/30151/tango-weather-showers-scattered-by-warszawianka. Copyright in the Public Domain.

Figure 76j: Copyright © Kyle R. Burton (CC BY-SA 3.0) at http://commons.wikimedia.org/wiki/File:Forest-Creek-Eagleville-PA-USA.jpg.

Figure 76k: jonphillips, "Fish from the market," https://openclipart.org/detail/182553/fish-from-the-market-by-jonphillips-182553. Copyright in the Public Domain.

Figure 76l: cwleonard, "Hamburger," https://openclipart.org/people/cwleonard/hamburger.svg. Copyright in the Public Domain.

Figure 76m: ernes, "Medicine - Drugs," Adapted from: https://openclipart.org/detail/27394/medicine--drugs-by-ernes-27394. Copyright in the Public Domain.

Figure 77a: http://www.blm.gov/ca/st/en/prog/aml/project_page/helen/helen_photo1.html

Figure 77b: "Stay Out, Stay Alive!," http://www.consrv.ca.gov/omr/abandoned_mine_lands/Pages/stay_out_stay_alive.aspx. Copyright in the Public Domain.

Figure 78: Copyright © Shizhao (CC by 3.0) at http://en.wikipedia.org/wiki/File:MercuryFoodChain-01.png.

Figure 79a: FishWatch, "Large bluefin tuna on deck," http://commons.wikimedia.org/wiki/File:Large_bluefin_tuna_on_deck.jpg. Copyright in the Public Domain.

Figure 79b: Copyright © Jackiemora01 (CC BY-SA 3.0) at http://commons.wikimedia.org/wiki/File:Stripe_marlin_right_off_the_coast_of_Carrillo.jpg.

Figure 80a: Copyright © Christian Rewitzer (CC BY-SA 3.0) at http://en.wikipedia.org/wiki/File:Greenockite-259580.jpg.

Figure 80b: Copyright © Christian Rewitzer (CC BY-SA 3.0) at http://commons.wikimedia.org/wiki/
File:Otavite-89481.jpg.

Review Questions

1. What kinds of elements are classified as heavy elements?

2. Why are heavy minerals important for industry?

3. In what different forms does mercury exist in the geological environment?

4. In what geological formations do heavy metal ores form?

5. Why are heavy metals dangerous to the environment and human health?

6. What kinds of heavy minerals, native metals, and elements do you know? Give at least ten of them.

7. What is native mercury?

8. What kinds of minerals contain mercury? Where do they form?

9. What is methylmercury, and how does it act in the environment?

10. How large are the global sources of mercury? What country is the leading producer of mercury?

11. Mercury in the environment: describe the natural geological sources and the anthropogenic (man-made) sources.

12. Where is mercury used?

13. What type of mercury compounds are the most toxic?

14. What continent produces the most mercury emissions?

15. Explain the mercury pathways to the environment and humans.

16. Would you recommend your sister to use mercury skin-lightening cosmetics?

17. What should you do if a small mercury spill happens in your house?

18. Why is mercury used for gold extraction? How does it work? What are the reasons that mercury is still used for gold mining?

19. Industrial extraction of mercury: roasting, condensation, and refining. What kinds of chemical reactions can extract mercury from cinnabar and other mercury-bearing ores?

20. What is dental amalgam?

21. What kinds of environmental and health hazards are associated with the mining of cinnabar ores?

22. What is acid mine drainage? Would you use water from a main drainage reservoir to wash your hands?

23. Explain how bio-accumulation of mercury occurs.

24. Explain human ingestion of mercury through the food chain.

25. Could mercury fumes released from a refining factory into the atmosphere contaminate a rice crop grown in neighboring agricultural fields? If yes, how does it usually happen?

26. What mercury level is safe in food according to the EPA?

27. How does exposure to mercury affect human health? Who are at risk?

28. What is the Minamata disease? How long did it take to link the epidemic health problem of the Minamata citizens to the mercury contamination of the environment?

29. What was wrong in the Minamata ore-processing factory?

30. What kinds of symptoms help to identify people's overexposure to mercury?

31. If two types of fish were offered in a restaurant—shark and tilapia— which one would you choose to eat?

32. Would you prefer swordfish to salmon if your regular diet is a seafood diet? Explain your choice in the context of a possible mercury contamination.

33. What kinds of measures should be taken to prevent human exposure to mercury?

Quizzes (see answers on page 307)

1. In the environment, an inorganic form of mercury is transformed into an organic form, which is extremely toxic and easily absorbed by plants, trees, grass, and living organisms. This organic form, which builds up in the body, is called
   a. Aquatic mercury.
   b. Mercury vapor.
   c. Methylmercury.
   d. Mercury salt.

2. People are more likely to ingest significant quantities of mercury by eating
   a. Seaweed.
   b. Clams.
   c. King crab.
   d. Tuna.
   e. Tilapia.

3. In the U.S., the EPA maximum allowed level of mercury in fish is
   a. 0.1 ppm.
   b. 0.5 ppm.
   c. 0.001 ppm.
   d. >0.6 ppm.

4. What should you do if there is a mercury spill in your room?
   a. You have to vacuum up the mercury immediately!
   b. Don't worry, just pour the mercury down the drain.
   c. Be careful: neither a nor b.

5. What are tailings, and why are they potentially dangerous to human health?
   a. Tailings are large deposits of metal ores. They are dangerous because of the high levels of mercury that they contain; the mercury can contaminate water supplies.
   b. Tailings are the unused, broken, ground-up rocks and dust that are usually piled up in mining operations. If not properly contained, tailings are always dangerous because any toxic elements that originally occurred in the rocks before their excavation are easily spread around by wind and water.
   c. Tailings are the chemicals that are used to separate useful metals from the rocks they are found in. They pollute the nearby environment.
   d. Tailings are a type of naturally occurring dust particle. They can be dangerous because they are easily inhaled due to their small size, and they can cause silicosis.

6.3.1. Natural occurrences

Cadmium (Cd) is a chemical element (atomic number 48) that belongs to the heavy metals group. The estimated abundance of cadmium in the continental crust is 0.1–0.5 ppm, and in the ocean, it is 0.1 ppm. Minerals that contain significant amounts of cadmium, such as the native metal (Cd), greenockite (CdS), monteponite (CdO), and otavite ($CdCO_3$); (see Fig. 80), do not occur in rocks in high enough concentrations to be profitable for mining. In fact, the native metal Cd is a very rare mineral and has been reported in only one place in the world (Sakha Republic, Yakutia, Russia), where it associates with Cu-polymetallic deposits located in sandstone-carbonaceous formations (Fleischer et al., 1980). Cadmium also is known to be accumulated in sulfide minerals such as sphalerite (ZnS), galena (PbS), and covellite (CuS) with the range of Cd being 0.2–0.7% by weight. Such accumulation occurs because cadmium, zinc, lead, and Cu share similar chemical properties, which means that Cd ions may easily substitute for Zn, Pb, and Cu in the sphalerite, galena, and covellite crystal lattices.

Natural accumulation of cadmium in crustal rocks occurs through volcanic eruptions, the weathering of rocks and soils, biomass self-burning processes, sea-salt spray, and marine aerosols. The worldwide estimated emission of Cd into the atmosphere through geological processes is ~1400 t/year, and 60% of this emission is contributed by volcanic activities.

6.3.2. The world cadmium production and where it is used

Because cadmium minerals do not form large ore deposits, most of the natural cadmium is recovered as a byproduct of the mining and refining of copper, zinc, and lead sulfides. The world cadmium production was ~20 tons per year at the beginning of the twentieth century, and it was stabilized by 1990 at the level of 20000 tons (Fig. 81). In 2011, worldwide production increased to ~22000 tons, with more than 500000 tons remaining as total world reserves according to the USGS Mineral Commodity Report. The leading countries in global cadmium production are China, South Korea, and Japan, and in 2011, they produced approximately 3800 t, 2846 t, and 1939

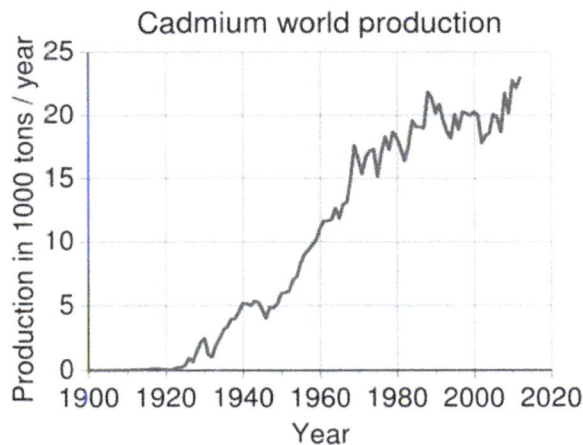

FIGURE 81 Growth of world cadmium production, from 1920 to 2020 (projected)

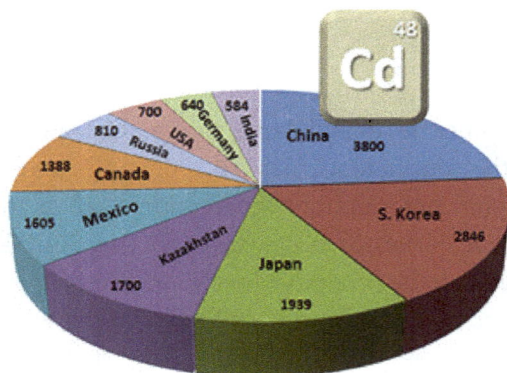

FIGURE 82 Ten top countries in world cadmium production in 2011

t, respectively (Fig. 82). About 80% of Cd is obtained as a byproduct of primary zinc extraction from ZnS ores, and the remaining 20% comes from recycled Cd products and natural PbS and CuS ores.

Cadmium is mainly used in rechargeable Ni-Cd batteries, for manufacturing pigments, in the metal-plating industry to protect other metals from corrosion, and as a stabilizer for plastic materials production (Fig. 83). Because cadmium easily absorbs neutrons, it is also used to make control rods for nuclear reactors. Other applications include manufacturing of low-melting alloys and many kinds of solder and bearing alloys because of its low coefficient of friction and metal-fatigue resistance. Hydrated cadmium sulfate ($3CdSO_4 \cdot 5H_2O$) is used in a special type of battery to calibrate medical and laboratory equipment. Cadmium is also added to phosphorous fertilizers.

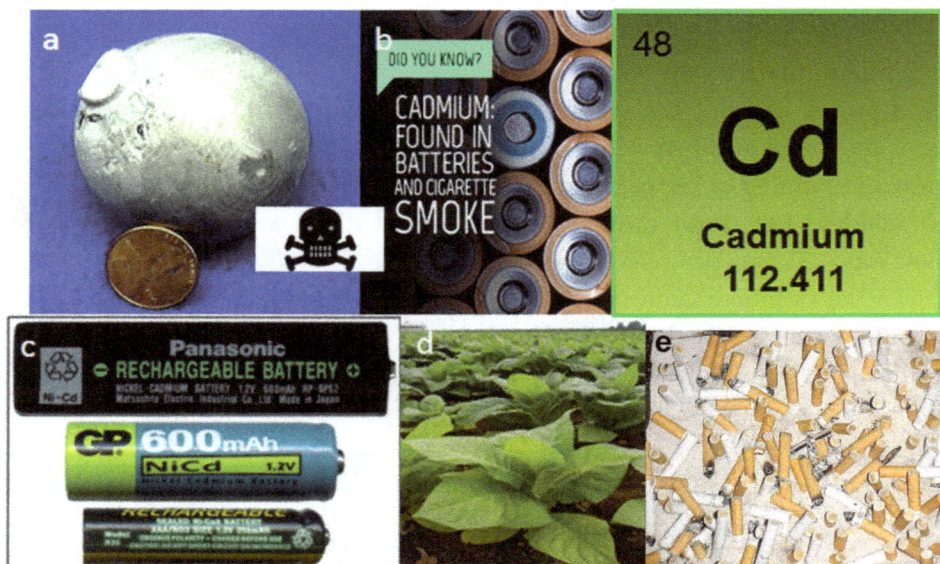

FIGURE 83 Cadmium in our life: (a) Cd crystal bar; (b) Cd found in batteries and cigarettes

6.3.3. Toxic effect and cadmium pathway to humans and the environment

Cadmium is a poisonous metal, and it is extremely toxic even in small concentrations. Concerns with regard to Cd toxicity have resulted in many regulations, and as a result, the industrial use of Cd has declined, which in turn has lowered hazardous emissions over the past forty years (Sigel et al., 2013). Although exposure to metals is mostly occupational, the general population can be easily exposed through inhalation of particulate matter (metal dust and aerosols) and ingestion of contaminated drinking water and food.

Cadmium in the atmosphere. The main sources of cadmium in the atmosphere are the burning of coal or oil and the incineration of municipal waste, which may contain uncounted numbers of expired Ni-Cd batteries and other Cd-bearing technological products. In mining and metal-refining industrial areas, large amounts of cadmium are emitted from zinc, lead, or copper smelters. Workers in the smelting and metal electroplating industries are still at high risk, and their safety depends on how exposure in the workplace is controlled through personal protective equipment and clothing, industrial hygiene regulations, and the monitoring and reduction of Cd emissions. Children can be exposed to cadmium through contact with the clothing and shoes of parents who work in cadmium-emitting industries, and thus taking a shower and changing clothes and shoes before returning home may reduce the risks of cadmium transportation from the workplace to home.

Cadmium in soil, food, and cigarettes. Cadmium easily accumulates in soils treated with phosphorous fertilizers to maintain agricultural productivity. The inorganic phosphate fertilizers are manufactured from natural phosphate rocks, which usually contain cadmium. Cadmium can also be found in animal manures, which are used as fertilizer as well. Application of both non-organic and organic fertilizers over long periods accumulates the cadmium in the soil and thereby creates imbalances in the nutrient uptake by plants. This causes the potential risk for high cadmium levels in crops grown in fertilized soils.

Food products such as rice, lettuce, spinach, and some other vegetables grown in fertilized soils have been found to have Cd concentrations as high as 0.01 mg/kg (Das et al., 1997). It has also been observed that older leaves of lettuce and spinach have higher accumulations of Cd than younger leaves, and Cd concentration is higher in roots than in shoots. In general, leafy vegetables (e.g., lettuce, spinach), potatoes, grains, peanuts, soybeans, and sunflower seeds contain high levels of Cd—up to 0.05–0.12 mg/kg. If domestic animals eat Cd-contaminated plants, their meat, and especially their kidneys and livers, can contain 0.1–1 mg/kg Cd. These numbers are significantly higher than the EPA's recommended dosage of 0.001 mg/kg/day for human dietary exposure to cadmium.

Observations of tobacco plants have shown that their leaves may accumulate rather high Cd concentrations, so that one cigarette produced from such a contaminated plant will contain

0.001–0.002 mg of Cd, of which 2–10% is transmitted to cigarette smoke. Therefore, the content of cadmium in tobacco is a warning sign to cigarette users (see Fig. 83). The EPA has reported that direct measurements of cadmium levels in human tissues shows that people who smoke have about twice the level of cadmium in their bodies as non-smokers.

Cadmium in drinking water and aquatic reservoirs. Cadmium accumulation in water may originate from corrosion of galvanized pipes, erosion of sulfide ore deposits, accidental industrial discharge from mines and metal refineries, aerosol precipitations, and runoff from leaking wasted Ni-Cd batteries, paints, and fertilizers that have been added to the soil. Cadmium pollution of water reservoirs has been considered to be a possible cause of physiological anomalies in fish and kidney damage in seabirds. Because fish accumulate Cd, people who fish from local waters as a means of food supplies should take seriously any advisories from their government environmental organizations. The EPA and the FDA suggest that their reference dose for cadmium in drinking water is 0.005 mg/kg/day.

6.3.4. Adverse effects of cadmium on human health

Cadmium is one of the extremely toxic environmental pollutants produced by anthropogenic activities. Transported to human bodies through inhalation and ingestion, cadmium can damage the liver, the kidneys, and bones, as well as cause renal dysfunction and kidney stones. Cadmium also has an adverse effect on the central nervous system (the brain and spinal cord),

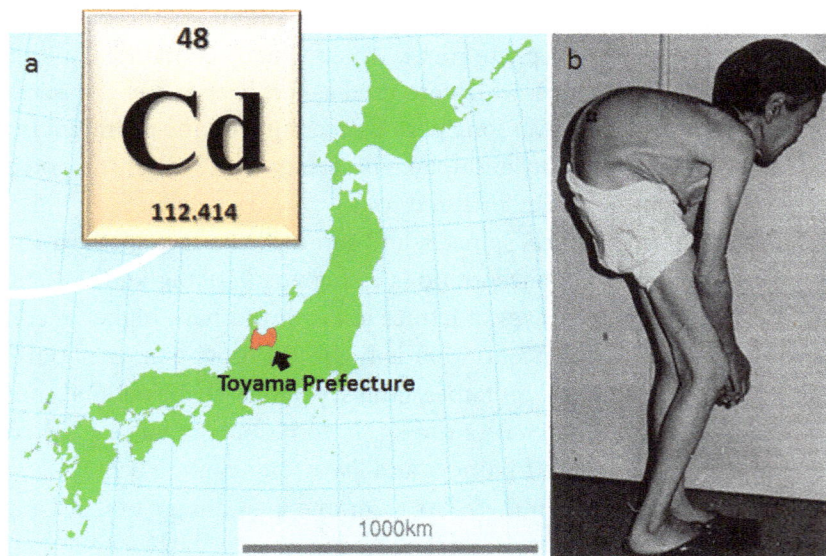

FIGURE 84 Cadmium pollution in (a) Toyama Prefecture, Japan, shown in red highlight; (b) a patient showing signs of itai-itai disease caused by cadmium pollution

and it is not transmitted through the skin. Cadmium is classified as a Group 1 human carcinogen (Sigel et al., 2013).

It is worth considering that if cadmium is constantly ingested through food, it will be progressively accumulated in the human body with age. This is because the human organism is capable of excreting only about 0.001% of accumulated cadmium per day. Whether cadmium can be transformed by human organisms into biologically acceptable and non-harmful compounds still remains a subject of intensive study. Cadmium is more actively absorbed and accumulated during certain periods or physiological states, such as pregnancy and iron deficiency in adults (Akkeson et al., 2002).

Acute effects. The acute effects of Cd poisoning through inhalation of a high level of Cd airborne particles (the most dangerous are PM_{10}) or fumes may result in bronchial and pulmonary irritation and impairment of lung function.

Chronic effects. The chronic effects of long-term accumulation of the Cd in such important internal organs as the liver and kidneys can result in increased frequency of kidney stone formation and damage to the lungs, immune system, and central nervous system. Lung cancer has been found in workers who were exposed to cadmium in the air and in experimental studies of rats that were forced to breathe air that contained powdered cadmium.

6.3.5. The Itai-Itai epidemic disease from Cd contamination

Serious kidney and bone disorders (weakening of bones), accompanied by severe pain were recognized in the Toyama Prefecture in Japan within local populations during 1945–1950 (Fig 84). This prefecture is situated in the western central part of the main island of Honshu, and since the eighteenth century, it has been a mining area for Au and associated metal sulfides containing cadmium. The mining activity was restarted in 1910, and the symptoms of kidney malfunction and weakening and softening of bones, especially in middle-aged women, became noticeable and attracted medical attention as early as 1912. However, the toxic Cd source was not recognized at that time, and it was thought that the illness was caused by a kind of bacterial infection.

In the 1950s, it became clear that the disease was caused by a lifetime of drinking Cd-contaminated water and eating rice harvested in the fields with Cd-contaminated soils. The Japanese women were ingesting up to 37.5 μg/day of cadmium, whereas in Europe, women of the same age category were ingesting about 6.3–27 μg/day, and in Asian regions free from severe anthropogenic contaminations, they were ingesting 5–15 μg/day. In Toyama, the rice fields situated along the Jinzu and Kakehashi river basins were irrigated and flooded with metal-laden industrial wastewater from the Mitsui Mining and Smelting operations (Shigematsu, 1984).

The polymetallic (ZnS, PbS, and CuS enriched in Cd) ore mining and smelting operations lasted about 40 years (from 1910 until 1950) were continuously discarding wastewater in significant quantities into the Jinzu and Kakehashi rivers. Cadmium that was deposited in sediments in the beds of the rivers killed fish and contaminated plants and drinking water reservoirs. However, only in recent decades has it become clear that there are links between the mining-smelting source of Cd contamination, local waters, rice, and the Itai-Itai disease, which has been directly related to the high mortality rate in the Toyama area (Ichihara et al., 2001).

In other words, even though the first record of the Itai-Itai symptoms appeared in 1912—just two years after the discharging of toxic wastes to the rivers began—it was only in 1968 that the Ministry of Health and Welfare of Japan accepted that the symptoms were the result of cadmium poisoning. However, by 2012, the cadmium-polluted areas along the Jinzu and Kakehashi rivers had been cleaned up at a total cost of ¥40.7 billion from the Japanese government, Mitsui Mining, and the governments of the Gifu and Toyama prefectures.

References

Åkesson, A., M. Berglund, A. Schütz, P. Bjellerup, K. Bremme, and M. Vahter. 2002. Cadmium exposure in pregnancy and lactation in relation to iron status. *American Journal of Public Health* 92:284–287.

Das, P., S. Samantaray, and G.R. Rout. 1997. Cadmium toxicity in plants: A review. *Environmental pollution* 98:29–36.

Fleischer, M., L. J. Cabri, G. Y. Chao, and A. Pabst. 1980. New mineral names. *American Mineralogist* 65:1065–1070.

Ishihara, E. Kobayashi, Y. Okubo, Y. Suwazono, T. Kido et al. 2001. Association between cadmium concentration in rice and mortality in the Jinzu river basin. *Japan, Toxicology* 163:23–28.

Shigematsu, I. 1984. The epidemiological approach to cadmium pollution in Japan. *Annual Academic Medicine* 13:231–236.

Sigel, A., H. Sigel, and R. K. O. Sigel. 2013. Cadmium: from toxicity to essentiality. Springer, 596 p.

Web resources

http://minerals.usgs.gov
http://safecosmetics.org/article.php?id=223

Image Credits

Figure 81: United States Geological Survey, "Cadmium - world production trend," http://en.wikipedia.org/wiki/File:Cadmium_-_world_production_trend.svg. Copyright in the Public Domain.

Figure 83a: United States Geological Survey, "CadmiumMetalUSGOV," http://commons.wikimedia.org/wiki/File:CadmiumMetalUSGOV.jpg. Copyright in the Public Domain.

Figure 83b: U.S. Department of Health and Human Services, "Cadmium: Found in Batteries and Cigarette Smoke," http://therealcost.betobaccofree.hhs.gov/facts/did-you-know/index.html. Copyright in the Public Domain.

Figure 83c: Copyright © Boffy-b (CC BY-SA 3.0) at http://en.wikipedia.org/wiki/File:NiCd_various.jpg.

Figure 83d: Copyright © 2006 Derek Ramsey, GNU Free Documentation License at: http://commons.wikimedia.org/wiki/File:Nicotiana_Tobacco_Plants_1909px.jpg. A copy of the license can be found here: http://en.wikipedia.org/wiki/GNU_Free_Documentation_License

Figure 83e: Copyright © Ben Schumin (CC BY-SA 3.0) at http://commons.wikimedia.org/wiki/File:Cigarette_ashtray.jpg.

Figure 84a: Copyright © Lincun (CC BY-SA 3.0) at http://commons.wikimedia.org/wiki/File:Map_of_Japan_with_highlight_on_16_Toyama_prefecture.svg.

Figure 84b: Copyright © XingXiong (CC BY-SA 3.0) at http://2009.igem.org/File:Itai_itai.png.

Review Questions

1. In what forms does cadmium exist in the geological environment?

2. What are the levels of cadmium in the continental crust and the oceans?

3. How do people get exposed to cadmium? How many sources of human cadmium exposure do you know?

4. What is the main route by which people are exposed to cadmium in industrial settings?

5. How can rice be contaminated with cadmium?

6. Which type of exposure is known to be most hazardous?

7. What health problems does exposure to cadmium create?

8. What is cadmium used to make?

9. In what kinds of rocks do cadmium-bearing minerals occur on the Earth?

10. What cadmium minerals do you know, and in what geological formations does native cadmium metal exist?

11. How is cadmium accumulated in geological formations?

12. Which are the top countries in Cd production?

13. Describe the toxic effects and cadmium pathways to humans and the environment.

14. What is the EPA recommended dose for human dietary exposure to cadmium?

15. Is it safe to eat lettuce and spinach grown in agricultural fields that are located within one mile of a smelting factory that extracts Cd from ores?

16. How small is the Cd concentration in drinking water that is recommended by the EPA as a safe level for human health?

17. Is it OK to smoke, eat, or chew gum while working with cadmium in a regulated area?

18. What are the symptoms of acute and chronic cadmium poisoning?

19. Would you eat rice grown near a river in which Cd wastewaters were accidentally discharged from an industrial operation?

20. How large is the worldwide estimated emission of Cd to the atmosphere through geological processes? How much do volcano eruptions contribute to that emission?

21. Cigarettes contain cadmium. Explain how they are exposed to cadmium during a technological process required for their preparation or during their delivery to consumers.

22. What is the Itai-Itai epidemic disease? Where and when was this disease first recorded?

23. What was a primary source of the Itai-Itai disease? Who were at risk?

Quizzes (see answers on page 307)

1. Cadmium exposure in the workplace occurs during:
   a. Mining and smelting work with ZnS, PbS, and CuS ores.
   b. Manufacturing of cadmium-bearing products such as paints.
   c. Industrial activities such as metal plating, soldering, and welding.
   d. All of the above.
   e. Only b and c.

2. Non-occupational exposure to cadmium occurs through which of the following?
   a. Eating cadmium-contaminated foods such as rice and cereal grains.
   b. Inhaling cigarette smoke since tobacco plants take up cadmium from the soil avidly.
   c. Drinking water from a private well that is about one mile from a neighboring smelting plant.
   d. All of the above.

3. Eating food contaminated with high levels of cadmium can cause short-term effects such as
   a. Abdominal cramps.
   b. Flu-like symptoms.
   c. Tingling hands.
   d. All of the above.

4. What processes are responsible for the natural accumulation of cadmium in crustal rocks?
   a. Volcanic eruptions.
   b. Weathering of rocks and soils.
   c. Biomass self-burning processes.
   d. Sea-salt spray and marine aerosols.
   e. Earthquakes.
   f. All of the above.
   g. All except e.

5. The EPA's recommended dose for dietary exposure to cadmium is:
   a.   0.001 mg/kg/day.
   b.   0.005 mg/kg/day.
   c.   >0.2 mg/kg/day.
   d.   None of the above.

6.4.1. Natural occurrences

Lead (Pb) belongs to the heavy metals group, and although it is extremely toxic, it has many industrial and technological applications. Lead minerals include ninety-four species, within which galena (PbS), anglesite (PbSO$_4$), cerussite (PbCO$_3$), and menium (Pb$_3$O$_4$) are the most important minerals for lead production (Fig. 85). Galena is a primary mineral that is formed by hydrothermal processes and that is mostly found in hydrothermal veins in igneous rocks, pegmatites, and contact-metamorphosed sedimentary rocks. Anglesite, cerussite, and menium are secondary minerals that are formed during oxidation and other chemical reactions of galena governed by weathering. Galena can be found in association with silver, zinc, iron, and copper-bearing ore deposits, and in non-industrial concentrations, it often associates with such common minerals as quartz, calcite, and fluorite in silicate-carbonate sedimentary rocks. The average concentration of lead in the Earth's crust is 10–14 ppm.

FIGURE 85 Lead minerals: (a) galena (PbS); (b) anglesite (PbSO4); (c) cerussite (PbCO3); (d) lead oxide, also known as minium, (Pb3O4)

6.4.2. The world production of lead and its use

The mineral galena contains about 86.6% lead and 13.4% sulfur by weight, and it serves as the ore for most of the world's lead production. China is the world's leading lead producer with 3,000 × 10^3 tons per year, followed by Australia (690 ×10^3 tons per year), the U.S. (340 × 10^3 tons per year), Mexico (220 × 10^3 tons per year), Peru (250 × 10^3 tons per year), and other countries (Fig. 86), according to the 2012 USGS Mineral Commodities Report (for details, see http://minerals.usgs.gov/minerals/pubs/mcs/2012/mcs2012.pdf).

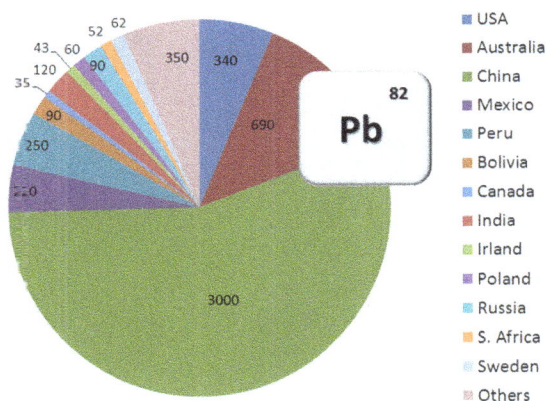

FIGURE 86 World mine production of lead in 2013, given in thousands of tons

The total world lead production in 2012 had reached 5400 x10³ tons. Significant amounts of lead are recovered as a by-product or co-product of zinc and copper mining and silver deposits.

In the U.S., lead is mined primarily in Missouri, Alaska, Idaho, and Montana. However, many mines are already closed, and the few that are still in operation are under consideration for closing because lead mining and extraction areas are identified as the most serious hazardous-waste sites in the nation. These sites are included on the National Priorities List (NPL), and they are undergoing long-term federal clean-up activities. Now, the U.S. has become the leading world producer and consumer of recycled scrap lead metal and "secondary" lead metal obtained from used lead and lead-acid batteries.

Lead is in a huge industrial and technological demand due to its unique physical and chemical properties. It resists corrosion, it is easily molded, shaped, and alloyed with other metals, and it does not transmit radiation. The majority of the lead is used to make vehicle storage batteries, ammunition, electrical cable covers, emergency power batteries for electronics and communications, television glass, construction materials, protective coatings, and sheets/aprons to shield the body from excess radiation exposure during X-ray examinations (Fig. 87). In the U.S., lead-enriched gasoline was slowly phased out starting in the 1970s and was banned completely for use in any type of vehicles in 1995. Lead compounds were used as colored pigments for painting furniture, wood fences, and many other products for domestic utilization, including toys for

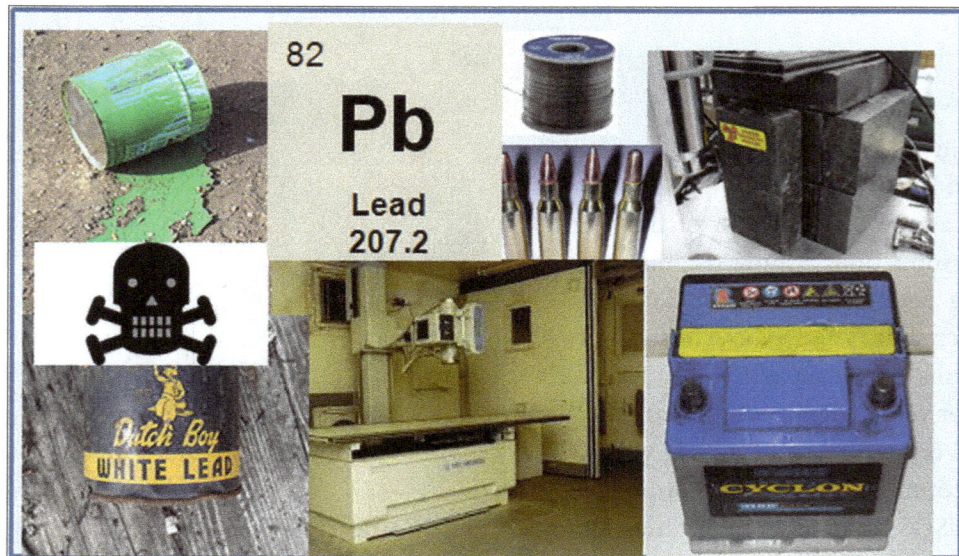

FIGURE 87   Lead usage in industrial and domestic needs

children. In the U.S., lead-containing paint and toys and furniture coated with such Pb-bearing paint were banned in 1978 to reduce the risk of lead poisoning, especially in children, who might ingest paint chips or peelings.

6.4.3. Industrial lead smelting and refining

The industrial extraction of lead from its primary mineral galena (PbS) is carried out by smelting processes that release a large amount of toxic pollution into the atmosphere and the environment, and the emissions from lead smelting are a big contributor to global lead pollution, including particulate matter, toxic effluents, and other solid wastes. Two types of smelting—primary and secondary—are used for extraction of lead metal.

Primary smelting is a technology developed for metallic lead extraction from raw metal-sulfide ores, and it is based on a combination of three processes—sintering, smelting, and refining—to extract lead of metallurgical quality. The sintering stage includes a hot-air combustion process that decomposes the mineral galena and removes sulfur from the lead ore. The resulting Pb reacts with oxygen to produce lead oxide (PbO), and this intermediate by-product is sent into a smelter, where it is heated to high temperatures (~1400°C) in order to isolate the pure lead from the other compounds. After the smelting is completed, the slag contaminated with lead particles is left over. Unnecessary minerals or chemical byproducts that are still in the lead ore after the smelting are removed by the refining process. Lead smoke and dust are usually released during all three stages, and slag contaminated with lead particles is usually stored in the vicinities of the smelting-refining factory sites.

Secondary smelting of lead includes two stages—smelting and refining—and does not require sintering. The secondary smelting is mostly used for extraction of pure lead metal from recycled lead-acid batteries and other electronic products. The used batteries and devices are heated with coke and/or charcoal to purify the lead from other compounds. This process also produces toxic fumes, dust, and slag, which are supposed to be released into the environment in small quantities if the smelting factory is properly constructed and the safety of the technological process is carefully controlled.

6.4.4. Adverse effects of lead on human health

Lead is a toxic substance that has adverse effects on human health and in extreme cases can cause irreversible neurological damage and death. The UNEP classified Pb as a "potent neurotoxin" and a "nerve poison" that affects every organ in the human body and destroys health and intellectual development of adults and especially children. Children are more susceptible to long-term lead exposure because their bodies can excrete only 33% of accumulated Pb, and therefore, if exposure continues, the primary Pb will be retained, and new portions of Pb will continue to accumulate in their blood and body tissues. Although international and national standards for Pb exposure have been established, numerous studies suggest that there is no amount of Pb small enough to avoid adverse biological reaction. Even such small concentrations as <50 µg/L in the blood can lower children's cognitive and academic skills, and Pb level of <100 µg/L in the blood can cause irreversible intellectual impairment and organ failure (Haefliger et al., 2009). Two types of lead poisoning—acute and chronic—are recognized. Acute poisoning happens immediately after inhalation of large quantities of lead dust or lead-containing fumes, and symptoms of poisoning will be visible soon. Chronic lead poisoning usually starts from a very low concentration of Pb through inhalation or ingestion. Initially, the lead poisoning effect is not visible, and it can be difficult to distinguish it from many other illnesses. The symptoms will appear after accumulation of a dangerous concentration of Pb >50 µg/L due to persistent inhaling or ingestion over longer periods.

Lead poisoning symptoms in adults include declines in mental functioning, pain and numbness or tingling of the extremities, high blood pressure, muscular weakness, headache, abdominal pain, memory loss, mood disorders, and miscarriage or premature birth in pregnant women. Lead poisoning in children causes irritability, loss of appetite, sluggishness and fatigue, weight loss, abdominal pain, vomiting, nose bleeding, and learning difficulties.

6.4.5. Levels of dangerous exposure to lead

People are usually exposed to lead through the air they breathe and through ingestion of water and food. Those who are at the greatest risk of lead poisoning, in addition to young children, are lead mining and smelters workers and people living in close proximity of mining and smelter operations.

The EPA permissible standard level of lead in air is 1.5 µg/m^3, and 1.5 µg/L in drinking water. The EPA permissible standard level of lead in human blood was 100 µg/L, but in 2012, it was lowered to 50 µg/L. Many studies, however, have arrived at the conclusion that *there is no safe dose of Pb exposure*, especially for children. Even very low levels of exposure in children (>50 to 100 µg/L) are associated with neuro-developmental deficiencies (Bellinger, 2008). Recent studies based on the modeling of the benchmark dose of lead exposure showed that lower confidence limits of about 1–10 µg/L for the dose resulted in a loss of one IQ point (Burtz-Jogerson et al. 2013). Primary prevention of exposure is the best hope of mitigating the impact of Pb poisoning in children.

6.4.6. Lead pathways to humans and the environment

There are many different routes of human exposure to lead, but most often it happens through:

1. Inhaling lead particulates in airborne emissions from uncontrolled smelters and, in some countries, from cars using lead-enriched gasoline;

2. Being exposed to airborne dust in lead-contaminated soil;
3. Ingesting lead particles/dust through normal hand-to-mouth activities;
4. Picking up lead-dust-covered objects and taking them home;
5. Drinking water contaminated with lead from groundwater and/or airborne dust entering water supplies;
6. Eating food contaminated by lead dust accumulating on locally grown rice, fruits, and vegetables;
7. Using lead-painted toys, lead-bearing cosmetics (e.g., lipstick, etc.).

6.4.6.1 Lead poisoning through industrial processes

In most cases, human lead poisoning is caused by anthropogenic activities. Mining processes, for example, leave behind large tailings that usually contain substantial amounts of lead toxins. In economically developed countries, the environmental pollution regulations require the tailing systems to be covered with protective fabrics or other materials that prevent them from uncontrolled blowing by wind and leaching by rain. If abandoned tailings are uncontained, lead-contaminated dust can be blown and precipitated into the surrounding environment, and lead and other toxic metals can leach out and contaminate groundwater systems.

Today, the highest concentrations of lead in the air and the environment are usually found near mining and smelting facilities. In modern lead-smelting facilities, there is infrastructure in place that allows monitoring the level of pollutions, controlling and keeping it within the environmental and health standards. However, modification of older smelters is expensive, and as a result, many smelting plants, especially in developing countries, forego safety measures. In places where the old lead-smelting operations have already contaminated large areas, the contaminated soil must be removed and disposed of to ensure that contaminated water can be returned to safe consumption levels. If the equipment of smelting plants is not properly constructed to control and minimize pollution, the lead toxins can enter the surrounding agricultural fields and underground water systems, which can contaminate livestock and crops. Recent studies show that rice easily accumulates lead from groundwater and soil, and therefore, the distribution and consumption of rice grown in fields close to mining and smelter operations need to be under control.

6.4.6.2. Industrial lead contamination in China

China is an unprecedented example of rapid economic and industrial development, including exploration of its geological resources followed by severe environmental degradation and pollution. Most mines of lead in China are located in Hunan, Yunnan, Sichuan, Guangdong, Guangxi, Henan, and Anhui provinces. By 2007, large-scale lead mining and processing operations in Tianying in the Anhui Province, which accounts for almost half of the country's total lead production, caused air and soil contamination eight to ten times above national and

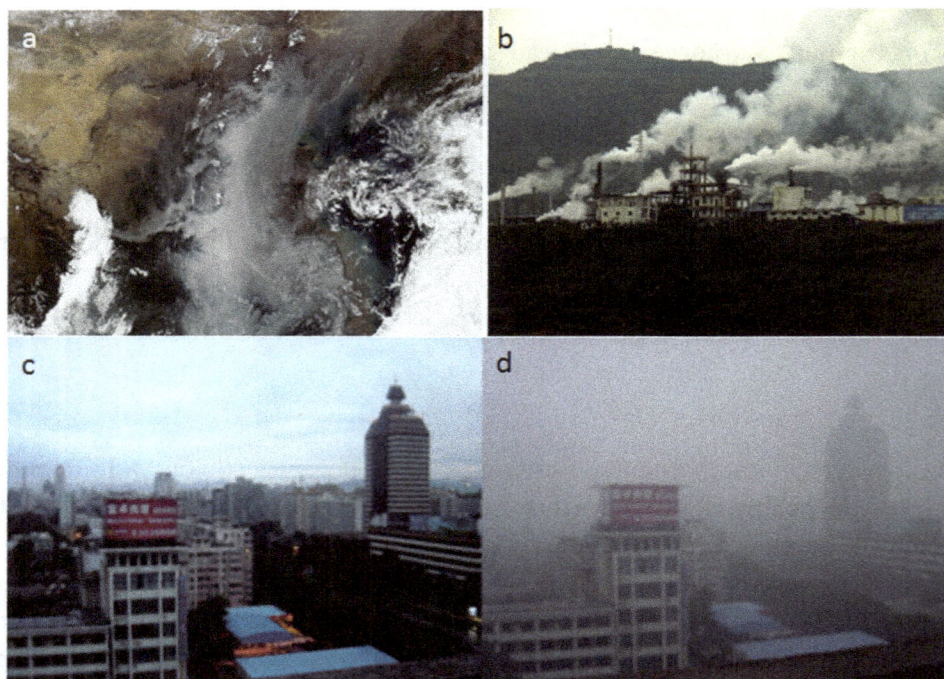

FIGURE 88 (a) Satellite image of a smog event in region of Tianying City, China: the milky white and gray covering the center, and the brighter whites at the left and right edges are clouds; (b) lead smelting factory in Tianying City, China; (c) clear sky in Beijing; (d) Beijing with smog cover

international standards (Fig. 88). Lead-dust contamination of locally grown vegetables and rice was twenty-four times higher than the national permissible level. Residents of Tianying and their children exposed to long-term lead contamination have suffered from mental deficiency, forgetfulness, irritability, loss of memory, and hallucinations. Tianying children poisoned with lead had lower IQ scores, short attention spans, learning disabilities, hyperactivity, impaired physical growth, and brain damage, and pregnant women have reported numerous cases of premature births and smaller and underdeveloped infants. The local administration suggested that all lead-processing firms must be shut down until their environmental impacts could be addressed.

Although the Chinese government has expressed concern over industrial pollution problems, high rates of childhood lead poisoning reflect the need to identify and control all other possible lead sources. In 2009, a lead-poisoning protest occurred in the town of Changqing, Shaanxi Province, where a Pb-Zn smelting factory of the Dongling Lead and Zinc Smelting Company had been in operation since 2003 (see Watts, 2009). During the six years of operation, the company produced more than 100000 tons of Pb and Zn and opened more than 2000 jobs for local residents. The Changqing residents, however, noticed that their children were having unexplained nose bleeding and memory problems. Eventually, it was recognized as Pb intoxication because in addition to those symptoms, 851 children from villages surrounding the factory had blood lead levels ten times higher than the standard level. Over 170 of those children were hospitalized with severe symptoms of lead poisoning. As a result of the protest, the Changqing factory was closed, but the ongoing fate of the poisoned residents and children remains unknown. Similar protests took place in Hunan Province, where children also were found to be Pb intoxicated, and more than 2000 children were lead poisoned in the Shaanxi and Hunan provinces (Watts, 2009).

The problem of Chinese children's exposure to lead is not over. Recent studies (e.g., van der Kuijp et al., 2013) have shown that 24% of the children in China were lead poisoned, with blood lead levels exceeding 100 mg/L during the period 2001–07, although many of them have never lived in the lead-mining regions. The children's blood lead level of 100 mg/L is 16% higher than the international standard. The research group concluded that China still has significant public health threats due to uncontrolled industrial pollution of the environment, with millions of children currently at risk of lead poisoning. They have also linked the persistently high level of children's lead poisoning to the rising growth of China's lead-acid and lead-battery industries, which are used in electric bicycle and automotive industry production. The increasing growth of lead industrial applications requires new regulations to create a safer environment and to protect the health of the next generation of China's citizens.

6.4.6.3. Other lead-related contaminations

Lead-enriched gasoline: past and present. From 1930 to 1970, lead additives to gasoline for obtaining higher octane ratings had raised the concern over vehicular pollutant emissions. At the same time, medical data and scientific studies indicated that lead exposure causes more neurotoxic and other adverse health effects than was previously thought. In 1973, the U.S. EPA announced a new phase-out program to gradually reduce the amount of lead in gasoline to 1.7 grams per gallon, and in 1985, "regular" gasoline contained only 0.5 grams of lead per gallon. In 1986, the U.S. banned leaded fuel altogether, and in 2000, it was also banned by the European Union. Statistical data showed that after leaded gasoline was banned in the U.S., the elevated blood lead levels decreased from 88% of children before the phase-out to approximately 1% in 2006. Now, in most parts of the world, the use of lead-enriched gasoline is forbidden, although in some developing and transitional countries, leaded gasoline is still being used. By 2009, only eleven countries from Africa, Asia, and South America still used lead-enriched gasoline, and among these, only three use it exclusively.

Cosmetics, paints, and other products. Special words should be said about lead-contaminated cosmetic products, particularly lipsticks. In 2009, the FDA released results of its studies of lipsticks produced by Procter and Gamble (Cover Girl brand), L'Oreal (L'Oreal, Body Shop, and Maybelline brands) and Revlon, which showed that they contain 0.09 to 3.06 ppm of lead. These levels of lead concentration appear to be four times higher than the levels found in their previous studies of the lipsticks from the same manufacturers in 2007; in that time, the maximum concentration of Pb was only 0.65 ppm. This situation shows that the FDA has not taken appropriate actions to protect consumers (for more information, see http://safecosmetics.org/article.php?id=223). The use of lead-based paints and lead-bearing cosmetics such as lipstick should be discontinued, and lead should not be used in the production of any food containers.

References

Bellinger, D. 2008. Very low lead exposures and children's neurodevelopment. *Therapeutics and Toxicology* 20:171–77.

Budtz-Jørgensen, E., D. Bellinger, B. Lanphear, and P. Grandjean. 2013. An international pooled analysis for obtaining a benchmark dose for environmental lead exposure in children. *Risk Analysis* 33:451–466.

Haefliger, P., Mathieu-Nolf, M., Laciciro, S., Ndiaye, C., Coly, M. et al. 2009. Mass lead intoxication from informal use lead-acid battery recycling Dakar, Senegal. *Environmental Health Perspectives* 117:1535–1540.

van der Kuijp, T.J., L. Huang, and C. Cherry. 2013. Health hazards of China's lead-acid battery industry: a review of its market drivers, production processes, and health impacts. *Environmental Health* 12:61–71.

Watts, J. 2009. Lead poisoning cases spark riots in China. *Lancet* 374:1596.

Web resources

http://minerals.usgs.gov/minerals/pubs/mcs/2012/mcs2012.pdf
http://safecosmetics.org/article.php?id=223

Image Credits

Figure 85a: Copyright © Rob Lavinsky (CC BY-SA 3.0) at http://commons.wikimedia.org/wiki/File:Baryte-Galena-Pyrite-203072.jpg.

Figure 85b: Copyright © Rob Lavinsky (CC BY-SA 3.0) at http://commons.wikimedia.org/wiki/File:Anglesite-113492.jpg.

Figure 85c: Copyright © Rob Lavinsky (CC BY-SA 3.0) at http://commons.wikimedia.org/wiki/File:Cerussite-18566.jpg.

Figure 85d: Copyright © Rob Lavinsky (CC BY-SA 3.0) at http://commons.wikimedia.org/wiki/File:Minium-232909.jpg.

Figure 87a: Copyright © Thester11 (CC by 3.0) at http://commons.wikimedia.org/wiki/File:LeadPaint1.JPG.

Figure 87b: Copyright © Snowjackal (CC BY-SA 3.0) at http://commons.wikimedia.org/wiki/File:100G115G130G150G.jpg?uselang=es.

Figure 87c: inductiveload, "60-40 Solder," http://commons.wikimedia.org/wiki/File:60-40_Solder.jpg. Copyright in the Public Domain.

Figure 87d: Copyright © Neep (CC BY-SA 3.0) at http://commons.wikimedia.org/wiki/File:GreenPaintBucketRome.jpg.

Figure 87e: "Sign Danger," https://openclipart.org/people/sheikh_tuhin/sign-danger.svg. Copyright in the Public Domain.

Figure 87f: Changlc, "Lead shielding," http://commons.wikimedia.org/wiki/File:Lead_shielding.jpg. Copyright in the Public Domain.

Figure 87g: Environmental Protection Agency, "Xray Machine," http://www.epa.gov/radtown/images/xray-machine.jpg. Copyright in the Public Domain.

Figure 87h: Shaddack, "Photo-CarBattery," http://commons.wikimedia.org/wiki/File:Photo-CarBattery.jpg. Copyright in the Public Domain.

Figure 88a: NASA, "2010 Smog over China," http://commons.wikimedia.org/wiki/File:2010_Smog_over_China.jpg. Copyright in the Public Domain.

Figure 88b: Copyright © High Contrast (CC by 2.0) at http://commons.wikimedia.org/wiki/File:Factory_in_China.jpg.

Figure 88c: Copyright © Bobak (CC by 2.0) at http://commons.wikimedia.org/wiki/File:Beijing_smog_comparison_August_2005.png.

Figure 88d: Copyright © Bobak (CC by 2.0) at http://commons.wikimedia.org/wiki/File:Beijing_smog_comparison_August_2005.png.

Review questions

1. What kind of lead-bearing minerals are used for Pb-metals production?

2. What is the chemical formula of the mineral galena? Based on its chemistry, describe why this mineral is the most useful for industry as well as the most toxic for the environment and public health.

3. What country is the world leader in lead production? List four more top countries in Pb production.

4. Why is lead metal in huge industrial and technological demand?

5. What is the majority of lead used for?

6. Explain primary and secondary industrial lead smelting and refining.

7. What kind of process is used for sintering during primary smelting?

8. Why should industrial slag storage be controlled by the smelting operation's authorities?

9. How toxic is lead?

10. Who are the most vulnerable to lead poisoning?

11. Why are children more susceptible to long-term lead exposure than adults?

12. How high is the EPA's maximum level of Pb in human blood? How high is the EPA blood Pb permissible for children?

13. After what level of Pb concentration will the signs and symptoms of lead poisoning appear in adults? In children?

14. What kind of symptoms will be developed in lead-poisoned adults?

15. What is the EPA's standard level of lead in the air? In drinking water?

16. How small is a safe dose of Pb exposure for children and for adults?

17. List the major routes of humans' exposure to lead.

18. Describe lead poisoning through industrial processes.

19. Describe the adult and child Pb poisoning in modern China. Why should we be concerned?

20. Discuss lead-enriched gasoline: past and present.

21. Is lead in cosmetics dangerous for health?

Quizzes (see answers on page 307)

1. Lead minerals include ninety-four species, within which galena is the major mineral for lead production. The chemical formula of galena is:
   a. Pb_3O_4.
   b. PbS.
   c. $PbCO_3$.
   d. None of the above.

2. Through which of the ways listed below can people get lead poisoning?
   a. Scraping lead paint off of a wall.
   b. Eating food stored in lead-glazed pottery.
   c. Trying to repair lead-acid batteries at home.
   d. None of the above.

3. Which of the following is not a step that the U.S. has taken since the 1970s to reduce humans' exposure to lead?
   a. Removed lead from use in gasoline.
   b. Removed lead from use in car batteries.
   c. Advising against using lead-containing paint.
   d. All of the above.

4. Which of the following symptoms are related to children's lead poisoning?
   a. Low IQ.
   b. Weight loss.
   c. Sluggishness.
   d. Loss of appetite.
   e. Irritability.
   f. All of the above

5. The EPA recommends the following Pb level in water and air:
   a. $1.5\ \mu g/m^3$ in both.
   b. $5\ \mu g/m^3$ in both.
   c. $1.5\ \mu g/m^3$ in water and $7\ \mu g/m^3$ in air.
   d. None of the above.

Asbestos—Hazardous Minerals of the Serpentine and Amphibole Groups

Chapter 7

What is asbestos? 7.1.

"Asbestos" is a commercial name for some industrially used minerals of the serpentine and amphibole groups, which have a fibrous texture with a 3:1 length-to-width ratio of each fiber. Altogether, there are only six asbestos-type minerals: chrysotile in the serpentine group and crocidolite, amosite, anthophyllite, tremolite, and actinolite in the amphibole group. These minerals are classified as hazardous because exposure to them may have irreversible adverse effects on human health, including various fatal lung diseases and a special form of lung cancer—mesothelioma. However, these six minerals do not always exhibit fibrous structure. For example, non-fibrous actinolite, tremolite, and anthophyllite of the amphibole group are found in many rocks, and therefore, they are not considered to be hazardous, asbestos-like substances.

Minerals of the serpentine group 7.2.

7.2.1. Chemistry and structure

The serpentine group includes three major minerals—chrysotile, antigorite, and lizardite—which have the same chemical formula [$Mg_3Si_2O_5(OH)_4$], where Mg may be substituted for Fe, but have different structures (Fig. 89). All of them are secondary minerals that are formed

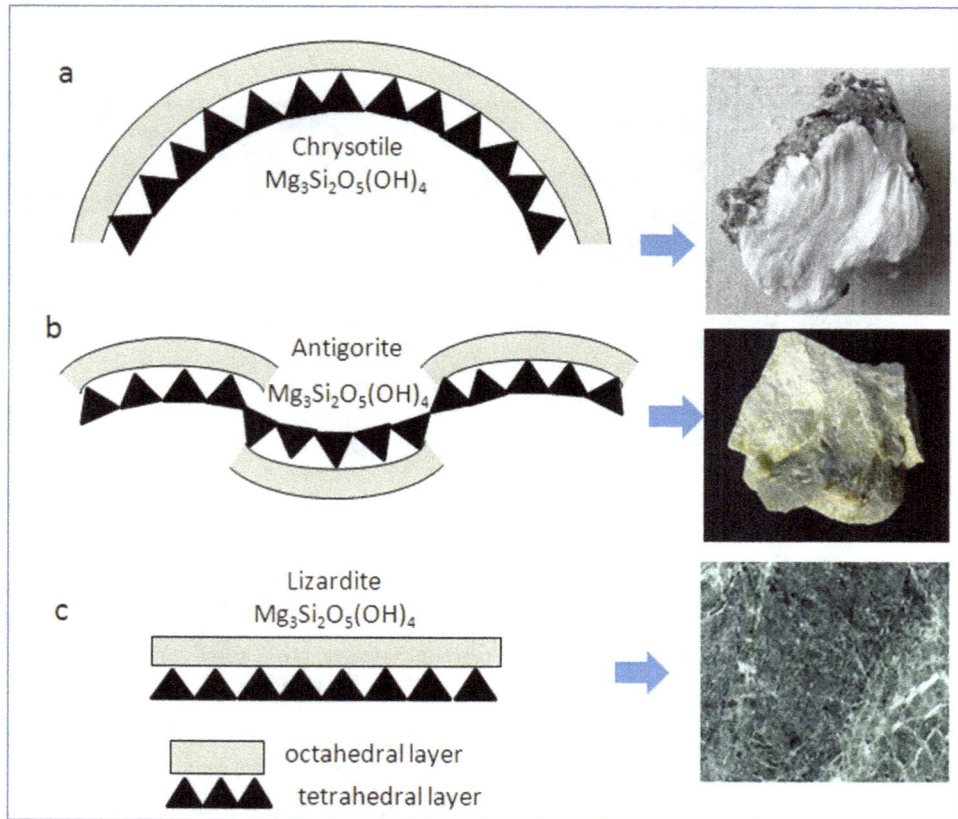

FIGURE 89 Minerals of serpentine group: chrysotile, antigorite, and lizardite

by replacement of olivine (Mg_2SiO_4) during metamorphism of the peridotites and pyroxenites, the main rocks of the oceanic crust and the Earth's mantle. Three basic chemical reactions may transform olivine to serpentine minerals:

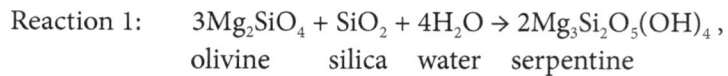

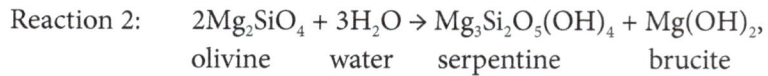

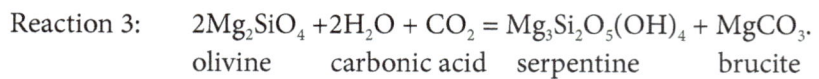

Reaction 1: $3Mg_2SiO_4 + SiO_2 + 4H_2O \rightarrow 2Mg_3Si_2O_5(OH)_4$,
olivine   silica   water   serpentine

Reaction 2: $2Mg_2SiO_4 + 3H_2O \rightarrow Mg_3Si_2O_5(OH)_4 + Mg(OH)_2$,
olivine   water   serpentine   brucite

Reaction 3: $2Mg_2SiO_4 + 2H_2O + CO_2 = Mg_3Si_2O_5(OH)_4 + MgCO_3$.
olivine   carbonic acid   serpentine   brucite

7.2.1.1. Why does chrysotile belong to the asbestos category, while antigorite and lizardite do not?

Chrysotile—white asbestos. Chrysotile is the only mineral of the serpentine group that is considered to be asbestos. It has a fibrous structure and exhibits small, silky fibers grown parallel to each other in veins or as a matted mass within rocks (Fig. 89a). Chrysotile, like other minerals of the serpentine group, belongs to sheet silicates (see Chapter 2), and generally, such a structure would favor the formation of flakes or sheets rather than fibers. The fibrous texture of the chrysotile is explained by a mismatch in the positions of the cations and the anions that causes bending of the tetrahedral and octahedral layers (Fig. 89a). During growth, the curved layers roll themselves into hollow (tubular) fibers, which are elongated in the direction parallel to the chrysotile crystallographic a-axis (Fig. 90). The internal diameter of the chrysotile asbestos tube-like fibers is 5–6 nm, and its external diameter is up to 50 nm.

Antigorite and lizardite. Although in the crystal structure of these minerals, there is also a mismatch in the positions of the cations and the anions between layers, they are not rolled into the hollow fibers. In antigorite structure, the octahedral layers contain reversals in the facing of the tetrahedral layer, as shown in Fig. 89b, and lizardite has a distortion of the tetrahedral mesh from its ideal geometry to match the dimension of the octahedral layer, as shown in Fig. 89c. As a result of such structural arrangements, neither antigorite nor lizardite form fiber-like or any other distinguishing crystal shapes; they usually occur as fine-grained, greenish masses in serpentinite rocks.

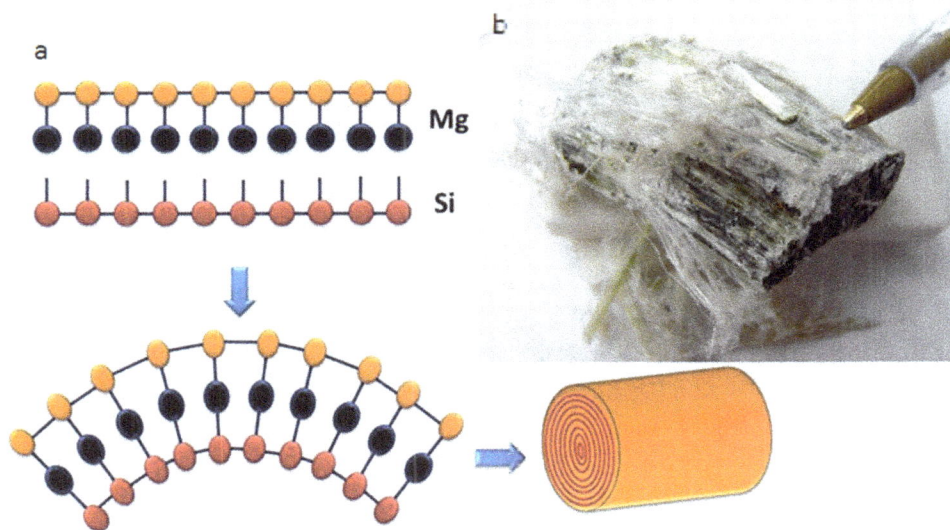

FIGURE 90 (a) Fibrous nature of chrysotile is due to mismatch in the position of the silica ions (tetrahedral layer) and magnesium ions (octahedral layer); (b) chrysotile asbestos

7.2.2. Geological occurrences of chrysotile asbestos

Serpentinite is a metamorphic rock that consists of minerals of the serpentine group, and it is a product of the recrystallization of the mantle rocks peridotite and pyroxenite at low pressures (<0.5–1 GPa) and temperatures (<500°C) in the presence of hydrous fluids. Serpentine minerals may also form at temperatures of 150–300°C by hydrothermal recrystallization of magmatic and metamorphic rocks of mafic and ultramafic bulk chemistry. The serpentinites consist of a green mass of antigorite and lizardite that is usually cut by veins of chrysotile asbestos. The mass of fibrous chrysotile is sometimes called "mountain leather" or "mountain cork" because of its appearance.

Sepentinites are found on all continents, mostly within ophiolite formations that represent fragments of an ancient oceanic crust. Other rocks that contain chrysotile asbestos are located within Archean greenstone belts and carbonate-bearing magmatic and metamorphic formations. Although chrysotile and the other varieties of serpentine minerals are ubiquitous, they are not necessarily present in quantities or the quality required for their commercial exploitation. The economically significant chrysotile asbestos deposits are known within ophiolite formations in Canada, Russia, Kazakhstan, Italy, Greece, Cyprus, and the U.S., while chrysotile asbestos deposits in carbonate rocks are found in South Africa. Serpentinites rich in chrysotile asbestos veins are abundant in greenstone belts of Canada. In the U.S., natural chrysotile asbestos deposits occur abundantly within serpentinites and sepentinized peridotites in northern Vermont and in Humboldt County and areas of San Benito, Monterey, and El Dorado counties in California.

7.3. Minerals of the amphibole group

7.3.1. Chemistry and structure

Amphiboles include an extensive and chemically complex group of hydrous silicate minerals with a general chemical formula:

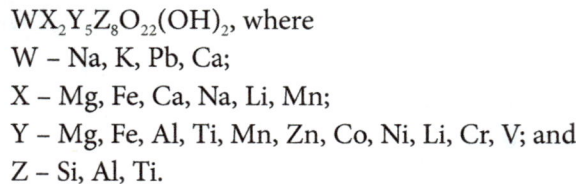

$$WX_2Y_5Z_8O_{22}(OH)_2, \text{ where}$$
W – Na, K, Pb, Ca;
X – Mg, Fe, Ca, Na, Li, Mn;
Y – Mg, Fe, Al, Ti, Mn, Zn, Co, Ni, Li, Cr, V; and
Z – Si, Al, Ti.

Most of the amphiboles are secondary minerals because they are formed during metamorphic recrystallization of pyroxenes, which are abundant in lower continental crust, oceanic crust, and upper mantle. Transformation of pyroxenes to amphiboles occurs at temperatures of ~500–700°C in the presence of fluids. The amphiboles may also be formed as a product of more complicated chemical reactions between different silicate minerals, but in all cases heating and fluid are required. Within the large family of amphiboles, only five exhibit asbestos-like textures: crocidolite, amosite, tremolite, actinolite, and antophylite, although there are many species of tremolite, actinolite, and antophylite that are not necessarily asbestos-like.

Crocidolite [$Na_2(Fe^{2+},Mg)_3Fe^{3+}_2Si_8O_{22}(OH)_2$] is one of the amphiboles of the riebeckite group that exhibits asbestos-like texture, and it is often called blue asbestos because of its distinctive blue color and straight, brittle fibers (Fig. 91a). The fiber dimensions are well recognized in the image (see Fig. 91b) obtained with the aid of a scanning electron microscope (SEM).

Amosite (brown asbestos) belongs to the amphiboles of the cummingtonite-grunerite solid-solution series:

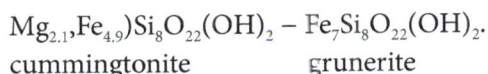

$$Mg_{2.1},Fe_{4.9})Si_8O_{22}(OH)_2 - Fe_7Si_8O_{22}(OH)_2.$$
cummingtonite   grunerite

Amosite has a light brownish to grayish color (Fig. 91c), and it consists of straight, needle-like, brittle fibers, which are well recognized in both hand specimens and with the SEM (Fig. 91d).

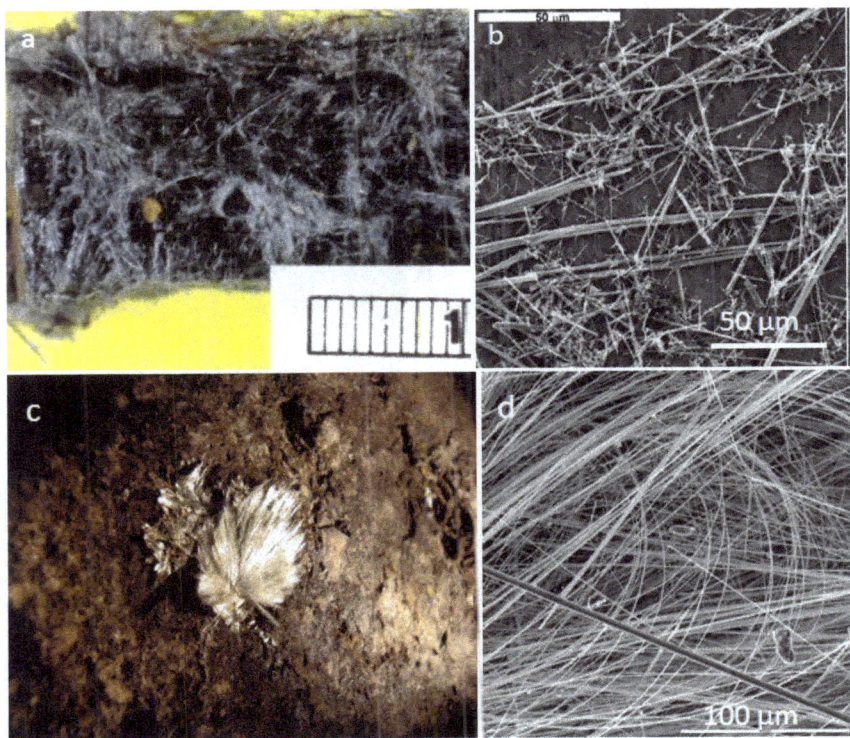

FIGURE 91   (a) Crocidolite, blue asbestos, fluffed to show the fibrous nature of the mineral (from mine at Wittenoom, Western Australia, photo by J. Hayman); (b) scanning electron microscope (SEM) image of standard crocidolite; (c) amosite, brown asbestos; (d) SEM image of standard amosite

Tremolite and actinolite are amphiboles that represent the end members of the tremolite-actinolite solid-solution series:

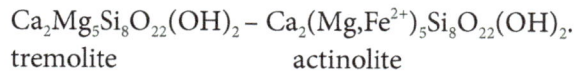

$$Ca_2Mg_5Si_8O_{22}(OH)_2 - Ca_2(Mg,Fe^{2+})_5Si_8O_{22}(OH)_2.$$

tremolite   actinolite

The asbestos-like tremolite and actinolite usually have a white color (Figs. 92a–d), which is reflected in their industrial name—"white asbestos"—similar to chrysotile asbestos, and this often causes confusion among non-geologists. Both tremolite and actinolite are frequently associated with talc, $[Mg_3Si_4O_{10}(OH)_2]$, and vermiculite, $[(Mg,Ca,K,Fe^{2+})_3(Si,Al,Fe^{3+})_4O_{10}(OH)_2O_4 \cdot H_2O]$, with the latter being a form of mica that changes its volume during heating.

Anthophyllite is magnesium-rich amphibole, $\{Mg_2\}\{Mg_5\}Si_8O_{22}(OH)_2$, and Mg may be substituted for Na, K, Ca, and Fe. Similar to tremolite and actinolite, anthophyllite (Fig. 92e) is often associated with talc and vermiculite. It has a dark green and brownish color, and when it is powdered, it is pale green. High-resolution SEM images clearly demonstrate its fibrous structure (Fig. 92f).

FIGURE 92   Amphibole asbestos: (a) silky fibers of tremolite developed on muscovite; (b) tremolite asbestos, Death Valley, California; (c) anthophyllite from Tuusniemi, Finland; (d) anthophyllite asbestos, Georgia. SEM images: (b) and (d)

Amphiboles belong to the group of double-chain silicates (Nesse, 2000), and their fibers in the asbestos-like species follow the elongation of their prismatic shape. The schematic structure of amphibole shown in Fig. 93 consists of two ribbons of silicate tetrahedrons placed back-to-back with a layer of cations (usually metals) between them. Such an arrangement allows the individual fibers to be strong and durable.

Asbestos-like amphiboles usually do not show any curved fibers except for those that are very long with a length/diameter ratio of 100:1. Observation of many fibers is often necessary to determine whether a sample of interest is an asbestos-like amphibole.

7.3.2. Geological occurrences of amphibole asbestos

Amphiboles, including amphibole asbestos, are major minerals of metamorphic rocks, amphibolites. Their protoliths (primary "parent" rocks) are volcanic and intrusive rocks of mafic and ultramafic composition (e.g., basalts, gabbro, some peridotites, and pyroxenites). These

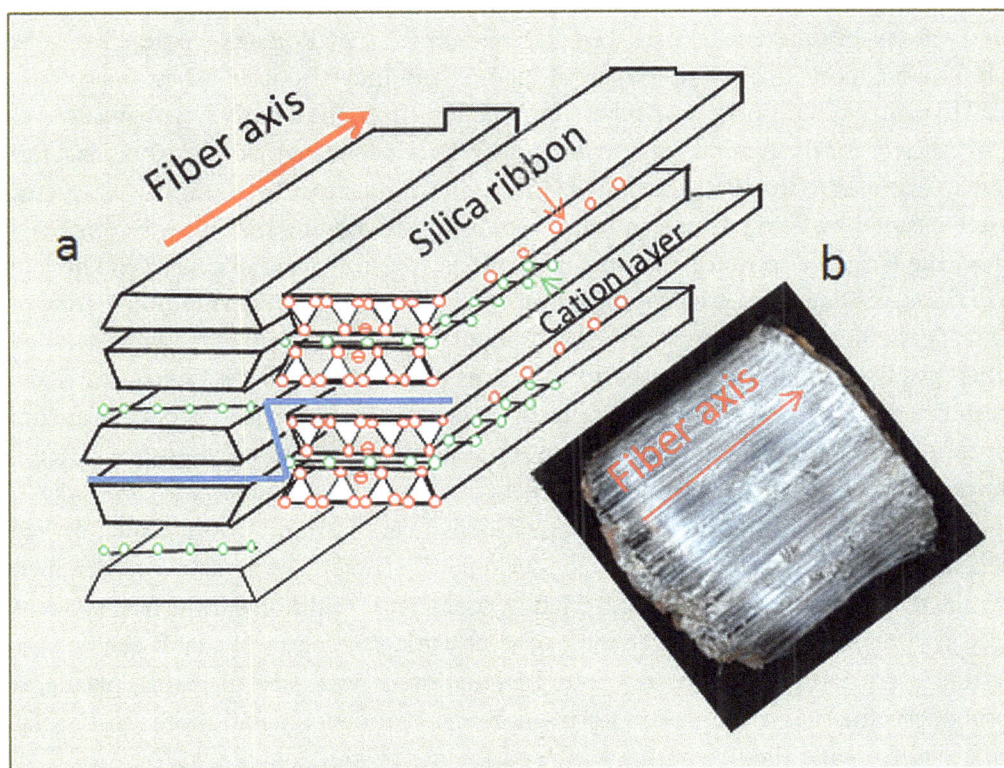

FIGURE 93   (a) Schematic model of amphibole structures showing series of fibers composed of SiO2 ribbons and cation layers; (b) amphibole crocidolite (blue asbestos) from Pomfret Mine, Vryburg, from Mineralogical Museum, Bonn, Germany

rocks compose the Earth's oceanic crust and upper mantle (chiefly peridotites and pyroxenites). Amphiboles are also common in magmatic rocks of intermediate to felsic composition (e.g., andesites, diorite, granodiorites, and dacites), which are abundant in the Earth's lower continental crust. Deposits of amphibole asbestos commonly occur in veins where fibers grow up perpendicular to the walls of the vein. Amphibolites are in abundance on all continents, and amphibole asbestos deposits of economically viable concentrations are found in in Russia, Canada, Australia, Italy, Bolivia, and South Africa. In the U.S., most natural amphibole asbestos deposits occur in the states of Alaska, California, Montana, Virginia, and Washington.

7.4. Asbestos uses and production

Asbestos was known to the ancient Greeks as a non-flammable natural material, and the name asbestos means "inextinguishable." They used asbestos to produce fabrics used in clothing and in a variety of different textiles. Legends mentioned that Romans would clean asbestos tablecloths after meals by throwing them in the fire, and they also used asbestos as a building material to protect their homes and other constructions from fires. Early Egyptians used cloths made of asbestos for the funeral dress to wrap their dead, believing that they would last forever. The Italian explorer Marco Polo, during his travels in China, described "miraculous garments" that were cleaned by being placed in fires—today, we can assume that they were made from asbestos. The Greeks also noted that this miraculous asbestos material caused sickness of the lungs in the slaves that wove asbestos into cloth, but the "magical" properties overshadowed the harmful effects, and nobody thought seriously about the adverse health effects.

Other physical properties of asbestos, such as the strength of its fibers, coupled with its resistance to heat made it one of the most important technological products during the Industrial Revolution in the nineteenth century. Asbestos began to be intensively used as an insulation material for steam pipes, boilers, ducts, and other high-temperature products. Later, many other amazing physical properties of asbestos fibers were discovered, such as its high tensile strength, resistance to corrosion, chemical inertia (except for acids), adsorption capacity, low electrical conductivity, and high friction coefficient. Additional industrial applications included its use as a spray for the coating of steel work, concrete walls, and ceilings for fire protection, corrosion protection, and sound insulation. It was used in paints, plastic, resin, and rubber products, as a foundation for many cement products to build walls and roofs, and as a material to make tiles. Asbestos was used for the production of textiles, thread, cloth, tape, and rope, and it was also in great demand for fabrication of wrapping, asphalt coating, electrical insulation, friction products in brakes and clutch pads, and in the electronic and printing industries. About 3,000 asbestos applications or products were used between 1930 and 1970—the peak period of world asbestos production and consumption—until it became clear that asbestos is a carcinogenic material.

Asbestos mining operations are of the open-pit type and use the bench drilling technique. After excavation from the pit, asbestos ore is usually delivered to special crushing facilities, where the ore is crushed to a nominal size and dried. The extraction of asbestos fibers goes through a series of additional crushing operations followed by separation of the individual fibers and collection of them in a vacuum system. Various secondary operations are used to collect fibers of specific lengths and to remove all non-fibrous mineral dust.

During the period from 1960 through 2000, Canada and Russia were leading countries in the world's asbestos mining production, with maximum 1.5 mln. tons by Canada in 1970 and maximum 2.4 mln. tons by Russia in 1900 (Fig. 94). During roughly the same period, from 1980 to 2003, many significant producers of asbestos such as Australia, Greece, Italy, Switzerland, and the U.S. ceased asbestos production because they recognized its adverse effects on human health. The U.S. completely stopped asbestos mining production in 2002, and now the U.S. imports only a limited volume of asbestos to satisfy some manufacturing needs. As a result of these changes, by 2003, the number of asbestos-producing countries had decreased significantly,

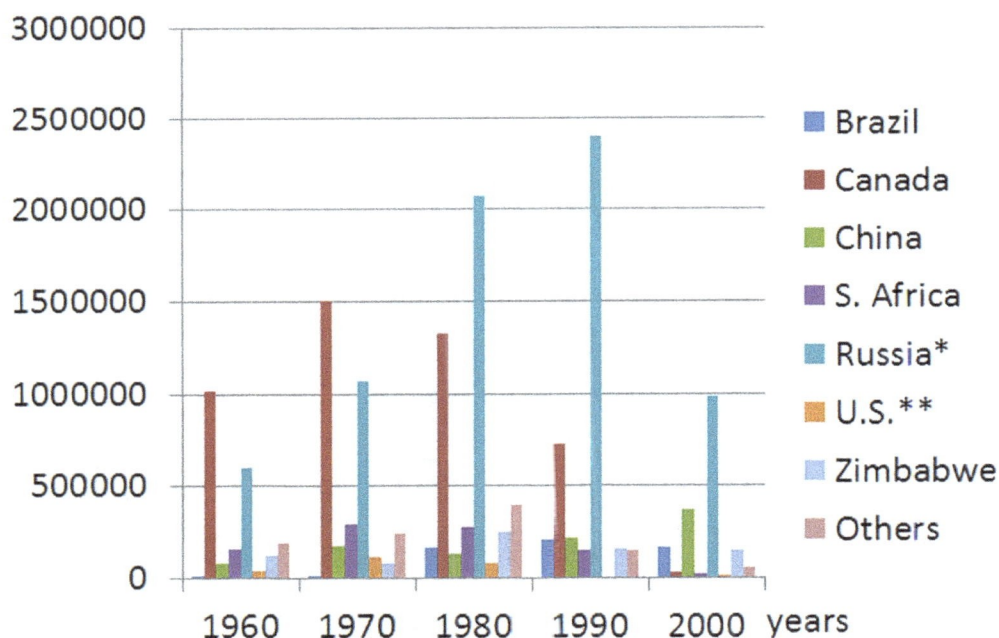

FIGURE 94 Asbestos mining production (in tons) from 1960 through 2000

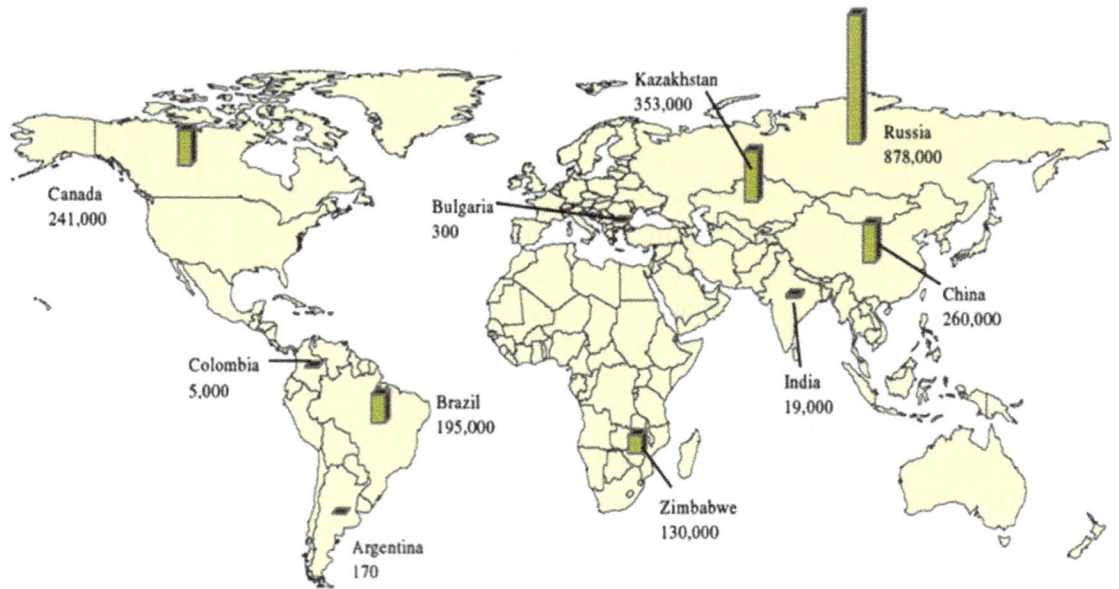

FIGURE 95 World map of asbestos mining and production in 2003; volume of asbestos is given in tons

and overall, asbestos production was also decreased (in mln. tons) to 0.88 by Russia, 0.35 by Kazakhstan, 0.26 by China, 0.24 by Canada, 0.19 by Brazil, and 0.13 by Zimbabwe, followed by others with production volumes as small as 0.003–0.019 (Fig. 95); (Vitra, 2005). During the last decade, Russia has remained a leading producer (1 mln. tons) of asbestos, followed by China (0.44 mln. tons), Brazil (0.3 mln. tons), and Kazakhstan (0.24 ml. tons); (see Fig. 96).

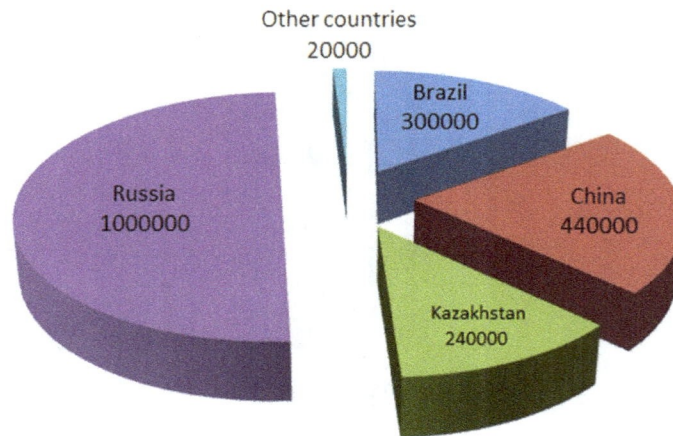

FIGURE 96 World asbestos mining production (in tons) in 2012

7.5.1. The adverse effects of asbestos and its pathway to humans

Although there were a lot of historical facts pointing toward possible connections between certain lung diseases and the inhaling of asbestos fibers, the first scientific publication confirming such relations appeared in 1924 (Cooke, 1924). Asbestos-caused death was confirmed by the autopsy of a 33-year-old worker in an asbestos factory, who had died from lung fibrosis caused by a prolonged occupational exposure to asbestos.

Later, studies of asbestos properties coupled with monitoring of different age groups of workers exposed to asbestos provided convincing information that inhalation of high doses of asbestos dust or fibers may cause a variety of functional impairments. By the 1940s, the risk of asbestos exposure was widely recognized by mining and manufacturing companies, but unfortunately, it was many years before they began to implement measures to prevent or minimize the exposure of workers to this deadly material.

Once inhaled into the lungs, tiny asbestos fibers stay in the body for a long time, eventually causing respiratory diseases such as asbestosis and the far more dreaded and incurable disease mesothelioma, which is a specific fatal form of lung cancer (Fig. 97). The time lag between significant inhalation of asbestos and the adverse health symptoms can be as long as fifteen to thirty or more years, so asbestos-related diseases tend to remain invisible and silent. Asbestos fibers do not dissolve in water, and they are resistant to heat and chemicals. When asbestos fibers are inhaled, they move through the air passages of human body, and finally, the smallest particles and fibers may be deposited in the deeper part of the lungs.

Asbestosis is a chronic, progressive disease that can eventually lead to disability or

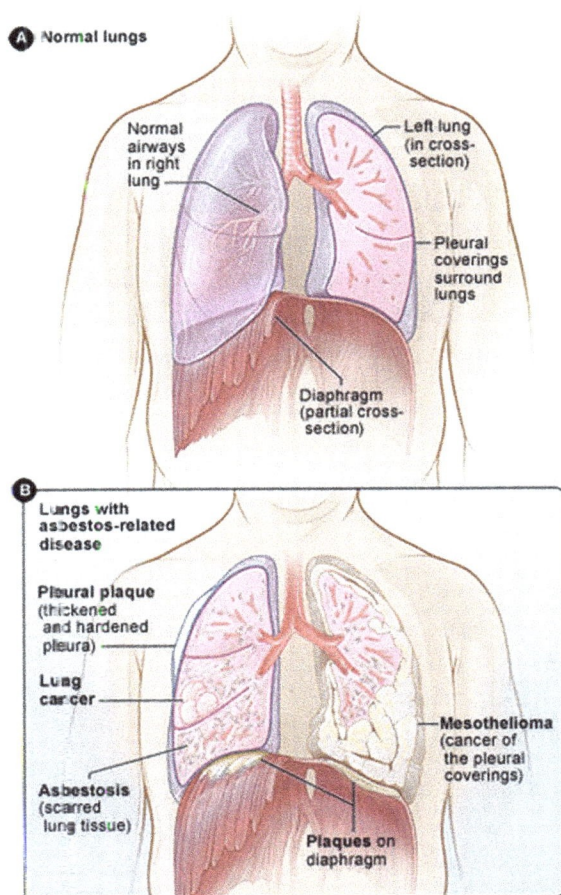

FIGURE 97 Asbestos-related diseases in the human body

death in people exposed to large amounts of asbestos, usually in their workplace, over a long period of time. Inhaled asbestos fibers scar the lung tissues, and when the scarring spreads widely, it causes a shortage of air to breathe. Symptoms include shortness of breath accompanied by coughing, chest pain, and a dry, crackling sound in the lungs when inhaling. In most known cases, asbestosis in workers has occurred at least fifteen years (in some cases more) after the person was first exposed to asbestos, and the time required for the manifestation of symptoms depends on many other factors—for example, the size of the fibers inhaled, tobacco smoking/ non-smoking, and other individual health conditions. There is no effective cure for asbestosis, and it usually leads to disability or death.

It is believed that after inhalation, the longer asbestos fibers are usually expelled, but some of them may be coughed out in a layer of mucus in the throat, where they are swallowed into the stomach (see Fig. 97). According to some existing concepts, asbestos fibers may penetrate the cells of the stomach and intestines, some may penetrate all the way through and enter into the bloodstream, and some may become lodged in the stomach-intestine system and in the alveoli (small air sacs or cavities) in the lungs, causing cancer of some internal organs and/or mesothelioma. However, this question is still under discussion between medical scientists and practitioners.

Cancers related to asbestos include cancer of the lung, gastrointestinal tract, kidney, and larynx. The most common symptoms of lung cancer are coughing, shortness of breath, persistent chest pains, hoarseness, and anemia. Many studies have concluded that the latency period of asbestos-related cancer is affected by the level of asbestos exposure coupled with other carcinogens such as cigarettes, suggesting that the risk multiplies in smokers. The U.S. Department of Labor (DoL) data show that asbestos-contaminated smokers and non-smokers have different levels of lung cancer risk (Fig. 98).

People who are exposed to asbestos and continue to smoke have fifty to ninety times higher risk of contracting lung cancer than people who have not been exposed to asbestos and do not smoke. Smokers (with no asbestos) have ten times higher risk, and asbestos-contaminated non-smoking individuals have five times higher risk than the general population.

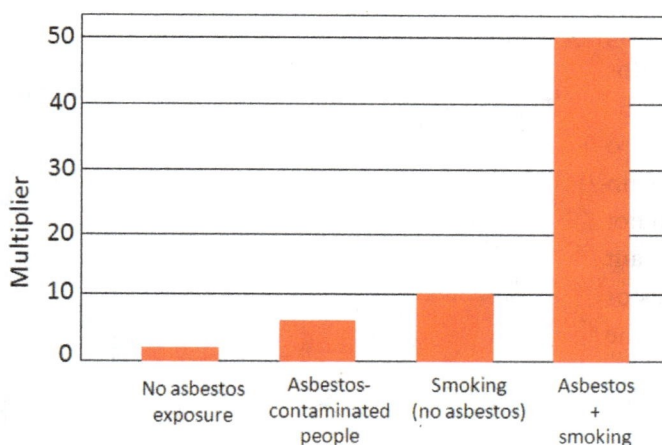

FIGURE 98 Diagram showing the risk of lung cancers within different groups of people

Mesothelioma is a specific (uncommon) form of cancer that involves the cells (called the mesothelium) that line the lungs. This cancer is extremely invasive and spreads quickly, crushing the lungs, and the process is painful and always fatal because the patient cannot breathe. Approximately 70–80% of people who are diagnosed with mesothelioma have been heavily exposed to asbestos at work. Usually, the disease shows its symptoms about thirty to fifty years after exposure, so most people with it are over the age of sixty to sixty-five. In the U.S., during World War II, asbestos was widely used in ships as insulation for pipes, boilers, steam engines, and turbines, and as a result, for every thousand former workers, about fourteen died of mesothelioma and an uncounted number from lung cancer and asbestosis. Currently, 2000–3000 new cases are diagnosed each year in the U.S. among former shipyard workers and those who worked in asbestos mines and mills, asbestos factories, and the asbestos-related heating and construction industries.

7.5.2. Asbestos in the environment

Asbestos is present everywhere in the air, natural rocks, soils, drinking-water reservoirs, and man-made products. Once released into the environment, small, airborne asbestos fibers can travel significant distances by wind or water before settling. Amphibole asbestos fibers are stable in water and can persist under typical environmental conditions, while chrysotile asbestos fibers may be subject to chemical alteration in acidic media, which leads to the dismembering of its shorter fibers. Asbestos minerals often occur in association with other hydrous silicates such as vermiculate and talc, and they often intergrow in such a way that it is difficult to distinguish one from the other. These complex mineralogical substances may be found in soils that were formed from erosion of asbestos-bearing rocks.

The level of asbestos particles in the air may be very high near naturally exposed asbestos-bearing rocks, open-pit mines, quarries and tailings, and nearby buildings (if they contain asbestos products in their walls and roofs) and especially those that are mechanically damaged. Asbestos is also prevalent near waste landfill sites if demolished asbestos materials and dust are not properly covered up or stored to protect from wind erosion. However, asbestos minerals in geological settings are not hazardous if they are left undisturbed, and if asbestos-containing materials are solidly embedded or contained in industrial products, exposure will be negligible.

Asbestos can contaminate aquatic reservoirs through deposition of the airborne particles, weathering of the natural rocks, industrial discharges, atmospheric pollution, and asbestos-containing pipes used in drinking water distribution systems. Usually, asbestos-containing pipes do not contribute very much asbestos into the water unless they are deteriorated. No convincing data have been collected to prove the carcinogenic effect of the ingested asbestos fibers in populations supplied with highly asbestos-contaminated drinking water (Polissar et al., 1984). Also, no convincing correlation has been found between cancer mortality and ingestion of asbestos fibers from drinking water within animals. Extensive experimental researches related to feeding animals with asbestos-contaminated food also have not shown consistently that ingested asbestos increases the incidence of malignant tumors of the animal gastrointestinal systems.

7.5.3. Asbestos in the home

Asbestos hazards may also be found in homes, but one cannot determine whether a material contains asbestos simply by looking at it unless she/he has been trained to do so. What people should know is that many houses built between 1930 and 1950 in the U.S. probably have attic or wall insulations made from vermiculite that contains asbestos. If this is the case, the vermiculite insulation should remain undisturbed, boxes or other items should not be stored in the near vicinity, children should not be allowed to play in an attic, and homeowners should not attempt to remove the vermiculite insulation by themselves.

Residential roofing and siding shingles also may be made of asbestos cement, and textured paint and patching compounds may have been used for the joints between walls and ceilings. Artificial ashes and embers that you may buy for use in a gas-fired fireplace to make it look more natural may also contain asbestos. Finally, asbestos may be found in some vinyl floor tiles and in the backing on vinyl sheet flooring and adhesives.

If people are planning to remodel their homes, they should keep in their minds that remodeling can disturb building materials and release airborne asbestos. It is recommended that homeowners enlist the services of an accredited asbestos professional who knows where to look for asbestos and who will take samples for analysis. Taking samples yourself and sending them to an asbestos laboratory is not recommended because if the sampling was done incorrectly, it can be more hazardous than leaving the material intact. Further in this chapter, it will be shown that there is no known safe level of exposure to asbestos, and most of the vermiculite used for attic insulation in the United States originated from a mine near Libby, Montana, where vermiculite was naturally contaminated with tremolite asbestos.

7.6. Is there any "safe level" of asbestos exposure?

Typical concentrations of asbestos fibers with lengths ≥5 µm in ambient air, where there is no asbestos mining, milling, or production, may be about 0.0001–0.00001 fiber/mL. According to regulations issued by the U.S. Occupational Safety and Health Administration (OSHA), asbestos concentration in workplace air is limited to 0.1–0.2 fiber/mL to protect against the development of pulmonary fibrosis and cancer (for details, see https://www.osha.gov/SLTC/asbestos/index.

html). A study of the indoor air of homes, schools, and other buildings that contain asbestos materials suggests that the average asbestos concentration is ~0.0001 fiber/mL (Lee et al., 1992).

Despite intensive studies and regulations related to asbestos, in reality, there is no convincing data that could be used to determine a "safe level" of asbestos exposure. However, the more an individual is exposed to asbestos and the more fibers that are inhaled and remain in the body, the more likely that this individual will develop asbestos-related diseases. Thus, people who have been exposed to asbestos frequently over a long period of their lives have a higher probability of becoming ill with asbestosis, lung cancer, and mesothelioma. Factors involved with asbestos exposure may be formulated as follows:

1. How much asbestos was an individual exposed to?
2. How long was the individual exposed?
3. What were the size, shape, and chemical composition of the asbestos fibers?
4. Was there any presence of other lung diseases before asbestos exposure?
5. Did the individual smoke during the period of exposure to asbestos?

Although no clear "safe" asbestos exposure dose has been established by scientific studies, the International Labor Organization (ILO) has accepted that an exposure limit may be defined as an average value measured by the number of asbestos fibers per-unit-volume, with only fibers of a length >5 μm being recommended for measurement. Exposure limit varies considerably between different countries and may range between 0.1 fiber/mL to 2 fibers/mL for four to eight hours of exposure.

Who is at risk?   7.7.

Although asbestos is now largely banned in many countries, currently about 125 million people in the world are exposed to asbestos at their workplaces, and at least ~90000 people die each year from occupational asbestos-related cancer and asbestosis that they developed through their exposure in workplaces (see for details www.mesotheliomhelp.org/mesothelioma/statistics). In general, those workers who are involved in mining and quarrying, the manufacturing of asbestos products, asbestos spraying and insulation, demolition of old buildings, construction, shipbuilding, the heating trades, pipefitting, and sheet-metal work are at risk of definite exposure. The workers in transport and railway systems, ship engine crews, and firefighters, as well as workers in the oil refining, chemical, paper, and metal industries and workers in car repair and general maintenance are at risk of so-called "probable and possible" exposure. People who work in offices, agriculture, forestry, health care, education,

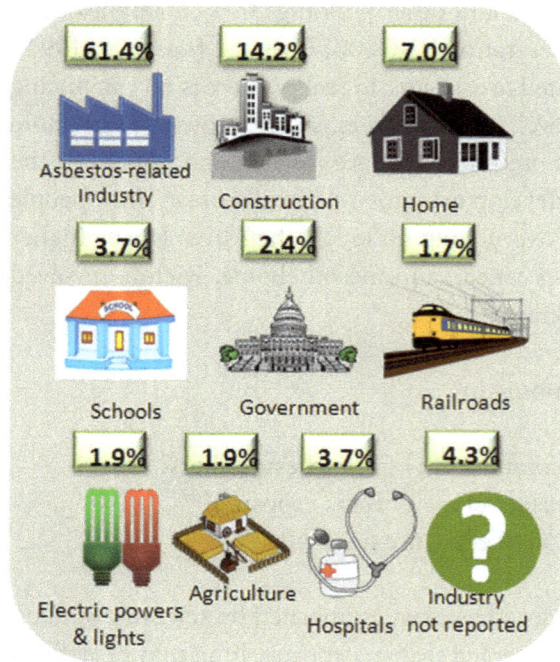

FIGURE 99 Most frequently recorded places of work on death certificates with cause of death attributed to mesothelioma

and telecommunications are classified as a group unlikely to experience exposure.

Statistical data (www.mesotheliomhelp.org/mesothelioma/statistics) have shown that the most frequently recorded occupations with regard to mesothelioma death are asbestos-related industries (61.4%), construction (14.2%), asbestos exposure at home (7%), and others ranging between 1.7% and 4.3%, including school and government buildings, railroads, electric power and light, agriculture, hospitals, and non-reported industries (see Fig. 99).

The greatest concerns are smokers and children exposed to asbestos. Smokers should immediately give up smoking if they are exposed to asbestos because it might reduce their chances for developing cancer. Although children have lower chances than adults to be exposed to asbestos, mesothelioma has occurred in young people, and the histories of the disease have often shown that these young people had parents who worked in asbestos mining or milling factories. Therefore, it is possible that the young people were exposed in their childhood to asbestos dust brought home on the working clothes and shoes of family members. This is why one of the important regulations for asbestos-industry workers is that they should not wear their work clothing outside of their workplaces. In cases where they do not change their clothing, they must clean the clothing by vacuuming or any other method that removes dust without causing the asbestos fibers to become airborne.

7.8. Controversial discussions about the adverse effects of asbestos on human health

Although asbestos toxicity was well recognized many decades ago and many countries have banned asbestos mining and some asbestos products, there is still a lot of discussion claiming that some forms of asbestos are not dangerous for human health. One group of scientists has suggested that amphiboles have more potential for mesothelioma risk than chrysotile because amphiboles have longer and more durable fibers and tend to stay in the lungs longer, while chrysotile has only curly and short fibers. Another group of scientists concluded that such parameters as the size of fibers are important for evaluation of the carcinogenic potency, but their chemical compositions should not necessarily be considered as a carcinogenic factor. This group also insists that all types of asbestos may be equally carcinogenic for human beings and that there is no known exposure threshold. The main parameters of asbestos fibers are discussed below to provide better understanding of whether there are "non-harmful" types of asbestos and why the discussion about "non-harmful chrysotile" versus "harmful amphibole" is still continuing.

7.8.1. Does the size of the asbestos fibers create a carcinogenic risk?

There are two approaches in understanding how to compare the cancer-risk potency of the different mineralogical types of asbestos (e.g., chrysotile vs. amphibole) by using their fiber dimensions (e.g., the ratio of length to diameter) and their chemical compositions.

Approach #1 indicates that for causing cancer, the length of asbestos particles is a more "aggressive" parameter than thickness since thick fibers are considered to have low pathogenic potential because of their low rate of deposition and retention (Lippmann, 1988). It was noticed that among the workers exposed to chrysotile asbestos the risk of lung cancer increases with exposure to longer fibers (>5 μm), and there is some evidence showing that these effects are most pronounced if these fibers are ~0.25–0.1 μm in diameter (Loomis et al., 2010, 2012). Other studies have also emphasized that the size of the particles is a crucial component of the carcinogenicity of any mineral fibers, no matter if they are chrysotile or amphibole asbestos (Stanton et al., 1981).

Approach #2 is based on the conclusion that there is a difference between chrysotile and amphibole asbestos with respect to the kinetics of their reactions with chemical and biological substances (Bernstein et al., 2013). The authors suggested that longer fibers of chrysotile asbestos fall apart into smaller particles that are destroyed by acidic environments produced by macrophages—e.g., "defense" cells of the human immune system. By contrast, amphibole fibers resist the acidic actions and remain intact for a long time in the human body, which means that they are capable of damaging and penetrating the cells and eventually causing a carcinogenic response. The authors also concluded that the risk of carcinogenic diseases is very low even with a high level of exposure to chrysotile asbestos over a short period time and that a low dose of exposure to chrysotile does not provide any detectable risk of cancer.

7.8.2. Is chrysotile asbestos safer than other types of asbestos?

There were long-term discussions on whether chrysotile is safer than crocidolite or any other type of amphibole asbestos, and the conclusion was that chrysotile asbestosis associated with neither mesothelioma nor other types of cancer (e.g., Hein et al., 2007). Chrysotile asbestos historically has been used for more applications than amphibole asbestos, making up over 95% of all asbestos used. The opinions about the safety or unsafety of chrysotile asbestos have often been extremely opposite and controversial. For example, a different group of activists and media (mostly those who were financed by commercial asbestos companies) say that "Chrysotile asbestos saved our lives," while scientists and health-and-safety organizations have clearly demonstrated connections between the rate of mortality and exposure to chrysotile in workers from the asbestos industries (Wang et al., 2013). As was pointed by Wang et al. (2013) in scientific studies, there were many controversies because some of them have been sponsored by industries whose economic interests collided when adverse health effects became more visible.

Some controversy derived from non-precise and often confusing formulations used by asbestos-industry propagandists such as the Chrysotile Institute (see http://www.chrysotile.com), a non-profit organization that was established in 1984 by the asbestos industries and the federal government of Canada. The Chrysotile Institute tries to convince the public that chrysotile asbestos is safe and does not cause any hazardous effects if it is used within the framework of the permissible level of exposure. To support its point of view, the Chrysotile Institute often used the WHO old standards established in 1985 that prohibited the amphibole asbestos but allowed mining and use of chrysotile asbestos for manufacturing at exposure levels as small as 0.01–0.001 fiber/mL in workplaces. However, since 1985, the WHO standards have been changed, and in its most recent statement, the WHO says: "…there is no evidence for a threshold for the carcinogenic effect of asbestos…" and "…the most efficient way to eliminate asbestos-related diseases is to stop the use of all types of asbestos" (see WHO, 2006).

However, no matter what the Chrysotile Institute and other media say, scientific studies and peer-reviewed publications show that occupational exposure to chrysotile asbestos is similar to exposure to amphibole asbestos and that they both lead to the development of life-threatening diseases—asbestosis, cancer, and mesothelioma (e.g., Stayner et al., 1996; Kanarek, 2011; Wang et al., 2013).

7.9. Asbestos ban

The earlier public awareness of asbestos toxicity began in 1900–1910, followed by widespread recognition in 1924–1930, and it has continued to grow until the present, with emphasis on the occupational nature of asbestos-related diseases and high mortality. There was great resistance by asbestos companies to accept the adverse effects of asbestos, and some companies even tried to remove all references to asbestos-cancer connections before allowing publication of research that they had sponsored.

In 1952, Dr. K. Smith, the medical director for the Johns-Manville company (at that time a leading supplier of roofing products in the U.S. located in Denver, Colorado) suggested placing warning labels on the company's products that contained asbestos, but his suggestions were not accepted. Although by that time the Johns-Manville company and other corporations were already familiar with the facts that asbestos is toxic, they denied the application of a caution label identifying a product as hazardous because such labeling would cut into sales and cause serious financial loss. The U.S. government has also been criticized by the media and activists for not acting quickly enough to inform the public about asbestos danger and to reduce work-place exposure, although some measures to eliminate asbestos exposure were undertaken. For example, in the late 1970s, the U.S. Consumer Product Safety Commission (CPSC) banned the use of asbestos in wallboard patching compounds and gas fireplaces because the asbestos fibers in these products could be released into the environment.

In 1989, the manufacturing, processing, importation, distribution, and use of asbestos was banned in the U.S. based on the document "An Asbestos Ban and Phase-out Rule" issued by the EPA. However, many products in which asbestos fibers were well bound and contained were not banned. In addition, in 1991, the document was modified, and although the U.S. no longer mines asbestos, the import of asbestos ore continues and reached ~1060 tons in 2012. According to the USGS data, 67% of the asbestos imported in 2012 was used for the chloralkali industry (mainly to produce chlorine and sodium hydroxide for industry), 30% for roofing products, and the remaining 3% for different compounds, coatings, and plastics. Thus, in the U.S., the chance to be contaminated with asbestos has not been completely ruled out.

The European Union had completely banned the use of any asbestos in most applications by 2005, and other countries that have banned asbestos either completely or with some exemptions include Argentina, Australia, Chile, Finland, Norway, Sweden, Japan, Saudi Arabia, South Africa, South Korea, and Swaziland. Some other countries are still in a stage of asbestos "phase-out" or are developing some new rules or regulations. However, such countries as Brazil, Canada, China, Russia, Kazakhstan, Zimbabwe, and India continue to mine asbestos, and according to the USGS, they produced ~2 million tons of asbestos in 2012.

Specific words should be addressed to the Canadian asbestos issue. Canada is enormously rich in natural asbestos deposits, and asbestos mining has historically taken place in Québec, Newfoundland, British Columbia, and the Yukon, with the largest mining activity being in Québec, including the Jeffrey Mine, which is the largest open-pit asbestos mine in the world. From 1950 through the 1960s and early 1970s, asbestos mining in Canada was enormously large, with domestic production of up to 1.7 million tons. After the 1970s, the Canadian production

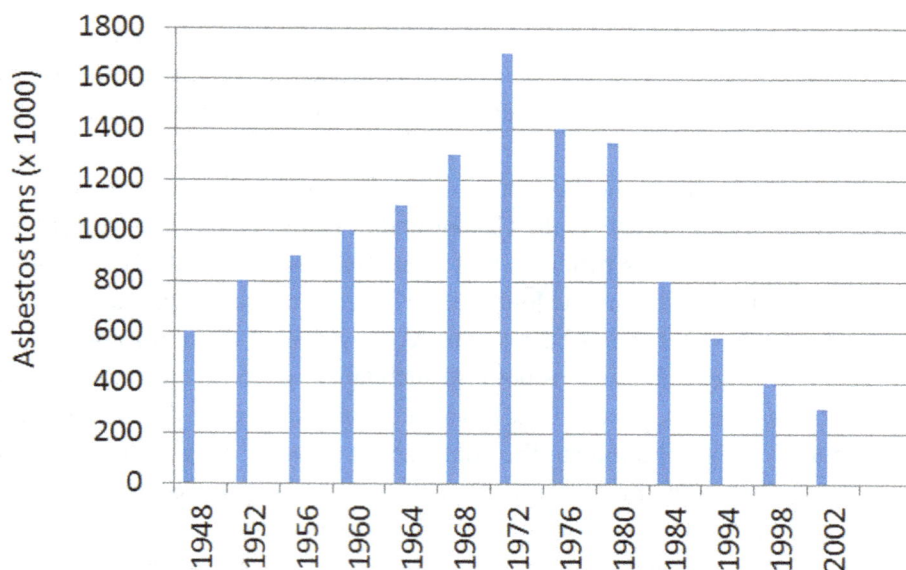

FIGURE 100 Asbestos production and export vs. asbestos-related mortality in Canada, from 1948 through 2002

and export of asbestos declined (Fig. 100), but it has not completely ceased despite the known risks of asbestos for human health.

The Canadian government and corporations have for a long time resisted the asbestos ban, and Canada remained for a considerable time the second-largest exporter of asbestos to developing countries, behind only Russia. They sold their raw asbestos ores to developing countries with labeling that claimed chrysotile asbestos is not as carcinogenic as amphibole and that Canadian chrysotile asbestos and its products are therefore "safe." The asbestos industries also extended their influence to scientific researchers by providing them with financial support in exchange for their conclusions that asbestos exposure can be completely controlled and that chrysotile asbestos has no harmful effects on human health. A truthful analysis of how the asbestos industries tried to extend their myth of asbestos safety through their influence on Canadian researchers was published in the *American Journal of Industrial Medicine* (Egilman et al., 2003). The authors of this paper evaluated studies conducted by researchers at McGill University who had received grants from the Quebec Asbestos Mining Association, and they concluded that:

> "The Canadian asbestos mining industry has a long history of manipulating scientific data to generate results that support claims that their product is innocuous. Researchers complicit in this manipulation seem to be motivated by a variety of

interests, including a desire to support an important national industry and a pre-existing ideological commitment to support corporate interests over worker or community interests. Conducting industry-friendly research can also anchor an academic career by guaranteeing the steady stream of funding necessary to stay afloat in the published or perished environment of the university" (Eligman et al., 2003).

The last asbestos mines in Canada were shut down only in 2011, but since that time, the production of asbestos has decreased to 50,000 tons. In 2012, the Canadian federal government finally announced that it would add asbestos to the list of hazardous minerals. Indeed, in 2012 and 2013, according to the USGS commodity database, Canada was not in the list of top countries that mine asbestos. The world's leading asbestos producers, with total production of ~1.94 million tons in 2013, were Russia, China, Brazil, and Kazakhstan (Table 16), where asbestos is still not banned.

TABLE 16   Worldwide Asbestos Production in 2012 and 2013 (the USGS Asbestos Commodities)

Country	2012 Asbestos Production (Tons)	2013 Asbestos Production (Tons)
Russia	1000000	1000000
China	420000	400000
Brazil	307000	300000
Kazakhstan	241000	240000
Others	300	300
Total	1970000	1940000

In China and India, where the mining and use of asbestos is not banned, there are practically no regulations, except for some small oversights, and these two countries lead the world in the use of asbestos (Table 17) and have a high percentage of asbestosis, lung cancer, and mesothelioma mortality.

TABLE 17   Top Five World Users of Asbestos in 2012 (Data of the USGS)

Country	2012 Asbestos Consumption (tons)
China	530834
India	493086
Brazil	167602
Indonesia	161824
Russia	155476

The fact is that in developing countries, asbestos is not considered a toxicological threat to humans. As a result of decreased asbestos mining in industrial countries, the multi-national asbestos corporations are progressively moving the asbestos markets (and therefore health hazards) to the developing world, and it is predicted that a huge hike in asbestos-caused diseases will take place in the coming decades (Harington and McGlashan, 1998; Jamrozik et al., 2011).

7.10.   How to keep asbestos safe

In most developing countries, asbestos is typically disposed of as hazardous waste in land-fill sites, where it becomes easily airborne and therefore hazardous. In the U.S. and other countries where asbestos is banned, there are new technologies that can transform asbestos into non-hazardous materials. Moreover, most products made today in asbestos-banned countries do not contain asbestos. However, asbestos still may be present in previously installed products in older homes and commercial buildings, and therefore, it still poses a hazard to workers because asbestos fibers can be released during repair work, demolition, and renovation. In addition, in the U.S., for example, asbestos is still used in brake pads, clutches, some roofing materials, vinyl tiles, and some cement pipes, and it can be released when these materials or goods are mechanically disturbed. Any debris, scrap, waste, or clothing contaminated with asbestos dust must be collected in sealed, impermeable containers or bags followed by further recycling or disposal while avoiding any disturbances that would release airborne fibers.

Thermal technologies of asbestos recycling. New recycling technologies have been created to decompose asbestos at temperatures of 1000–1200°C with production of a mixture of non-hazardous silicate phases. Another modification of the thermal recycling is the melting of asbestos at temperatures of >1250°C followed by fast cooling and production of non-harmful ceramic material—e.g., silicate glass (Gualtery and Tartaglia, 2000). Another method, a microwave radiation treatment, is used in an industrial manufacturing process to transform asbestos and asbestos-containing waste into porcelain stoneware tiles, porous, single-fired wall tiles, and ceramic bricks (Leonelli et al., 2006). These technologies lead to destruction of the crystal structure of any asbestos minerals and transformation of them into amorphous matter, multi-component synthetic compounds, or a non-asbestos mineral, forsterite, which is an iron-magnesium silicate [$(Fe,Mg)_2SiO_4$]. These new byproducts are not harmful to humans or the environment.

The case of the asbestos exposure disaster at the Libby mine, Montana, U.S.

7.11.

Beginning in 1900, in the U.S., asbestos mines were in operation in 14 states—Alaska, Arizona, California, Georgia, Maryland, Massachusetts, Montana, North Carolina, Oregon, South Carolina, South Dakota, Vermont, Virginia, and Wyoming. One of the mines, the vermiculite-tremolite-asbestos mine near Libby, Montana, became one of the world's largest asbestos-contaminated sites and a classical example of the national environmental and public health disaster caused by asbestos. This case was and remains instructive because of the high rate of mortality among workers in the Libby mine, because it did not gain any attention for a long time either from asbestos corporations or the government, and because the toxic asbestos products were distributed without any warning over the whole U.S. until 1999.

The town of Libby is situated near the Rainy Creek alkaline-ultramafic igneous complex, which occupies an area as large as 14230 m² (Boetcher, 1967). Biotite-pyroxenite rocks, which comprise 40% of this igneous massif, were subjected to a low-temperature hydrothermal alteration, during which the mineral biotite was transformed into vermiculite (Fig. 101). Another rock-forming mineral, pyroxene, was transformed into amphiboles of various compositions, and part of those amphiboles (winchite and richterite) had a fibrous morphology—e.g., asbestos (Gunter et al., 2003). In the non-geological literature, the winchite and richterite asbestos from the Libby mine were mistakenly called tremolite. In the mine, the vermiculite and amphibole asbestos minerals were so closely intergrown that it was practically impossible to distinguish them with the naked eye, and therefore, nobody planned to separate them during mining. The Libby mine vermiculite was used for attic insulation, construction building materials, soil conditioning, and as a bulk

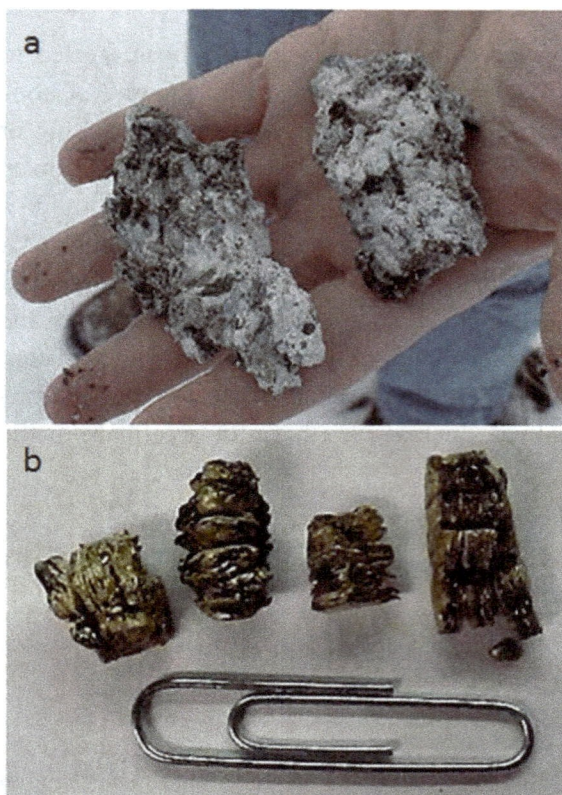

FIGURE 101 Asbestos production and export vs. asbestos-related mortality in Canada, from 1948 through 2002

carrier for agricultural chemicals and other related products. The most famous product was "Zenolite," a kind of cement containing crushed vermiculite ore that was widely used for loose-fill insulation.

During the period from 1963 to 1990, the Libby mine was operated by the W.R. Grace and Company, with over 200 employees who—without any special precautions or protections—continued to mine and mill vermiculite ore enriched in amphibole asbestos. It was a huge and extremely profitable operation: the W.R. Grace and Company mined and processed about 80% of the world's supply of vermiculite at the Libby mine. For decades, asbestos-tainted vermiculate products were distributed nationally and sold to countless countries around the world. At the Libby mine and processing, many workers who spent hours each day around the deadly substance eventually were contaminated with asbestos and suffered from lung diseases. Ironically, the W.R. Grace and Company had distributed the leftover vermiculite for use in playgrounds, backyards, gardens, roads, and a number of other popular locations in the town of Libby. Thus, in addition to the asbestos circulating around the mine and milling, asbestos was released into the air from vermiculite that was disturbed in baseball fields and other areas where children played and adults spent time.

More than 400 deaths were attributed to asbestos-caused diseases in the Libby mining region, and since 1979, more than 1500 people who used to live and work in Libby have developed similar illnesses. Although it has been concluded that pure vermiculite has no adverse effect on human health, the health effect of the amphibole asbestos exposure observed in the Libby mine workers was similar to many other cases of exposure to asbestos in the world (Ross et al., 1993).

The high rate of asbestosis and related lung diseases and the high mortality from lung cancer and mesothelioma among the Libby mining workers finally came to national attention in 1999 due to a series of articles published by the *Seattle Post-Intelligencer*. The newspaper published an article titled "Uncivil Action: A Town Left to Die," in which mortality from asbestos in the Libby mine was clearly formulated and supported by medical statistics and facts. The paper wrote:

"From 1924 until 1990, miners extracted a large percentage of the world's vermiculite from a mountainside near Libby (Montana). As they mined and milled the ore, millions of tons of tremolite asbestos were released into the air... 192 people from Libby had died, and 375 were currently diagnosed with fatal asbestos-related disease, directly traceable to the mining operation ... . The W.R. Grace Co., which owned the mine for three decades, was well aware of the deadly asbestos being inhaled by the miners and their families, but for years did not tell its workers of the hazards... And doctors say the people of Libby will keep dying for decades" (Seattle Post-Intelligencer, 1999).

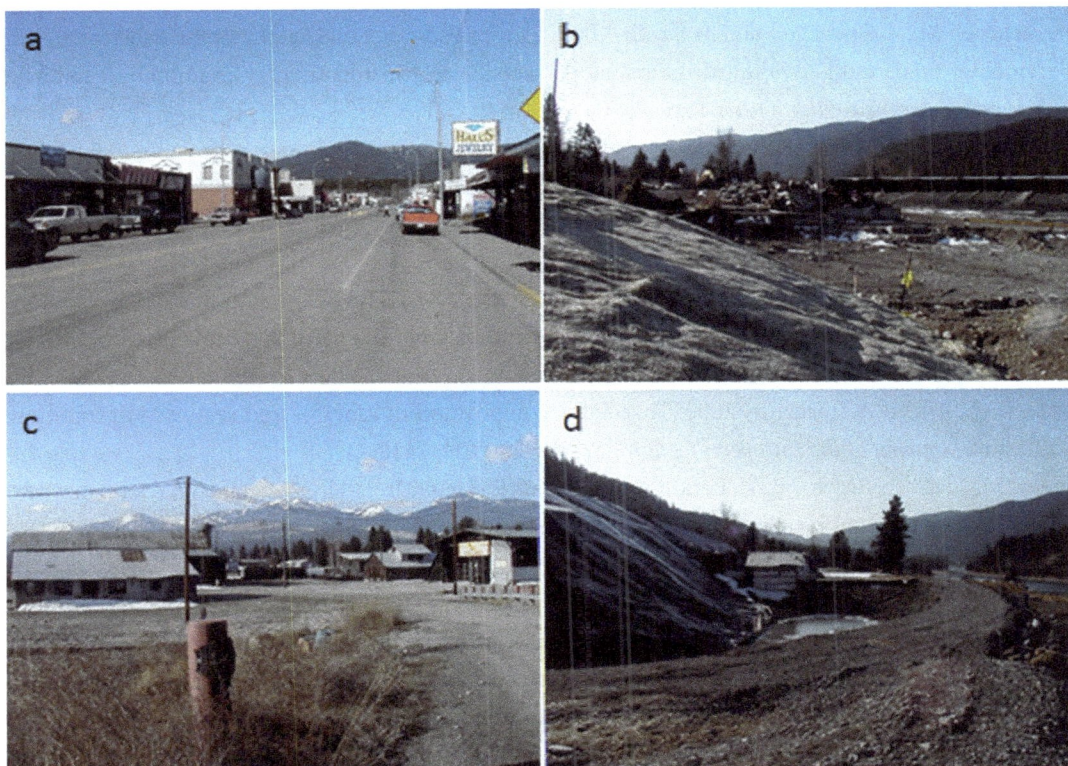

FIGURE 102 (a) City of Libby, Montana; (b) screening plant for vermiculite in Libby before clean-up; (c) vermiculite exfoliation facility in Libby; (d) contaminated soil is covered with a heavy tarp

After this publication, the EPA scheduled an investigation to identify the source of the diseases and mortality of the local population and organized an information center in Libby to address the asbestos issues. During this project, the EPA team removed vermiculite/asbestos from homes and business buildings in the town of Libby. The mine site and vermiculite screening, milling, and exfoliation facilities were closed, and the tails of vermiculite and contaminated soil were covered with heavy tarps (Fig. 102). The ATSDR determined that "persons living in Libby have a standardized mortality ratio for asbestos-related death of 40 to 60 times higher than a normal population." In 2002, the town of Libby was declared a Superfund site, which was the biggest in U.S. history, and in 2008, the W.R. Grace and Company was ordered to pay $250 million to cover clean-up costs in Libby.

References

Bernstein, D. M., J. Chevalier, and P. Smith. 2005. Comparison of Calidria chrysotile asbestos to pure tremolite: final results of the inhalation biopersistence and histopathology examination following short-term exposure. *Inhalation Toxicology* 17:427–449.

Boettcher, A. L. 1967. The Rainy Creek alkaline-ultramafic igneous complex near Libby, Montana. I: ultramafic rocks and fenite. *The Journal of Geology* 75:526–553.

Cooke, W. E. 1924. Fibrosis of the lungs due to inhalation of asbestos dust. *British Medical Journal* 2:147.

Egilman, D., C. Fenel, and S. R. Bohme. 2003. Exposing the "Myth" of ABC, "Anything But Chrysotile": A critique of the Canadian asbestos mining industry and McGill University chrysotile studies. *American Journal of Industrial Medicine* 44:540–557.

Gualtieri, A. F., and A. Tartaglia. 2000. Thermal decomposition of asbestos and recycling in traditional ceramics. *Journal of the European Ceramic Society* 20:1409–1418.

Gunter, M. E., B. R. Brown, B. R. Bandli, M. D. Dyar, F. F. Foit, Jr. et al. 2003. Crystal chemistry, crystal structure, and morphology of amphibole and amphibole asbestos from Libbi, Montana, U.S.A. *American Mineralogist* 88:1970–1978.

Harington, J. S., and N. D. McGlashan. 1998. South African asbestos: production, exports, and destinations, 1959–1993. *American Journal of Industrial Medicine* 33:321–325.

Hein, M. J., L. T. Stayner, E. Lehman, and J. M. Dement. 2007. Follow-up studies of chrysotile textile workers: Cohort mortality and exposure-response. *Occupational and Environmental Medicine* 64:616–625.

Jamrozik, E., N. de Klerk, and A. W. Musk. 2011. Asbestos-related disease. *Internal Medicine Journal* 41:372–380.

Kanarek, M. S. 2013. Mesothelioma from chrysotile asbestos: Update. *Annals of Epidemiology* 21:688–697.

Lee, R. J., D. R.Van Orden, M. Corn, and K. S. Crump. 1992. Exposure to airborne asbestos in buildings. *Regulatory Toxicology and Pharmacology* 16: 93–107.

Leonelli, C., P. Veronesi, D. N. Boccaccini, M. R. Rivasi, L. Barbieri et al. 2006. Microwave thermal inertisation of asbestos containing waste and its recycling in traditional ceramics. *Journal of Hazardous Materials* 135: 149–155.

Lippmann, M. 1988. Asbestos exposure indices. *Environmental Researches* 46:86–106.

Loomis, D., J. M. Dement, D. Richardson, and S. Wolf. 2010. Asbestos fibers dimensions and lung cancer mortality among workers exposed to chrysotile. *Occupational and Environmental Medicine* 67:580–584.

Loomis, D., J. M. Dement, L. Elliot, D. Richardson, E.D. Kuepmel et al. 2012. Increased lung cancer mortality among chrysotile asbestos textile workers is more strongly associated with exposure to long thin fibers. *Occupational and Environmental Medicine* 69:564–568.

Nesse, W. D. 2000. *Introduction to mineralogy*. Oxford University Press, p.442.

Polissar, L., R. K. Severson, and E. S. Boatman. 1984. A case control study of asbestos in drinking water and cancer risk. *American Journal of Epidemiology* 119:456–471.

Ross, M., R. P. Nolan, A. M. Langer, and C. V. Cooper. 1993. Health effect of mineral dust other than asbestos. In: Guthrie Jr. G.D., and B.T. Mossman. (Eds.) *Health effect of mineral dust*. Review in Mineralogy. Mineralogical Society of America, Washington D.C., p. 35–408.

Seattle Post-Intelligencer. 1999. "Uncivil action – a town left to die, asbestos: Forgotten killer." https://www.ire.org/resource-center/stories/16058.

Stanton, M. F., M. Layard, A. Tegeris, E. Miller, M. May, et al. 1981. Relation of particle dimension to carcinogenicity in amphibole asbestos and other fibrous minerals. *Journal of the National Cancer Institute* 67:965–975.

Stayner, L. T., D. A. Dankovich, and R. A. Lemen. 1996. Occupational exposure to chrysotile asbestos and cancer risk: A review of the amphibole hypothesis. *American Journal of Public Health* 89:179–186.

Vitra, R. L. 2005. Mineral commodity profiles–asbestos: U.S. Geological Survey Circular 1255–KK, 56 p.

Wang, X., E. Yano, S. Lin, I. T. S.Yu, Y. Lan, et al. 2013. Cancer mortality in Chinese chrysotile asbestos mines: Exposure-response relationships. *PLOS ONE* 8:e718999.

WHO. 2006. Elimination of asbestos-related diseases. World Health Organization. Public Health and the Environment, WHO/SDE.OEH/06.03, September.

Web resources

http://www.chrysotile.com
https://www.ire.org/resource-center/stories/16058

http://www.mesotheliomhelp.org/mesothelioma/statistics

http://www.nochrysotileban.com

https://www.osha.gov/SLTC/asbestos/index.html

Image Credits

Figure 99a: netalloy, "nettaloy industrial," https://openclipart.org/detail/37249/netalloy-industrial-by-netalloy. Copyright in the Public Domain.

Figure 99b: netalloy, "building," https://openclipart.org/detail/70195/buildings-by-netalloy-70195. Copyright in the Public Domain.

Figure 99c: netalloy, "NetAlloy Cape Code," https://openclipart.org/detail/38929/netalloy-cape-code-by-netalloy. Copyright in the Public Domain.

Figure 99d: netalloy, "school-building," https://openclipart.org/detail/70423/schoolbuilding. Copyright in the Public Domain.

Figure 99e: Gerald_G, "US Capitol Building," https://openclipart.org/detail/7869/us-capitol-building-by-gerald_g. Copyright in the Public Domain.

Figure 99f: rdevries, "Dutch train," https://openclipart.org/detail/170494/dutch-train-by-rdevries-170494. Copyright in the Public Domain.

Figure 99g: gsagri04, "Electric Bulb," https://openclipart.org/detail/161977/electric-bulb-by-gsagri04. Copyright in the Public Domain.

Figure 99h: nicubunu, "RPG map symbols: Farm," https://openclipart.org/detail/11443/rpg-map-symbols:-farm-by-nicubunu. Copyright in the Public Domain.

Figure 99i: metalmarious, "Medicine and a Stethoscope," https://openclipart.org/detail/5013/medicine-and-a-stethoscope-by-metalmarious. Copyright in the Public Domain.

Figure 100: "Asbestos production and export vs. asbestos-related mortality in Canada," http://www.ehatlas.ca/asbestos/trends/asbestos-production-canada-0. Copyright in the Public Domain.

Figure 101a: United States Geological Service, " Popped-out vermiculite," http://www.atsdr.cdc.gov/asbestos/more_about_asbestos/health_consultation/. Copyright in the Public Domain.

Figure 101b: Environmental Protection Agency, ""Popped" vermiculite," http://www2.epa.gov/region8/abcs-asbestos. Copyright in the Public Domain.

Figure 102: Center for Disease Control, "Pictures of Libby, Montana," http://www.atsdr.cdc.gov/asbestos/sites/libby_montana/libby_pictures.html. Copyright in the Public Domain.

Review Questions

1. What is asbestos?

2. What minerals are included in the asbestos group?

3. Where does chrysotile asbestos occur in the Earth?

4. Where do amphibole asbestos minerals occur in the earth?

5. What is the mineral name for blue asbestos? For white asbestos?

6. What are some common properties of asbestos that make it attractive for industrial use?

7. What is the difference between chrysotile asbestos and the minerals of the amphibole asbestos group?

8. Define the major crystalline structure features of chrysotile and amphibole asbestos minerals.

9. Why is chrysotile considered asbestos while antigorite and lizardite are not?

10. What has asbestos been used for historically? In ancient times? Since the Industrial Revolution?

11. What are the kinds of diseases caused by asbestos inhalation?

12. What are the asbestos pathways to humans and to the environment?

13. Is there any "safe level" for asbestos exposure?

14. What has caused the controversial discussions about the adverse effects of asbestos on human health?

15. Is chrysotile asbestos safer than other types of asbestos?

16. Does a carcinogenetic risk depend on the size of the asbestos fibers?

17. Are asbestos-related diseases curable?

18. What can be done to reduce exposure risk at abandoned asbestos mines?

19. Has asbestos been banned from all products in the U.S.?

20. What manufactured products containing asbestos can still be found in the U.S.?

21. Who is at risk for asbestos exposure? What industries are the most vulnerable?

22. How can adults prevent children's exposure to asbestos?

23. What mineral was being mined in Libby, Montana?

24. How did exposure to asbestos occur in the town of Libby?

25. When was the vermiculite danger known?

26. What was done to prevent exposure and stop the deaths?

27. What kinds of people outside of Libby appeared to be potentially at risk to asbestos exposure as well?

28. What are the ways in which asbestos can be safely disposed of and recycled?

29. What are the main factors affecting the likelihood of developing diseases related to asbestos exposure?

30. How often does drinking water contaminated with asbestos cause asbestos-related disease?

Quizzes (see answers on page 307)

1. Given unlimited resources, what would be the best way to get rid of asbestos removed from old buildings?
   a. It should be treated as hazardous waste and dumped in a safe landfill.
   b. It should be burned to make sure it cannot re-enter the environment, but the CO_2 produced should be captured so as not to enter the atmosphere.
   c. It should be recycled by heating to produce harmless glass, ceramic, and porcelain products.
   d. It should be reused by cleaning and used as insulation in new buildings.

2. Three asbestos-related diseases are:
   a. Colon cancer, asbestosis, and malaria.
   b. Lung cancer, mesothelioma, and AIDS.
   c. Lung cancer, asbestosis, and mesothelioma.
   d. Hepatitis B, mesothelioma, and asbestosis.

3. If you accidentally knock off a chunk of sprayed-on asbestos insulation, you should:
   a. Leave it for a custodian to clean up.
   b. Immediately get a dusk mask, then carefully sweep it into a Ziploc bag, zip and label it, and deliver it to an asbestos recycling facility.
   c. Don't touch it and report it immediately.
   d. Get a vacuum.

4. Asbestos fibers are most hazardous if they are:
   a. Inhaled.
   b. Touched.
   c. Swallowed.
   d. Smelt.

5. Asbestos-related diseases have:
   a. No latency period.
   b. About a fifteen- to thirty-year latency period.
   c. About a five- to twelve-year latency period.
   d. About a one- to ten-year latency period.

Coal—a Fossilized Geological Fuel and its Hazardous Effects

Chapter 8

What is coal? 8.1.

Coal is a sedimentary rock that consists mainly of carbon (~50–98 wt%) that originates from biological matter and various quantities of other organic components, minerals, native elements, and volatile matter. Coal has an organic origin—e.g., it was formed from vegetal matter that was buried at a certain depth and underwent complex chemical reactions under conditions requiring a deficiency of oxygen and elevated pressures and temperatures. Coal formation has taken millions of years and has occurred in humid climates and in swampy regions, where dead trees, leaves, and other plant debris remained wet most of the time and were substantially converted first to peat and then to more dense substances—e.g., coal deposits. Evidence that coal is derived from plants is supported by the presence of recognizable fossilized leaves or other plant remains, and its sedimentary origin is recorded by occurrences of layer-like coal deposits in close association with limestone, sandstone, and other sedimentary formations (Fig. 103). Usually coal beds, which are also called "seams," are presented by layers of coal from a few centimeters thick to meters or even hundreds of meters thick.

FIGURE 103 (a) Coal seam near Hartley Bay, UK (courtesy of Peter Cowling); (b) fragments of fossilized fern in coal from Morocco

8.2. Coal formation in the Earth

Most of the world's richest coal deposits were formed during the Paleozoic era, in the Carboniferous period, from about 360 to 286 million years ago. Rich coal deposits are also found within the rocks formed during the late Cretaceous period through the early Tertiary period—e.g., from 130 to 50 million years ago (see Chapter 1, Table 1).

The Carboniferous period was characterized by a more uniform, tropical, and humid climate, with no seasons being distinguished, and during this time, all continents were covered with huge trees, ferns, and other plants. The climate of the Cretaceous through the early Tertiary period continued to be favorable for coal formation, and studies have shown that there were significant coal accumulations during this time.

In the course of geological time, shallow marine water often flooded the continents, and because it did not evaporate quickly due to the high humidity, the land was covered with swamps and stagnant pools. The latters were eventually filled with the fallen dead leaves and plants, algae, and anaerobic bacteria (i.e., living without oxygen). The bacteria partly decayed the biological matter, which was further transformed into a spongy, slightly compacted material—peat—and new sediments were deposited over the peat layers and buried them. Eventually, the heat and pressure created in such burial reservoirs transformed peat into coal (Fig. 104). During this process, which is called "coalification," some isolated "domains" filled with plant remains may not be decayed by microorganisms and converted to coal, and thus, the overprints of the fossilized plants are sometimes preserved inside the coal layers. The fossils are used by paleontologists for reconstruction of the history of a paleoclimate, a biological and geochemical evolution, and interactions among oceans, land, and the atmosphere over long geological time periods.

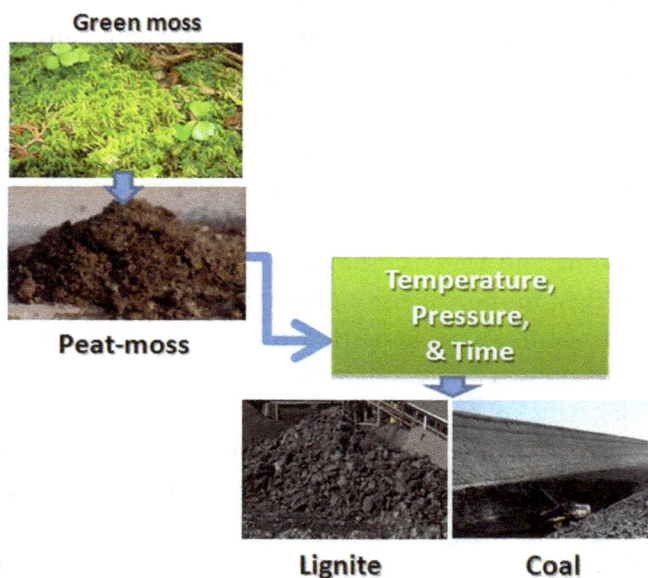

FIGURE 104 Schematic diagram showing stages of coal formation

8.2.1. Coalification, coal composition, and rank

The formation of coal includes several major stages, which start with a transformation of organic matter to peat and extend through lignite to subbituminous and bituminous coal and through the transformation of bituminous coal to anthracite (Fig. 105). In geological terminology, the coalification term encompasses chemical, physical, and biological processes that lead to the decay of organic matter; diagenesis—i.e., lithification, (from the Latin word *Lithos*, or stone); and metamorphism (a recrystallization requiring heat and pressure). Volatile phases rich in hydrogen and oxygen (e.g., water, carbon dioxide, and methane) produced during coalification will substantially escape from the buried plant remains, and hence, the organic matter becomes progressively denser, with the coal substances becoming richer in carbon form. The most important factors in coalification are the burial processes, with the implication of higher temperatures and pressures increasing with the depth (O'Keefe et al., 2013).

FIGURE 105   (a) Peat, a precursor of coal; major types of coal: (b) lignite, (c) bituminous, and (d) anthracite

The composition of coal varies widely from one geographic place to another. In general, coal consists of carbon (50–98 wt%), volatile matter (2–50 wt%), other impurities such as S (~1–2 wt%), Cl (~1 wt%), P (<1 wt%), N (~1–3 wt%), H (~1–6 wt%), trace elements, and some dirt. An approximate molecular formula of coal is: $C_{135}H_{96}O_9NS$.

The main types of coal and products of coalification are peat, lignite, and different varieties of subbituminous, bituminous, and anthracite coal.

8.2.1.1. Peat

Peat is not coal, but rather a precursor for coal (see Fig 105a) that usually accumulates in mires and swampy and boggy wetlands with stagnant water full of algae and anaerobic bacteria. Peat formation is controlled by two important factors:

  i.   The accumulation rate of organic matter must exceed the rate of its decay.
  ii. The deposition of sediments (gravel, sands, etc.) into the peat-forming reservoirs must be minimal.

Peats are formed at shallow depths, with the temperature inside peat-forming reservoirs not higher than 25–30°C, and negligible pressure ~0.05 kbar. As a result of such conditions, peat generally contains many compacted fragments of plants, about ~70–75% moisture and ~15–30% carbon, which is less than the ~50% carbon of higher-ranked coals such as subbituminous, bituminous, and anthracite (~90% carbon) (Fig 106). Though ancient peat formations are favored by warm and moist climates, but modern peat formations are found in the cold regions of Canada, the Scandinavian countries, and Siberia in Russia.

The air-dried peat burns for a long time, and thus, it was used as a source of local heat by prehistoric people and is still used as a fuel in Ireland and England despite the fact that its burning produces a lot of smoke. Fresh peat

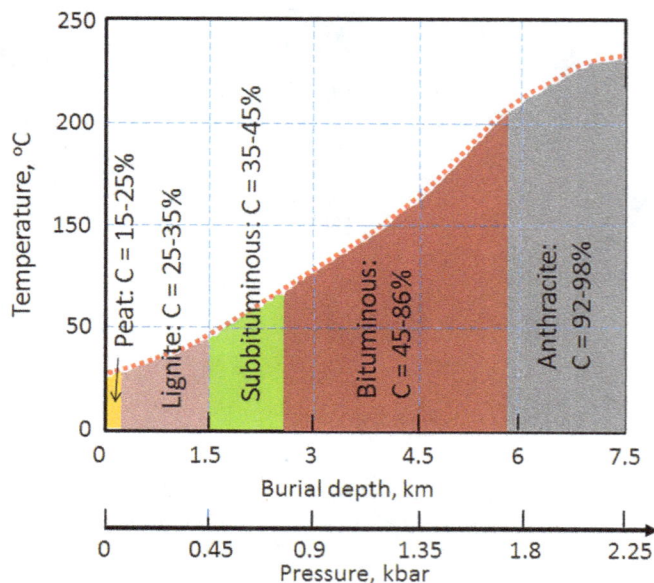

FIGURE 106 Schematic diagram showing that coal ranks and carbon content increase with the temperature, pressure, and burial depth

is easily cut by spade or shovel in the form of blocks, which are usually left on the neighboring land to dry, as shown in Fig. 107. Under conditions of continuing air-drying by the wind and sun, the peat blocks shrink and become compressed, and these denser blocks can be either stored or transported as a domestic fuel material.

The mining of peat for energy purposes has increased in developing countries as an alternative to relatively expensive, imported, high-quality coal, oil, and natural gas. In addition, peat can be also added to soils for increasing their moisture and acidity, which are important for specific plants. Peat can also store nutrients in soils, even though peat is not fertile itself. In many developing countries, air-dried peat is used in buildings as an insulator because of its poor heat-conducting properties.

Notable peat deposits are found in Russia, Ireland, Scotland, Poland, Germany, the Netherlands, the Scandinavian countries, and North America. Worldwide, approximately 60% of modern wetlands have peat deposits.

FIGURE 107 Peat mining and air-drying: (a) in Lewis; (b) the peat stack in Ness, Outer Hebrides, Scotland

8.2.1.2. Lignite

Lignite, often called "brown" coal (see Fig. 105b), is a product of the second step of coalification of the organic matter after peat formation. During this stage, the sediments accumulated above the peat layers create inside the burial peat reservoir pressure from >0.05 to 0.45 kbar, and such a load raises the temperature up to ~50°C (see Fig. 106). Such conditions cause the compaction of peat and loss of the small amount of water and volatiles such as methane (CH_4) and carbon dioxide (CO_2). As a result, the peat is eventually transformed to lignite, which contains ~25–35% carbon, ~66% moisture, and 6–19% other organic solid and volatile impurities, some minerals, and some rare earth elements. Because of the high content of moisture, lignite burning does not produce high heat energy, and so it is ranked as an uneconomical fuel. The heat energy density (content) of lignite ranges from 10–20 MJ/kg (MJ—micro

Joule, an energy unit based on the International System of Units [SI]), which is very small compared with bituminous coal and anthracite, which have average energy densities of 24–35 MJ/kg.

Lignite is mined in Australia, Bulgaria, Canada, China, Germany, Greece, India, Kosovo, Poland, Russia, Serbia, Turkey, and the United States, where Texas, Montana, and North and South Dakota are the main producers. Lignite is used almost exclusively as fuel for steam-electric power generation.

8.2.1.3. Bituminous and subbituminous coal

Bituminous coal, a dense black to dark-brown geological material (see Fig. 105c), is formed in the third stage of the coalification process, accompanied by increase of pressures up to 0.7–1.6 kbar and temperatures up to >80–200°C inside the coal-forming reservoir. Ryer and Langer, (1980) calculated that in the burial peat or peat-lignite reservoirs at depths of ~2.2–5.8 km (see Fig. 106), the compaction and water loss lead to a reduction of ~3.5 vertical meters of original peat to ~0.32 vertical meter of bituminous coal. This means that the transformation of peat to bituminous coal requires ratio of volume change as 11:1. Under these conditions, primary peat becomes completely metamorphosed to that level when virtually all traces of the organic life (plants, leaves, etc.) have disappeared.

Bituminous coal consists of carbon (45–86 wt%) and impurities (14–35 wt%) that include S, N, H, O, and other occasional elements and minerals that may be specific for certain geological localities. Bituminous coal is energy efficient, producing ~24–28 MJ/kg of heat.

The intermediate fraction of coal between lignite and bituminous is called subbituminous, which is formed at temperatures of >50–100°C and pressures of ~0.5–0.7 kbar at depths of ~1.5–2.7 km, and it contains about 35–45 wt% carbon (see Fig. 106). Subbituminous coal produces ~20–28 MJ/kg heat energy, and in this aspect, it is of poorer quality than bituminous and anthracite coal.

The availability of subbituminous coal in the world is moderate, with notable resources in Brazil, Indonesia, and Ukraine. Bituminous coal, however, is abundant on all continents, with the largest reserves being in the U.S. and notable reserves in Africa, Australia, Canada, China, India, and Russia, among other countries (Fig. 108). The U.S. is the world's leading producer of both subbituminous coal (Alaska, Colorado, Illinois, Montana, New Mexico, and Wyoming) and bituminous

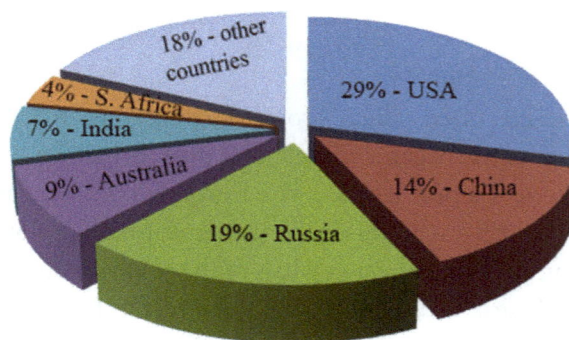

FIGURE 108 Diagram showing coal reserves of the world as of 2010

one (Alabama, Arkansas, Kentucky, Maryland, Oklahoma, Pennsylvania, and West Virginia), with estimated total reserves of approximately 300000 million tons (Mt). More than 90% of all coal consumed in the U.S. belongs to the bituminous and subbituminous grades, with the remaining 10% including mostly anthracite, the highest coal grade, and rarely lignite.

8.2.1.4. Anthracite

Anthracite is a "black" coal (see Fig. 105e) with a shiny luster and higher hardness and density than all other types of coal. It is formed under conditions of increasing pressures from ~1.6–2.25 kbar and temperatures from 200–250°C during the final stage (#4) of coalification (see Fig. 106). This stage is characterized by the complete transformation of subbituminous and bituminous coal to dense, fossilized, amorphous carbon matter—e.g., anthracite. Anthracite consists of 92–98% carbon, low volumes of impurities, and traces of moisture, and therefore, it burns far more efficiently, with less smoke than bituminous and subbituminous coal. The heat energy produced by anthracite ranges from 28–35 MJ/kg, which is twice as high as lignite.

The world's largest deposits of anthracite occur in Russia, Canada, China, Ukraine, North Korea, and Vietnam. In the U.S., anthracite deposits are rare compared with subbituminous and bituminous coal, with the largest reservoir being located in Pennsylvania and several smaller deposits being known in a historical mine in Colorado.

8.2.2. Minerals and trace elements in coal

Coal of all ranks contains small quantities of mineral impurities and trace elements that usually originate from neighboring geological formations. The minerals usually are products of the weathering or have crystallized due to chemical/biochemical reactions that occurred during coalification and metamorphism. More than 150 mineral species have been identified in coals from different geographic localities (Ward, 1986). These minerals occur in coal seams as tiny dispersed grains from–millimeter to micron in sizes, tiny nodules, thin lenticles, and microbands.

The minerals in coals are divided for three groups: abundant, common, and rarely occurring (Table 18). Abundant minerals include silicates (kaolinite, feldspars) and carbonates (calcite and siderite), and the group of common minerals consists of sulfides (pyrite and marcasite), oxides and hydroxides (quartz, limonite, and gibbsite), silicates (illite and montmorillonite), and carbonates (dolomite and ankerite). The group of rarely occurring minerals includes sulfides (galena and sphalerite), oxides (hematite and chalcedony), silicates (chlorite, muscovite, halloysite, and dickite), phosphates (apatite), and sulfates (barite and gypsum).

Heavy metals and trace elements are also abundant in coal, and their diversity and concentrations depend on the geochemical characteristics of the geological settings in which the

TABLE 18 Examples of Abundant, Common, and Rarely Occurring Minerals in Coal (Data Summarized from Xu et al., 2003; Ward, 1986)

Name of the Mineral Group	Abundant Minerals	Common Minerals	Rarely Occurring Minerals
Sulfides		Pyrite (FeS), Marcasite (FeS)	Galena (PbS), Sphalerite (ZnFe)S
Oxides		Quartz (SiO_2)	Hematite (Fe_2O_3), Chalcedony (SiO_2)
Hydroxides		Limonite $FeO(OH) \cdot nH_2O$, Gibbsite $[Al(OH)_3]$	
Silicates	Kaolinite $Al_2(Si_2O_5)(OH)_4$, Feldspars $NaAlSi_3O_8 - CaAl_2Si_2O_8$, $KAlSi_3O_8$	Illite $(K,H_3O)(Al,Mg,Fe)_2(Si,Al)_4O_{10} \cdot [(OH)_2,(H_2O)]$, Montmorillonite $(Na,Ca)_{0.3}(Al,Mg)_2Si_4O_{10}(OH)_2 \cdot n(H_2O)$	Chlorite $(Mg,Fe)_3(Si,Al)_4O_{10}(OH)_2 \cdot (Mg,Fe)_3(OH)_6$, Muscovite $KAl_2(Si_3Al)O_{10}(OH,F)_2$, Halloysite $Al_2Si_2O_5(OH)_4$, Dickite $(Al_2Si_2O_5(OH)_4)$
Carbonates	Calcite $(CaCO_3)$, Siderite $(FeCO_3)$		Dolomite $(CaCO_3)$, Ankerite $Ca(Fe^{2+},Mg,Mn)(CO_3)_2$
Phosphates			Apatite $Ca_5(PO_4)_3(OH,F,Cl)$
Sulfates			Barite $(BaSO_4)$ Gypsum $CaSO_4 \cdot 2(H_2O)$
Metals, Metal-loids, and Trace Elements		Cu, S, As, Hg, F, Cl, N, H, Bi, Cr, Pb, Cs, K, Li, Mg, Ni, P, Cd, Sb, Co, U, Th, V, Fe, Rb, Sr, Se, Ta, Mn, and others	

coalification process took place. Some of the trace elements may originate from clay minerals, sulfides, and silicates, and others are due to the chemical reactions that occurred during coalification and metamorphism. For example, clay minerals and feldspars release Pb, Rb, Cs, Ti, and other elements, while sulfides mostly release As, Hg, Cd, Fe, Cu, and other metals and metalloids in different concentrations. All these elements may be considered as both essential and non-essential.

Knowledge of the composition of the minerals and of how the trace elements behave in coal are extremely important for understanding their adverse toxicological and environmental effects because coal is the most sustainable world energy source, with relatively abundant reserves. Many heavy metals, metalloids, and other elements are already known as having adverse impacts on human health and the environment, and their emission from coal combustion into the atmosphere is an additional component of both air pollution and land contamination.

World coal production and use 8.3.

8.3.1. Coal deposit exploration and how much coal is left

Coal deposits of economically profitable concentrations are discovered in the Earth through geological and instrumental exploration techniques. The national geological surveys of each individual country use satellite imaging coupled with geochemical and geophysical surveys accompanied by rock sampling in order to create a geological map of areas that may have rich coal deposits, and this is followed by drilling to determine the thickness of coal seams. The combination of these techniques with modern instruments and technologies allows the development of an accurate picture of the area for the future mining operations. The area will become a mine if the discovered coal has a sufficient quality and if the coal seams are large enough and accessible enough for mining operations to be economically viable. Many environmental and societal restrictions such as neighboring towns, wetlands, national parks and preserves, and other sensitive ecosystems should be also taken into account before mining activity is begun.

Coal plays a leading role among other geological fuels because it is safer than gas and oil and is widely available in the world at relatively low prices that are both stable and predictable. In addition, there are new technologies that improve coal efficiency and make it environmentally cleaner than oil and gas.

Coal belongs to the category of non-renewable geological resources. It means that the coal deposits, formed in the Earth millions of years ago, are not being formed any longer, and therefore, all coal reserves are available only for the foreseeable future. Calculations, though they may vary, show that the total world coal reserves are enough for 112–150 years at current rates of their production and consumption (Fig. 109).

To understand such calculations, one should take in account that the total reserves of coal in the world as of 2011 were 891850 Mt, while in 1993, they were as large as 1031610 Mt (for details on reserves of coal as of 2011, see http://www.worldenergy.org/data/resources/resource/coal/), with the consumption of 139760 Mt over the eighteen–year period.

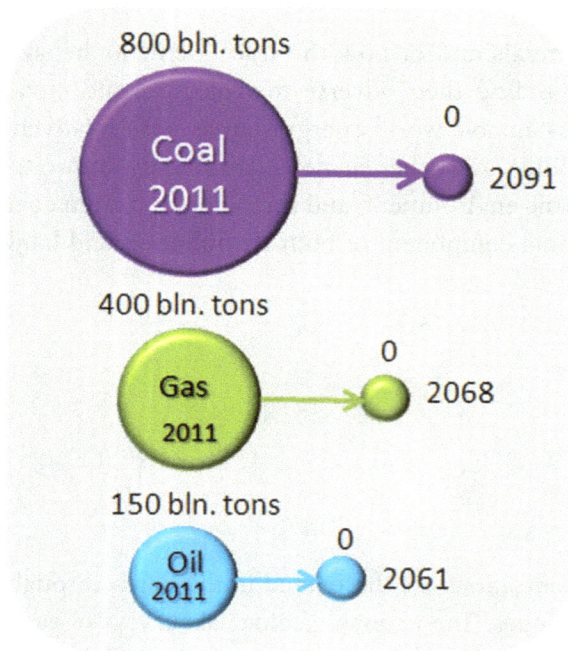

800 bln. tons

Coal 2011 → 0 2091

400 bln. tons

Gas 2011 → 0 2068

150 bln. tons

Oil 2011 → 0 2061

FIGURE 109 Projection of the world energy reserves from 2011 to 2091 (coal, gas, and oil in billions of tons [oil equivalent])

Equation 1 shows that world coal reserves will last for 114 years:

$$W = R^n \frac{T}{R}, \qquad \text{[Eq. 1]},$$

where W is the years of world coal reserves available for the future, R^n is world coal reserves (in Mt) as of 2011, R is world coal reserves (in Mt) as of 1993, and T is the time gap (in years) between 2011 and 1993.

The reserves of oil and natural gas in the Earth are not as large as the coal reserves, and the oil and gas deposits are restricted to certain regions in the world, many of them politically unstable. Calculations forecast that oil reserves in the world will be enough only for the next forty to fifty years, and natural gas for about sixty years (see Fig. 109), if no new deposits are discovered and the oil consumption and production rates remain the same as in 2011.

8.3.2. Coal mining

Coal mining usually takes place either by open-pit operations if the coal layers are situated at a shallow depth or by underground methods to extract coal from deeply buried seams.

Open-pit operations are applied to those seams of coal that lay close to the Earth's surface, and this method allows mining coal at a low cost (Fig. 110a). The soil and sedimentary rocks overlying the seams of coal are mechanically removed with blasting and excavation by bulldozers to make it as easy as possible to access the pure coal layers. Huge power shovels scrape out coal and load it into huge cars for transportation. Pits remain under operation during long periods of time—sometimes years.

Underground mining methods are applied when the coal seams are at depths of several hundreds of meters below the Earth's surface. Underground mining includes complicated engineering construction that consists of a series of vertical and horizontal tunnels (shafts) built in a way allowing transportation back and forth by mechanical devices and instruments (Fig. 110b).

All, including miners, move directly to the area of the coal extraction. One of the vertical tunnels has a coal elevator to deliver the extracted coal to the Earth's surface. Finally, there are series of vertical tunnels built specifically for air ventilation in shafts.

In both open-pit and underground mine operations, the extracted coal is delivered outside of the mine by cars or on conveyor belts, and it is temporarily stored there until it is shipped by trucks, railroads, or ships and barges to the industrial users, buyers, and local consumers. Crushed coal can also be delivered through pipelines as a slurry substance, which is made of mechanically crushed small-to-tiny particles and pieces of coal mixed with oil or water in a special proportion.

8.3.3. Uses of coal

Historically, coal was used to produce heat in homes, and later in business buildings, factories, and plants. Though

FIGURE 110 (a) Open-pit coal mining; (b) underground coal mining

coal's use for indoor heating was eventually replaced by natural gas, coal is still used for home heating in small towns and villages around the world, especially in developing countries. Presently, coal is the largest source of energy used by the electrical power industry. Other coal-related industries include paper, cement, and ceramic production, manufacturing of iron and steels, and industrial steam generation. In turn, steam produced by coal-burning power-generating plants operates turbines and generators. To make it usable for these and many other industries, raw coal must be re-modified to eliminate gases, ash, and other undesirable microelements and components. The oldest coal-upgrading technologies include "coking" (a high-temperature treatment), and newer technologies include coal gasification and liquefaction, which produce synthetic gas ("syngas") and liquid fuel, respectively. They are considered below.

8.4. Coal-upgrading technologies

8.4.1. Coking

Because coals of low and regular ranks contain ~12–56% volatiles, they must be "cleaned" to remove dirt (ash), volatiles, and moisture before they can be used for industrial processes. Such a cleaning procedure is called "coking," and it requires the heating of coal at high temperatures in the absence of oxygen. Coking is carried on inside a special furnace at temperatures of 1000–1100°C, and for extremely "wet" and volatile-rich types of lignite and subbituminous coals at 2000°C. The heating with no oxygen eliminates volatile hydrocarbons (propane, benzene, and others) and some other pollutants, and it drives out some sulfur and nitrogen-bearing gases and water. After coking, due to process of devolatilization, coal loses ~30% of its weight, so that, for example, 1 ton of raw coal will produce 0.7 ton of coke. A variety of byproducts such as hydrocarbon gases, sulfur-bearing gases, nitrogen-bearing gases, carbon dioxide and carbon monoxide gases, H_2O-vapor, and other impurities are formed in the course of coking. The final task of the coking treatment is to increase coal efficiency and reduce emissions of pollutants into the atmosphere when the coke is burned.

8.4.2. Coal gasification

Coal gasification is a technology that allows the transformation of coal to gas "in situ" while coal seams are still underground in their primary position—i.e., enveloped by other rocks. In simple words, a gasification technique may be described as a use of two vertical wells drilled to the depth of the coal seam, as shown in Fig. 111. One well is for the injection of oxygen into the coal seam to trigger its ignition, which produces carbon monoxide (CO) and hydrogen (H_2), called "syngas." Another well is constructed to collect and transport gases and byproducts of coal combustion to the surface, where they will be refined by gas-waste-CO_2 separation facilities (Fig. 111). Gasification of coal may be conducted at varying pressures and temperatures, and the choice of the condition depends on the coal quantity and quality, the thickness of coal reservoirs, the mechanical and physical properties of the surrounding rocks, hydrogeology, and surface conditions. Underground coal gasification is usually applied to thin coal seams or to coal layers that are situated too deeply or have quantity and/or quality that are not high enough to be economically recoverable by other technologies. The man-made transformation of coal to gas "in-situ" is an alternative to natural gas production because it has a small cost technology. Also, it does not require mining, destruction of the land and ecosystems, on-ground cars or railroads, or sea transportation, and it does not produce solid waste.

It is believed at the present that the gasification of coal is one of the most promising "clean coal" technologies. The Tampa Electric Polk Power Station in Florida is an example of an advanced, "clean" coal power, and it uses a coal gasification process. Being in operation since 1996, it combines two different technologies: (1) mixing coal and oxygen to create a clean-burning

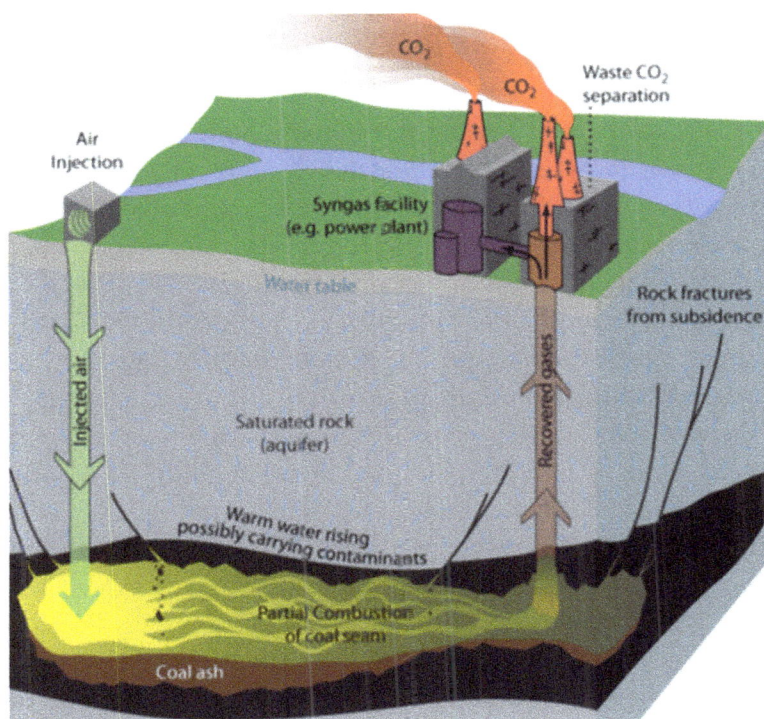

FIGURE 111 Scheme of underground coal gasification that resumed with the production of synthetic gas (syngas), which is a mixture of carbon monoxide (CO) and hydrogen (H)

gas (coal gasification) at the first stage, and (2) reusing exhaust heat to produce more electricity (e.g., "combined-cycle") at the second stage. This state-of-the-art power station is efficiently producing enough energy to provide electricity to 75000 homes in Florida.

8.4.3. Coal liquefaction

Coal liquefaction converts solid coal of subbituminous and bituminous grades to a petroleum-like liquid substance rich in hydrocarbon, and this technology operates in two different ways—indirect and direct (e.g., Williams and Larson, 2003).

Indirect coal liquefaction (ICL) is carried out in two stages: (1) coal gasification, which produces mainly "syngas," and (2) methanol synthesis that produces hydrocarbon-rich liquid fuel which may substitute oil and its refined products. The ICL technology is capable of producing synthetic gasoline, synthetic diesel fuel, or other types of the oxygenated fuels. These challenges are addressed using a variation of the H/C ratio, which is governed by the following reaction:

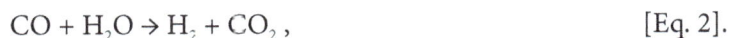

$$CO + H_2O \rightarrow H_2 + CO_2 , \hspace{3cm} [Eq. 2].$$

The byproduct of this reaction (CO_2) and other impurities such as H_2S and NO_x are usually removed from the system by a specific waste-gas filtering process, and as a result of this, the liquid fuel will be much cleaner than its precursor (coal) and will be more enriched in hydrocarbons, which increases the fuel energetic capacities.

A direct coal liquefaction (DCL) technology does not have any intermediate stages, but instead, it directly applies solvents or catalysts to mechanically crushed coal under high-pressure and high-temperature conditions. The solvents/catalysts break down the organic structures of coal, converting it to a liquid rich in hydrocarbons. Though the DCL technology is more advanced for clean energy production than the ICL, its results in liquid-fuel production have not yet been established on the commercial scale.

The final products of both the DCL and ICL coal liquefaction technologies generally have a higher hydrogen-carbon molar ratio than solid coal, and from this point of view, both are considered as alternative energy sources, which will keep the environment less polluted than the use of natural coals. However, coal liquefaction on industrial scales consumes a lot of energy, and therefore, its production as an alternative energy source is extremely expensive. At present, oil-based energy production is much less cost-effective than the synthetic liquid fuel made from coal. Thus, coal liquefaction could be economically viable only when the oil price reaches its historically highest level. Moreover, though the synthetic products are indeed cleaner than the raw coal, the processes of both coal gasification and liquefaction need further improvements in order to make "greenhouse gases" emission levels significantly less than those derived from crude oil products.

8.5. Why is coal used more than any other energy source?

There are six factors that make coal in the highest demand in comparison with other available fossil fuels or alternative energy sources:

1. *Abundance of coal deposits in the world.* Coal deposits are in abundance on all continents, and coal is mined in the territories of almost all countries, while the occurrences of oil and gas are restricted to certain geological and geographic regions. Historically, many of the richest oil and gas deposits have been found in politically unstable countries, which make oil and gas extraction and distribution to consumers difficult.

2. *Ease of mining and use.* Coal has inexpensive and relatively simple technological processes of extraction. It is a direct fuel that can be widely used from heating homes/buildings to the production of electricity.

3. *Ease of transportation.* Coal can be easily transported from the place of mining to distribution and consumption areas.

4. *Relatively cheaper transportation.* Most coal deposits occur in areas that are connected to industrial regions through railroads and sea shipping networks.

5. *More reliable source of energy than any others.* Many alternative energies such as solar and wind depend on the weather conditions, and they may not be available to a great extent in some populated regions of the world.

6. *Less cost and higher economic benefits for nations.* The lower cost of coal production is due to all of the factors above, and moreover the coal industries give opportunities for employment, they provide significant economic benefits to nations.

Adverse effects of coal on human health and the environment   8.6.

A chain of different processes, from coal mining to refining and consumption to the production of heat and electricity, may create physical, chemical, and epidemic hazards of different magnitudes for humans and the environment.

8.6.1. Physical disasters from the coal industries

8.6.1.1. Active coal mines

The coal-mining industry is one of the most dangerous on the Earth no matter if coal is mined by the open-pit or underground methods. In underground mining, injuries may be caused by roofs collapsing, seismic tremors caused by mining-induced seismicity, and explosions of gas and coal dust. These and other factors put miners' lives in danger, especially within the confined spaces of mines.

Roofs collapse. Injuries of miners may be caused by mechanical instabilities of the mine's construction because series of underground shafts/tunnels and passages remain there for long periods of time, during which they are under the load of heavy rocks overlying the mine construction. The collapsing of mine roofs may be due to human errors or engineering miscalculations—for example, if the walls and ceilings of the coal shafts have not been properly secured, or if the shafts were excavated too deeply and this caused cracks in their walls, thus weakening the structure.

Mining-induced seismicity. Seismicity of very low magnitude can be caused by mining because tunnels and "voids" created underground may alter the balance of forces in the rocks, causing rock failure, which is called "rock bursts." The deeper mining progresses, the larger is the stress in the rock, which could eventually lead to collapse of the tunnels and shafts and formation of "seismic" waves causing damage similar to small earthquake events.

Gas and coal dust explosions. Gas explosion originating from a buildup of methane gas is a frequent hazard in coal mines. Thus, ventilation of the underground work area is essential to prevent collections of gas. Coal dust explosions in mines are often caused by the ignition of coal dust particles triggered by friction, hot surfaces, electrostatic discharge, or fire. In the U.S., one of the largest and most devastating coal mine explosions occurred in West Virginia at the Upper Big Branch coal mine in 2010, which took the lives of twenty-nine of the thirty-one miners who were at work that day. The mine was operated by the Massey Energy Co., which, as investigations revealed, had mismanaged the proper maintenance of the underground ventilation systems, which caused high methane levels followed by explosion. The concentration of toxic gases in the mine remained high for over two months after the explosion, which prevented national Mine Safety and Health Administration (MSHA) investigators from entering the mine. The MSHA found 505 violations of the 2009 underground mining protocol, and although within those, many issues were related to the poor ventilation system, the Massey Energy Co. had never acted on any of them. In the final report of the MSHA investigation, the Massey Energy Co. was accused of intimidating miners and state officials and was ordered to pay a $10.8 million civil fine plus $209 million for the settlement (for full report, see http://www.msha.gov/Media/PRESS/2011/NR111206.asp).

8.6.1.2. Abandoned coal mines

Various adverse environmental impacts may occur at abandoned coal underground and open-pit mines. The Office of Surface Mining Reclamation and Enforcement (OSMRE) of the U.S. Department of the Interior is a responsible organisation for reclamations of abandoned coal mines. Reclamation is the restoration of land that has been mined to its natural pre-mining conditions, or so that it has some economically useful purpose. The OSMRE's responsibilities include establishing an inventory of high-priority sites posing health and safety hazards, collection of the reclamation fees from the coal mining industry, and other regulatory issues between the coal mining industry and local administrations and citizens.

When the mining operation is over, both abandoned underground mine sites and open-pits can cause some environmental problems. Although mining companies are obliged to restore the excavated lands and disturbed ecosystems to their original state, such processes are usually long-lasting because of their high cost, and sometimes they are never achieved because many companies declare bankruptcy. The major environmental impacts from abandoned mines include: (1) acid drainage, (2) toxic metal (As, Pb, Cd), metalloid (Hg), and halogen (Cl, F, etc.)

contamination of ground and surface waters and soils, (3) air emission and deposition of dust, and (4) physical impact. As open-pits remain "open" for a long period of time, wind blows out the coal dust into the atmosphere, creating particulate matter and aerosols that increase air pollution. Rain water easily gains access to the open-pits and washes the exposed coal remains, dissolving trace elements, including those that are toxic. When the water eventually evaporates, hazardous materials are left behind, and they continue to spread toxic components further into the environment. Local administrations and private companies responsible for recreation activities, tourism, fisheries, agricultural resources, and watersheds should be concerned about toxic pollutions originating from abandoned coal mines.

8.6.2. Spontaneous coal seam fires

A spontaneous coal seam fire is a process that starts "by itself"— e.g., independently of human-related external sources. Many scientists agree with the concept that the fire is caused by the heat accumulation in the course of oxidation processes. The latter may be accelerated by catalysts, such as an excess of water or mineral pyrite (FeS_2), which are available in coal, and other parameters such as coal mining methods, types of ventilation systems in underground mines, airflow rate, rank of coal, and ratio of organic and non-organic components in coal (Mao et al., 2013). Coal oxidation is an exothermic reaction producing heat, which eventually initiates ignition. Coal fires are observed in all parts of the world where coal reserves exist. Underground coal fire typically releases gases that contain up to fifty toxic compounds, and some are carcinogenic. These compounds include benzene (C_6H_6), methylene chloride (CH_2Cl_2), hydrogen sulfide (H_2S), sulfur oxides (SO_x), carbon monoxide (CO), arsenic (As), selenium (Se), mercury (Hg), lead (Pb), and fluorine (F), to name just a few. They are all extremely toxic and represent a great threat to human health, causing pulmonary heart disease, bronchitis, chronic pulmonary disease, lung cancer, and stroke. In addition, heat from the fire leads to deterioration of the ecosystem, and burning releases large amounts of greenhouse-relevant gases such as CO_2 and NO_x into the atmosphere, which contribute to global warming.

Another side of destruction caused by spontaneous coal fires is that they usually occur close to active mines, occupying large areas and lasting for several decades. For example, the South Canyon underground coal fire near Glenwood Springs, Colorado, has been burning about a century (starting in 1910), and the technology to stop this fire has not yet been found. Meanwhile, this underground coal fire has already burned the surface of >50 km^2 and destroyed the ecosystem, the forest, twenty-nine homes, and fourteen other buildings around Glenwood Springs. The underground coal seam fire is still continuing in Colorado with no sign of extinction.

Coal fires may also be caused by the ignition of shallow coal seams exposed to the Earth's surface, as well as ignition of coal dusts and debris by forest fires, lightning, technical accidents, and other problems that could be aggravated by people. One example is a famous coal fire that was ignited in 1962 in Centralia, Pennsylvania, in the course of a landfill reduction by burning to control rodents. This underground coal fire was not suppressed, and it continues to burn at present. In the 1980s, the government declared Centralia as an unsafe territory for human life, and over 1000 residents have been relocated. Another example is in China. Each year, coal mine fires burn about 200 million tons of coal, releasing ~360 million tons of carbon dioxide and other greenhouse gases into the atmosphere.

8.6.3. Air pollution from coal-burning plants and local home heating

8.6.3.1. Sulfur-bearing components, acid drainage, acid rain, and fly ash

There are considerable concerns over air pollution with uncontrolled, sulfur-bearing chemical compounds, gases, and other toxic elements released from coal combustion, which produce acid drainage, acid rain, and fly ashes.

Many minerals are burned together with coal and produce ash, but some of them, such as pyrite, break down with the production of Fe and S, which are eventually oxidized, producing ash consisting of small particles of FeO and/or Fe_2O_3 and sulfur dioxide (SO_2) gas. Sulfur dioxide is an extremely toxic gas, and if it is inhaled in a high concentration, it can be fatal for people. In other cases, the health effect is attributed to a non-fatal concentration of sulfur in aerosols, where it reacts with other substances, forming sulfates. In addition, remaining parts of SO_2 gas are absorbed by fine particles of dust, mostly $PM_{2.5}$ (e.g., particulate matter of 2.5 micron in diameter). Inhalation of such fine particles can cause severe irritation of the nose and throat, coughing, shortness of breath, difficulty with breathing, and tightness in the chest. A single exposure to a high concentration of SO_2 gas can cause long-lasting asthma. Sulfate aerosols may be transported for long distances before their precipitation takes place.

According to the EPA, in the U.S., about 60% of air pollution that comes from coal-burning power stations is composed of SO_2 (Fig. 112). SO_2 gas emitted into the atmosphere forms acid rain and can lead to widespread acidification of soils and ecosystems. Another product of SO_2 interaction with the atmosphere is

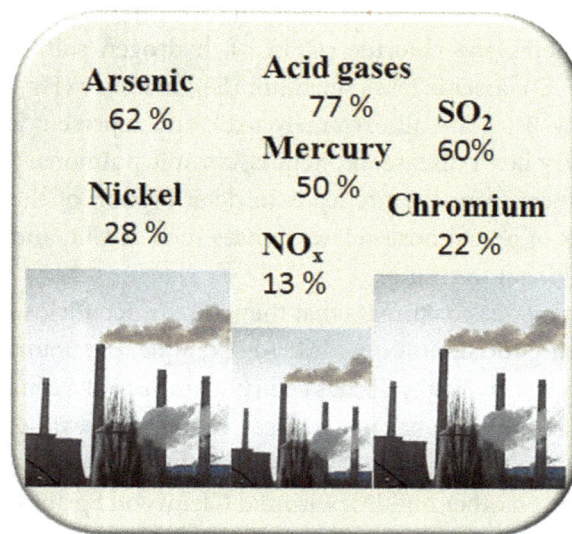

FIGURE 112 Portions of toxic gases and compounds contributed to US air pollution by coal-burned power plants (for example, if we assume that all sources of arsenic pollution equal 100%, then 66% of the arsenic pollution originates from the coal-burning plants)

the H_2SO_2-gaseous phase, which scatters solar radiation. Such a scattering produces a kind of "false" cooling effect that camouflages the real warming caused by increased greenhouse gases. A total portion of acid gases (which also include H_2S and CO_2) emitted from the coal power stations is about 77%.

Both acid rains and acidic drainages formed near abandoned coal mines are environmental hazards because they are enriched with toxic elements. They contaminate waters, can kill fish and other aquatic life, dissolve surrounding rocks by washing out additional toxic elements, and can destroy the surface of building facades and monuments. In some areas near coal-burned power plants, the landscape looks environmentally dramatic, with no plants or trees around. The average coal-fired power plant produces about one million tons of ash per year, which is usually collected in special ponds in the vicinities of plants. Fig. 113 is a photograph of one of such area near the Kingston Fossil Plant in Tennessee, where on December 22, 2008, coal fly ash slurry (a coal-combustion waste product that is captured and stored in wet form) was catastrophically spilled around the ash retention pond. About 20441 m^3 of wet ash escaped through a breach in the wall. As a result, a high wall of ash was formed that extended to approximately 1.6 km (Fig. 113) around the pond. The spill also infiltrated the Emory River, burying about 120 hectares of land in sludge. The 20441 m^3 of this extremely toxic sludge contained about 20 tons of arsenic, 22 tons of lead, 64 tons of manganese, and 635 tons of barium compounds. About forty homes, buildings, and other structures were damaged or totally destroyed by the ash slurry flow, and many residents had to permanently leave their homes.

On December 23, 2008, the environmental organization Greenpeace, after conducting an independent investigation, declared that the spill could have been prevented and asked for a criminal investigation of this hazardous event. The land and house owners filed suit for $165 million against the Tennessee Valley Authority (TVA), which operated the Kingston Fossil Plant, but for a long time, the TVA tried to avoid its responsibility for the coal

FIGURE 113 View of the Kingston Fossil Plant, coal fly ash slurry spill (Tennessee, USA)

ash sludge spill disaster. Only on August 1, 2014, was it finally announced that the TVA had agreed to pay $27.8 million to settle claims from property owners who suffered damages due to the 2008 hazardous spill. The spill was eventually cleaned up by the efforts of the TVA. It took more than three years to completely eliminate the remains of dried coal dust mixed with toxic elements and heavy metals. However, these measures have not been able to restore the contaminated region to the formerly environmentally healthy conditions. As a positive feedback, the Kingston Fossil Plant spill prompted the TVA and other coal-burning power plants to re-evaluate their approaches to how they store coal ash and other hazardous byproducts released from burned coal.

The U.S. Congress held several hearings in the aftermath. Until the Kingston catastrophe, there had been no regulations with regard to coal ash remains, because the industry and some lawmakers opposed the classifying of coal ash as a hazardous material. Two years after the spill, in May 2010, the EPA proposed some coal ash regulation, but unfortunately it did not finalize the complete protection protocol. Though since 2008, there have been more than 200 nationwide sites contaminated with industrial coal ash byproducts, only recently the EPA has announced that the agency will finalize the first-ever federal regulations for the disposal of coal ash by December 19, 2014 (Saylor, 2014).

8.6.3.2. Chronic arsenic and fluorine poisoning from unprocessed coal combustion

Arsenosis. All natural coal contains arsenic in small concentrations, and the average content is different in coal extracted from different geological locations. For example, median concentrations of arsenic in coal mined in the U.S. vary from 1.4 to 72 ppm, whereas Chinese coal from Guizhou Province contains up to 35000 ppm arsenic (Finkelman et al., 1999). Hundreds of millions of people living in rural areas all over the world still use unprocessed coal for household heating and cooking in home stoves. In the U.S., domestic coal use for heating and cooking is very rare even in rural areas, and therefore, arsenic exposure through coal combustion is limited. Although arsenic poisonings of people through burning unprocessed coal at homes is rare in industrial countries, it happens frequently in developing countries. For example, it is very common in China, India, and South Africa, where in addition to cooking, the coal heat is used to dry food to make it safe for months and even years ahead. Diseases that are regularly found among particular groups of people in a certain area are called *endemic*. Fig. 114 shows the interior of a home (without any ventilation) in southwestern Guizhou Province, China, where arsenic-rich coal is used for heating home and cooking meals. It is also used to dry chili paper, resulting in high arsenic contamination. As the cuisine among local people traditionally includes spicy food and most such chili papers contain an average of ~500 ppm of arsenic, it is a major source of poisoning. Chronic arsenic poisoning has affected more than 3000 people in the Guizhou Province of China alone (Finkelman et al., 1999). People affected by arsenic poisoning typically exhibit symptoms of hyperpigmentation and scaly lesions on the skin of the hands and feet (see Fig. 70, Chapter 5).

Fluorosis. Another serious health problem caused by indoor combustion of unprocessed coal in China is endemic fluorosis, mostly occurring in southwestern Guizhou and scattered as well through Sichuan, Shaanxi, Yunnan, and Hubei provinces (Liu et al., 2007). According to the WHO standards, the average level of fluoride (F⁻) that is not dangerous for life is 1–4 mg/L, and a fluoride concentration of >4 mg/L is dangerous for health (fluorine, F, is a chemical element, and its anion, F⁻, or any of the compounds containing the anion F⁻, are called fluorides). Geological studies have shown that local coals used in areas with endemic fluorosis contain clay minerals enriched in fluorine. These local clay rocks, rich in fluorine, are used by local people as a briquette binder for fine coals. The concentration of fluorine in Chinese coal is not homogeneous, as it falls in the 82 mg/kg to 4400 mg/kg range, and the concentration level of

FIGURE 114 Interior view of a residence in southwestern Guizhou Province, China, where arsenic-rich coal briquettes are used to dry crops of chili peppers, resulting in arsenic toxicity

fluorine depends on geochemical conditions of the surrounding rocks, geological position, and the ages of the coal seams (Chen et al., 2013). In the case of raw coal and clay combustion, most of the fluorine is released as HF gas in the indoor area, followed by its absorption by house dust, which adheres to the surface of any food available in the room. If corn and chili paper are dried indoors, the concentration of fluorine in these products will increase up to 1.5–110 mg/kg, compared with a level of <1 mg/kg in fresh crops.

Many studies have unconditionally shown that dental fluorosis and skeletal fluorosis, which causes deformities, are exclusively attributed to excessive fluoride. Consumption of local peoples' favorite foods—e.g., roasted corn and chili—with elevated levels of fluorine is the main cause of endemic fluorosis in China, India, and other developing countries. In 2000, about 18 million

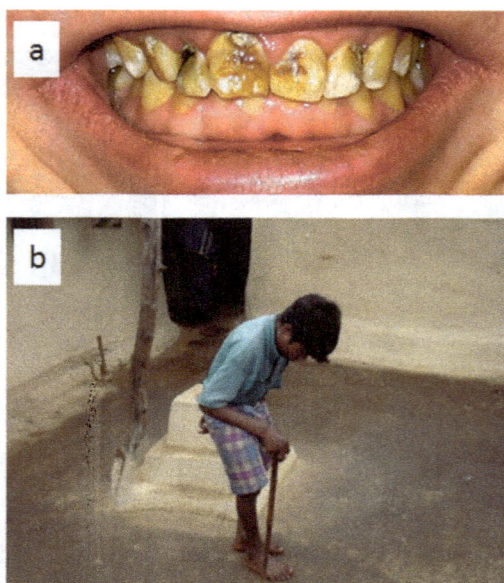

FIGURE 115 (a) Dental fluorosis (b) and skeletal fluorosis caused by nutritional deficiency combined with exposure to high levels of fluorine from domestic coal combustion (Chhattisgarh, India)

cases of dental fluorosis and 1.5 million cases of skeletal fluorosis were diagnosed in territories of China that have extensive use of unprocessed coal for cooking and drying corn and chili paper. The situation with endemic fluorosis in China is more severe than that of arsenosis. Endemic fluorosis is recognized by the presence of mottling of tooth enamel (dental fluorosis), deformity of the knees, abnormal spinal curvature, and limited movement of joints (skeletal fluorosis; Fig. 115). The WHO recently estimated that 2.7 million people in China have the crippling form of skeletal fluorosis, and in India, seventeen of its thirty-two states have been announced as endemic areas for fluorosis, with an estimated 66 million people at risk and 6 million people seriously afflicted.

In the U.S., the artificially fluoridated drinking water from the taps contains 0.7–1.2 mg/L of fluoride. The EPA has established 4 mg/L fluoride as the maximum contaminant level (MCL). Symptomatic skeletal fluorosis is almost unknown in the U.S., though about a dozen cases have been reported.

Other pollutants from raw coal. Mercury, selenium, thallium, and some organic compounds remain under consideration as extremely harmful toxic pollutants, although there is no direct evidence that these elements emitted by raw coal combustion in indoor domestic stoves have triggered endemic diseases. This is mainly because the many symptoms caused by the mentioned pollutants overlap with those related to other diseases of a non-environmental nature. Further studies of coal geochemistry, together with medicinal statistics of possible cases, may help to prevent some health problems caused by raw coal burning indoors.

8.6.4. Black lung—an occupational disease of coal miners

Due to long exposure to coal dust, all coal miners eventually suffer from one or more occupational pneumoconiosis diseases, which are also called "coal workers' pneumoconiosis" (CWP) or black lung disease (Fig. 116), or miner's asthma and miner's silicosis. These similar diseases also strike those working in coal-related industries such as coal loading and stowing coal for storage, milling graphite, manufacturing carbon electrodes, and carbon black compounds used

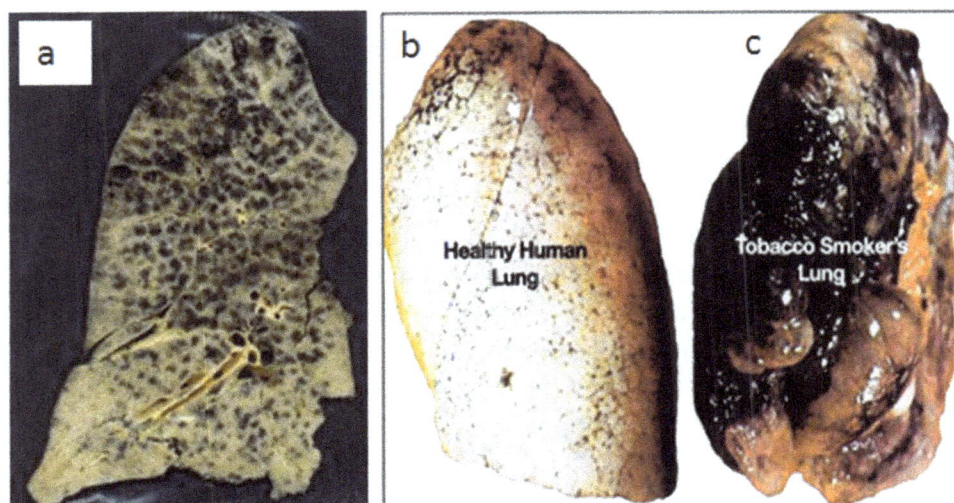

FIGURE 116 (a) Black lung of a coal miner suffering from progressive massive fibrosis; for comparison, (b) a healthy human lung and (c) a tobacco smoker's lung

for tires and other rubber product fabrication. Black lung disease is not curable because when coal dust is inhaled, the particles are accumulated in the lungs and cannot be removed by any human biological processes. The coal dust particles damage living tissues, causing inflammation, fibrosis, and sometimes necrosis, or premature death. In addition, the amount of the dust increases as exposure time increases. The severity of the black lung disease depends also on the type of coal dust and the amount of time the dust was in the air.

The National Institute for Occupational Safety and Health (NIOSH) shows that miners used to have to spend at least fifteen years underground before symptoms of the black lung disease appeared. Recently, it was found that the disease rose within miners of younger ages and is progressing more rapidly. NIOSH has estimated that black lung disease killed more than 76000 miners between 1968 and 2010.

Black lung disease is preventable, and since 1969, when the U.S. Federal Coal Mine Health and Safety Act became law, the number of cases of this disease has significantly decreased by 80–90%. However, according to reports by NIOSH, cases of black lung have been increasing since 1990 among coal miners (for full report, see http://www.msha.gov/S&HINFO/BlackLung. This tendency may be explained by several factors related to how coal is mined:

1. By extensive drilling through rock to reach coal seams;
2. Because drilling through rocks releases silica dust, which is more toxic than coal dust (silica dust is more dangerous for the respiratory tract than is coal dust);
3. Longer shifts and extended work weeks may increase exposure to coal dust.

The only real method of preventing black lung disease is avoiding inhalation of coal dust. In the U.S., there are standard procedures established by the OSHA for coal-related industries. Workers are required to wear a protective mask. If coal dust comes into contact with the skin, workers should wash the affected areas with soap and water, immediately remove their clothing (which is contaminated with coal dust), and refrain from eating, drinking, smoking, or taking medication in places where coal dust is handled, processed, or stored. There is no consensus between governmental organizations as to exposure limits in coal mines due to the lack of a suitable method to measure particle size in coal dust. In this situation, workers should follow the safety protocol suggested by their company.

8.7. Carbon dioxide emission and global warming

Coal-burning power plants and other industries related to consumption of refined coal, as well as the use of raw coal for indoor heating and food preparation, are increasing the levels of carbon dioxide and other gases in the atmosphere. This increase is causing an "enhanced greenhouse effect," which creates more heat on the Earth's surface. Although the subject of global warming is intensively debated among politicians, industry leaders, scientists, and environmentalists,

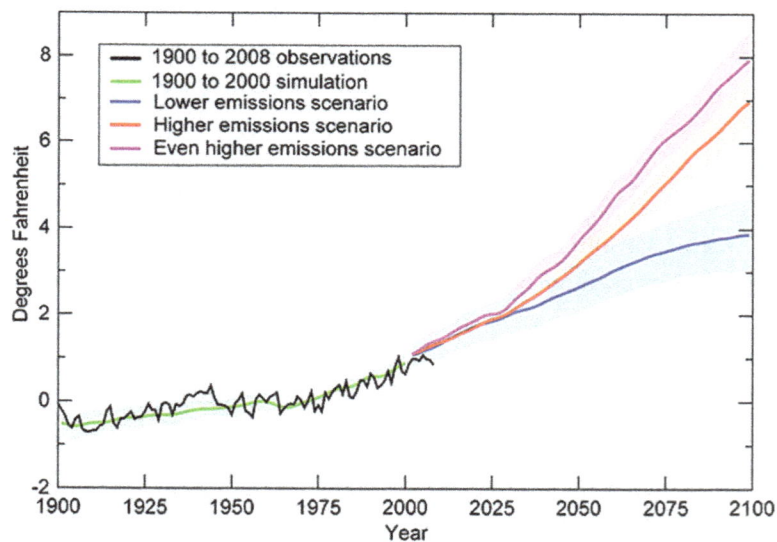

FIGURE 117   Past, modern, and projected changes in global warming according to three independent models. The shaded areas are the possible ranges; lines show the central projections.

the fact is that since pre-industrial times, the atmospheric concentration of greenhouse gases has significantly increased: CO_2 by 31%, CH_4 by 15%, and NO_x by 16% (Mitchel et al., 2001). The level of CO_2 in the atmosphere at present is 375 ppm (Fig. 117), which is the highest level of the past 20 million years. Data from seven different models, which are plotted in Fig. 117, show a coherently increasing concentration of CO_2 in the atmosphere by 2100 from 550 ppm (minimal—model B1) to 1000 ppm (maximal—model A1B).

The scientific community in general agrees that the enhanced greenhouse effect is a real phenomenon because industrial activity, transportation, and the use of fossil fuels additionally contribute to concentrations of greenhouse gases in the atmosphere. However, consequences that will take place from the long-term impact on human civilizations and the global climate are not known.

In order to minimize the anthropogenic contribution of CO_2 and other gases that cause an increase of the greenhouse effect, the United Nations Framework Convention on Climate Change (UNFCCC) has created the Kyoto Protocol—an international treaty that proposed binding obligations on industrialized countries to reduce greenhouse gas emissions. The Kyoto Protocol "… recognizes that developed countries are principally responsible for the current high levels of greenhouse gas emissions in the atmosphere as a result of more than 150 years of industrial activity, and it places a heavier burden on developed nations under the principle of common but differentiated responsibilities" (for details see http://unfccc.int/kyoto_protocol/

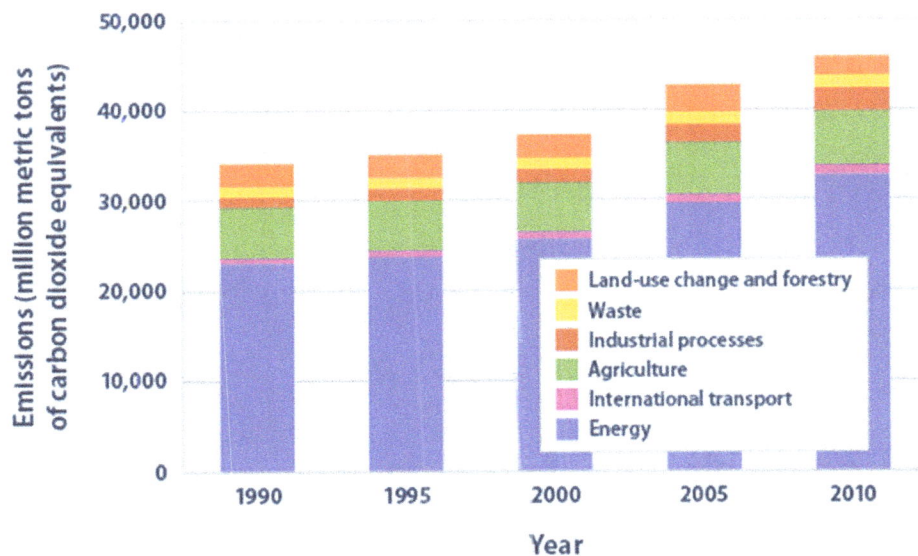

FIGURE 118   Global greenhouse gas emissions by sectors, from 1990 to 2010; emissions are expressed in million metric tons of carbon dioxide equivalents

items/2830.php). Since being enacted in 1997, the Kyoto Protocol has had a target of minimizing emissions of six greenhouse gases: carbon dioxide, methane, nitrous oxide (N_2O), hydrofluorocarbons (HFC), perfluorocarbons (PFC), and sulfur hexafluoride (SF_6). By ratifying the treaty, the developed countries pledged that in seven years (e.g., by the end of 2012), they would lower their greenhouse gas emissions to 5% below 1990 levels, but that target did not happen, as we now know. The global greenhouse gas emission diagram shows that by 2010, the energy-generation industry was responsible for increasing greenhouse gases up to 30% above the 1990 level, while other sectors have contributed only a few percent or have even slightly decreased their commitment (Fig. 118).

Though the Kyoto Protocol represented an important step forward in the effort to tackle global warming, its actions were not successful, mostly because there was no consensus between economically developed countries about the procedure. For example, the U.S. is not a member of the Kyoto Protocol, and Canada withdrew its participation in 2012 because of "enormous financial penalties" for exceeding the CO_2 limit suggested by the treaty. The U.S. Administration did not believe that the Kyoto Protocol was an effective tool for the reduction of greenhouse gas emissions because developing countries and those whose economies are rapidly growing were exempted from curbing their CO_2 and other gas emissions. Those countries included the world's largest greenhouse polluters—China, India, Thailand, Indonesia, Egypt, and Iran—that are rapidly increasing their CO_2 emissions by more than 10–20% per year. By 2011, the world total CO_2 emission had reached 34 billion tons, and more than 40 billion tons were forecasted for 2020, which is 20% more. The

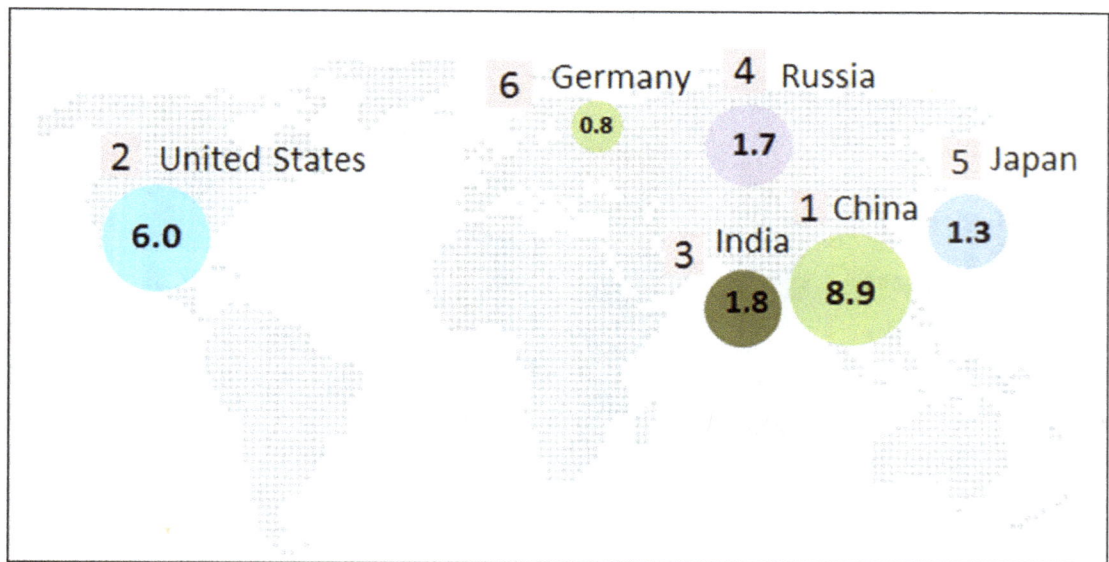

FIGURE 119   Top six nations that produce carbon dioxide, according to the World Resources Institute

leading country in CO_2 emission in 2011 was China, which emitted 8.9 billion tons; second in greenhouse emissions was the U.S., with 6 billion tons; third was India, with 1.8 billion tons; and Russia (1.7 billion tons), Japan (1.3 billion tons), and Germany (0.8 billion tons) followed (Fig. 119).

While the U.S. develops "clean coal" technologies and explores alternative energy sources, China, India, and the other countries mentioned above continue to produce pollution without trying to replace the older, less efficient coal-burning power stations with new clean coal technologies. There are also many other problems such as: (1) the fact that some countries do not accept the concept of global warming increasing through the influence of the anthropogenic emission of CO_2 into the atmosphere; (2) the financial aid shortage to developing countries to limit CO_2 emission; and (3) energy tax-adjustment uncertainties.

Review of publications (e.g., Biermann and Brohm, 2005) and data analysis show that the Kyoto Protocol has generated only a slight effect on curbing global emissions growth, and this is mostly due to the limited financial support to developing countries and the non-acceptance of the treaty by some economically developed countries. New efforts in the international harmonization of the environmental/energy tax-adjustment policies are required to boost the Kyoto Protocol actions.

References

Biermann, F. and R. Brohm. 2005. Implementing the Kyoto protocol without the USA: The strategic role of energy tax adjustments at the border. *Climate Policy* 4:289–302.

Chen, J., G. Liu, Y. Kang, B. Wu, R. Sun, et al. 2013. Coal utilization in China: environmental impacts and human health. *Environmental Geochemistry and Health* 36:735–753.

Finkelman, R. B., H. E. Belkin, B. Zheng. 1999. Health impacts of domestic coal use in China. *Proceedings of the National Academy of Sciences of the USA* 96: 3427–3431.

Liu, G, L. Zheng, C. Qi, Y. Zhang. 2007. Environmental geochemistry and health of fluorine in Chinese coals. *Environmental Geology* 52:1307–1313.

Mao, Z., H. Zhu, X. Zhao, J. Sun, Q. Wang. 2013. Experimental study of characteristic parameters of coal spontaneous combustion. *Procedia Engineering* 62:1081–1086.

Mitchell, T. D., M. Hulme, M. New. 2001. Climate data for political areas. *Observations* 109–112.

O'Keefe, J. M. K., A. Bechtel, K. Christianis, S. Dai, W. A. DiMichele, at al. 2013. On the fundamental difference between coal rank and coal type. *International Journal of Coal Geology* 118:58–87.

Ryer, T. A., and A. W. Langer. 1980. Thickness change involved in the peat-to-coal transformation for a bituminous coal of Cretaceous age in Central Utah. *Journal of Sedimentary Petrology* 50:987–992.

Saylor, J. EPA Agrees to Deadline for First-Ever U.S. Coal Ash Regulations. *Earthjustice.* http://earthjustice.org/news/press/2014/epa-agrees-to-deadline-for-first-ever-u-s-coal-ash-regulations.

The World Factbook 2013–14. Washington, D.C.: Central Intelligence Agency, 2013.

Xu, M., R. Yan, C. Zheng, Y.Qiao, J. Han, et al. 2003. Status of trace element emission in a coal combustion process: a review. *Fuel Processing Technology* 85:215–237.

Ward, C. R. 1986. Review of mineral matter in coal. *Australian Coal Geology* 6:87–110.

Williams, R. H., and E. D. Larsen.2003. A comparison of direct and indirect liquefaction technologies for making fluid fuel from coal. *Energy for Sustainable Development* 7:103–129.

Health effect of air pollution. Washington, D.C. USA:EPA. 2010. www.epa.gov/oar/caa/Healthslides/pdf.

Web resources

http://unfccc.int/kyoto_protocol/items/2830.php

http://www.msha.gov/S&HINFO/BlackLung

http://www.msha.gov/Media/PRESS/2011/NR111206.asp

Image Credits

Figure 103a: Copyright © Peter Cowling (CC BY-SA 2.0) at http://commons.wikimedia.org/wiki/File:Coal_seam,_Hartley_Bay._-_geograph.org.uk_-_571834.jpg.

Figure 103b: Copyright © Vzb83 (CC BY-SA 3.0) at http://commons.wikimedia.org/wiki/File:Fern_fossil.jpg.

Figure 104a: Copyright © Fungus Guy (CC BY-SA 3.0) at http://commons.wikimedia.org/wiki/File:Common_green_peat_moss_(Orphan_Lk)_3.JPG .

Figure 104b: Copyright © Ragesoss (CC BY-SA 3.0) at http://commons.wikimedia.org/wiki/File:Schultz_Sphagnum_Peat_Moss.jpg.

Figure 104c: Copyright © Edal Anton Lefterov (CC BY-SA 3.0) at http://commons.wikimedia.org/wiki/File:Lignite-coal.jpg.

Figure 104d: United States Geological Survey, "Coal Mine," http://www.usgs.gov/blogs/features/files/2013/02/PRB-Coal-Loading.jpg. Copyright in the Public Domain.

Figure 105a: United States Geological Survey, "Peat," http://www.usgs.gov/blogs/features/files/2011/12/Peat2_2009_dp.jpg. Copyright in the Public Domain.

Figure 105b: United States Geological Survey, "Lignite," http://www.usgs.gov/blogs/features/files/2011/12/ligniteB_2010_dp.jpg. Copyright in the Public Domain.

Figure 105c: United States Geological Survey. "Bituminous," http://www.usgs.gov/blogs/features/files/2011/12/bituminous1_2009_dp.jpg. Copyright in the Public Domain.

Figure 105d: United States Geological Survey. "Anthracite," http://www.usgs.gov/blogs/features/files/2011/12/Anthracite2_2009_dp.jpg. Copyright in the Public Domain.

Figure 107a: Copyright © Wojsyl (CC BY-SA 3.0) at http://commons.wikimedia.org/wiki/File:Peat_Lewis.jpg.

Figure 107b: Copyright © MacIomhair (CC BY-SA 3.0) at http://commons.wikimedia.org/wiki/File:Peat-Stack_in_Ness,_Outer_Hebrides,_Scotland.jpg.

Figure 110a: Copyright © Stephen Codrington (CC BY-SA 2.5) at http://commons.wikimedia.org/wiki/File:Strip_coal_mining.jpg.

Figure 110b: Tangopaso, "Underground mine Ales 2," http://commons.wikimedia.org/wiki/File:Underground_mine_Ales_2.jpg. Copyright in the Public Domain.

Figure 111: Copyright © Bretwood Higman (CC by 3.0) at http://commons.wikimedia.org/wiki/File:UCGprocessfigure-01.png.

Figure 112: Bogdangiusca, "Pollution de l'air," http://commons.wikimedia.org/wiki/File:Pollution_de_l%27air.jpg. Copyright in the Public Domain.

Figure 113: Copyright © Brian Stansberry (CC by 3.0) at http://commons.wikimedia.org/wiki/File:Kingston-plant-spill-swanpond-tn2.jpg.

Figure 114: United States Geological Survey, "Interior view of residence in southwestern Guizhou Province, China, where arsenic-rich coal and coal briquettes are used to dry crops (chili peppers) that are later consumed, resulting in arsenic toxicity," http://pubs.usgs.gov/fs/2005/3152/. Copyright in the Public Domain.

Figure 115a: Copyright © FinalGamer (CC BY-SA 3.0) at http://en.wikipedia.org/wiki/File:Severe_fluorosis.JPG.

Figure 115b: Copyright © Pankaj Oudhia (CC BY-SA 3.0) at http://en.wikipedia.org/wiki/File:Fluorosis_in_Chhattisgarh.jpg.

Figure 116a: Center for Disease Control, "Black Lung," http://www.cdc.gov/niosh/topics/pneumoconioses/. Copyright in the Public Domain.

Figure 116b: National Institute on Drug Abuse, "Healthy lung-smokers lung," http://commons.wikimedia.org/wiki/File:Healthy_lung-smokers_lung.jpg. Copyright in the Public Domain.

Figure 116c: National Institute on Drug Abuse, "Healthy lung-smokers lung," http://commons.wikimedia.org/wiki/File:Healthy_lung-smokers_lung.jpg. Copyright in the Public Domain.

Figure 117: Environmental Protection Agency, "Past, modern and future projected changes in global warming," http://www.epa.gov/climatechange/images/science/ScenarioTempGraph-large.jpg. Copyright in the Public Domain.

Figure 118: Environmental Protection Agency, "Global Greenhouse Gas Emissions by Sector, 1990-2010," http://www.epa.gov/climatechange/pdfs/climateindicators-full-2014.pdf. Copyright in the Public Domain.

Figure 119: "World Map," http://www.wagenborg.com/img/map-world.png. Copyright in the Public Domain.

Review Questions

1. What is coal?

2. In what geological era were the world's richest coal deposits formed?

3. What kind of climate was preferable for coal formation?

4. What is coalification?

5. What is the chemical composition of coal?

6. Rank of coal: from peat to anthracite.

7. How was peat formed? Where is peat used today?

8. Where are the world's richest peat deposits?

9. What is lignite? How is lignite formed?

10. What is the chemical composition of lignite?

11. Does lignite have a higher heat energy density than peat?

12. In what countries is lignite mined?

13. What is bituminous coal? How is it formed? What are the pressure and temperature conditions required for its formation?

14. What country has the largest reserves of bituminous coal?

15. What is anthracite, and what is its chemical composition? How does anthracite differ from other types of coal?

16. Explain the pressure and temperature conditions that are required for the formation of anthracite.

17. There are minerals and trace elements in coal. How does the rank of coal depend on these parameters?

18. What do we learn from minerals and trace elements in coal, and why is this important?

19. Is coal a renewable natural resource?

20. How much coal is left in the Earth?

21. How is coal mined?

22. What is open-pit mining?

23. What is underground mining?

24. Where was coal used historically?

25. Where is coal most used at the present time?

26. Define the coal-upgrading technologies: coking, coal gasification, and coal liquefaction.

27. Explain the principal chemical reactions that underlie the processes of "clean coal" production.

28. Is natural coal better than its synthetic product produced after coking?

29. Are bituminous coal and anthracite better than the final products of both DCL and ICL coal liquefaction technologies?

30. What are the six most important factors that make coal the highest in demand in comparison with any other energy sources?

31. Name one adverse effect of coal on human health and the environment.

32. Explain some of the physical disasters from the coal industries. Include active and abandoned coal mines.

33. What is a spontaneous coal seam fire? How long do such fires last?

34. Explain why sulfur-bearing gases, acid drainage, acid rain, and fly ash originated from coal combustion in electrical power-generating industries and home stoves are hazardous products.

35. Give examples of endemic diseases from combustion of unprocessed coals.

36. What is a standard average level of fluoride (F⁻) that is not dangerous for life according to the WHO?

37. In addition to arsenic and fluorine, what other pollutants are emitted from indoor combustion of raw coal?

38. "Black lung" is an occupational disease of the coal miners. How long does it take to develop?

39. How are carbon dioxide emission, global warming, and greenhouse gases connected with the coal industries?

40. What is the Kyoto Protocol?

41. What is status of the U.S. in the Kyoto Protocol organization?

42. What countries had leading roles in CO_2 emission in 2011?

43. Has there been any visible tendency in decreasing CO_2 emission since 1990?

44. Why did the Kyoto Protocol actions fail to minimize the global greenhouse gas emission?

45. Do you think that the greenhouse gas emission into the atmosphere can be controlled?

Quizzes (see answers on page 307)

1. Lignite is…
   a. A mineral containing arsenic.
   b. A mineral containing mercury.
   c. The lowest grade of coal, also called "brown coal."
   d. The highest grade of coal, also called anthracite.

2. Which of the following is NOT a problem associated with coal production?
   a. Because of the combustible nature of coal dust and gases, coal mines are prone to terrible fires.
   b. The U.S. has few coal reserves and is therefore dependent on the import of coal from politically unstable parts of the world.
   c. Coal mining produces toxic dirt, dust, and sludge that can pollute the environment.
   d. Inhalation of toxic dust produced by coal mining can lead to black lung disease in miners.
   e. Mining of coal can be very destructive to land by literally removing the tops of mountains.

3. Which one of the following belongs to a non-renewable energy source?
   a. Solar power.
   b. Coal.
   c. Wind power.
   d. Water power.

4. Adverse environmental impacts of coal mining do not include:
   a. Decreasing emission of greenhouse gases into the atmosphere.
   b. Acid mine drainage.
   c. Roofs collapse in the underground shafts.
   d. Coal seam fires.

5. The U.S. has the following largest resources of fossil fuel:
   a. Natural gas.
   b. Coal.
   c. Oil.
   d. Peat.

Abbreviations and Achronyms

AVHRR—Advanced Very High Resolution Radiometer
(http://noaasis.noaa.gov/NOAASIS/ml/avhrr.html)

ATCDR—Agency for Toxic Substances and Disease Registry (www.atsdr.cdc.gov)

BGS—British Geological Survey (www.bgs.ac.uk)

CDCMR—California Department of Conservation and Mine Reclamation
(http://www.conservation.ca.gov)

CDC-NCEH—Center for Disease Control and Prevention National center for Environmental
Health (www.cdc.gov/nceh)

CDPH—California Department of Public Health (www.cdph.ca.gov)

CNMNC—Commission on New Minerals, Nomenclature, and Classification
(http://www.ima-mineralogy.org/CNMNC_Strategy.htm)

DRC—Democratic Republic of Congo

DoL—U.S. Department of Labor (www.dol.gov)

EIA—U.S. Energy Information Administration (http://www.eia.gov)

EPA—U. S. Environmental Protection Agency (www.epa.gov)

EU-European Union (www.europa.eu/index_en.htm)

IARC—International Agency for Research on Cancer (www.iarc.fr)

ILO—International Labor Organization (www.ilo.org)

IMA—International Mineralogical Association (http://www.ima-mineralogy.org)

IMA-NA—Industrial Mineral Association of North America (www.ima-na.org)

IUGS—International Union of Geological Sciences (www.iugs.org)

FDA—Food and Drug Administration (www.fda.gov)

KPCS—Kimberley Process Certification Scheme (www.kimberleyprocess.com)

MSHA—Mine Safety and Health Administration (www.msha.gov)

NIEHS—National Institute of Environmental Health Sciences (http://www.niehs.nih.gov)

NIOSH—National Institute for Occupational Safety and Health (http://www.cdc.gov/niosh)

NRDC—National Recourses Defense Council (www.nrdc.org)

OSHA—Occupational Safety and Health Administration (https://www.osha.gov)

OSMRE—Office of Surface Mining Reclamation and Enforcement (http://www.doi.gov/ocio/information_assurance/privacy/osm_notices.cfm)

TOMS—Total Ozone Mapping Spectrometer (www.eospso.gsfc.nasa.gov)

TVA—Tennessee Valley Authority (www.tva.gov)

UNEP—United Nations Environment Programme (www.unep.org)

UNICEF—United Nations International Children Emergency Fund (www.unicef.org)

UNFCCC—United Nations Framework Convention on Climate Change (https://unfccc.int)

UNGA–United Nations General Assembly (www.un.org/en/ga)

UNSC–United Nations Security Council (www.un.org/en/sc)

USGS—U.S. Geological Survey (www.usgs.gov)

IUGS–International Union of Geological Sciences (www.iugs.org)

WB—World Bank Group (www.worldbank.org)

WHO—World Health Organization (www.whc.int)

WTC—World Trade Center

Quiz Answers

Chapter 1

1: b—Transform boundaries.
2: d—Convergent boundary.
3: a—Similar rocks and similar fossils on different continents.
4: d—The age of oceanic crust increases with distance from a mid-ocean ridge.
5: c—<45% SiO_2.
6: c—Hadean.

Chapter 2

1: b—It is not crystalline (e.g., there is no repetitively organized molecular structure).
2: e—All of the above.
3: b and c—Renewable energy sources (2.5%) and Fossil fuels (coal, gas, and oil – 97.5%).
4: c—Are mined in conditions of armed conflict and human rights abuses and may be used to support civil wars and conflicts between neighboring countries in Africa.
5: d—They consist of tiny needle-like fibers that, if you inhale them, become stagnated in the lungs for a long time, often causing cancer.
6: b—It consists of a single layer of carbon atoms that are packed in a honeycomb-like configuration.

Chapter 3

1: e—All of the above.
2: c—All of the above.
3: b—Weathering of the Se-rich rocks followed by formation of Se-rich soil.

Chapter 4

1: c—Hundreds of years.
2: e—b and c only.
3: b—Yes, there is plenty of dust that carries the fungus responsible for Valley Fever. In addition, there is the added danger of earthquakes loosening sediments and creating more dust.
4: e—On almost all continents except Africa and Australia.
5: c—The air at high altitudes is less compressed and is therefore "thinner," which means that in a given volume of air there are fewer oxygen molecules present.

Chapter 5

1: c—An element that is in very low concentrations (less than 0.1% by mass) in a rock or mineral.
2: a—An arsenic-bearing organic compound found in mushrooms and certain marine foods.
3: c—The practice of eating earthy or soil-like substances.
4: a—Organic or inorganic form.
5: c—The natural geological formations of Bangladesh are enriched in arsenic.
6: b—10 μg/L in any source of drinking water.

Chapter 6 Mercury

1: c—Methylmercury.
2: d—Tuna.
3: a—0.1 ppm.
4: c—Be careful: neither a nor b.
5: b—Tailings are the unused, broken, ground-up rocks and dust that are usually piled up in mining operations. If not properly contained, tailings are always dangerous because any toxic elements that originally occurred in the rocks before their excavation are easily spread around by wind and water.

Chapter 6 Cadmium

1: d—All of the above.
2: d—All of the above.
3: d—All of the above.
4: g—All except e.
5: a—0.001 mg/kg/day.

Chapter 6 Lead

1: b—None of the above.

2: a,b,c—Scraping lead paint off of a wall. Eating food stored in lead-glazed pottery. Trying to repair lead-acid batteries at home.

3: b—Removed lead from use in car batteries.

4: f—All of the above.

5: a—1.5 μg/m3 in both.

Chapter 7

1: c—It should be recycled by heating to produce harmless glass, ceramic, and porcelain products.

2: c—Lung cancer, asbestosis, and mesothelioma.

3: b—Immediately get a dusk mask, and then carefully sweep it into a Ziploc bag, zip and label it, and deliver it to an asbestos recycling facility.

4: a—Inhaled.

5: b—About a fifteen- to thirty-year latency period.

Chapter 8

1: c—The lowest grade of coal, also called "brown coal."

2: b—The U.S. has few coal reserves and is therefore dependent on the import of coal from politically unstable parts of the world.

3: b—Coal.

4: a—Decreasing emission of greenhouse gases to the atmosphere.

5: b—Coal.

www.ingramcontent.com/pod-product-compliance
Lightning Source LLC
Chambersburg PA
CBHW080929220326
41598CB00034B/5728